HYDRAULIC ENGINEERING V

PROCEEDINGS OF THE 5TH INTERNATIONAL TECHNICAL CONFERENCE ON HYDRAULIC ENGINEERING (CHE V), 15–17 DECEMBER 2017, SHANGHAI, P.R. CHINA

Hydraulic Engineering V

Editors

Guojun Hong & Gongxun Liu

CCCC National Engineering Research Center of Dredging Technology and Equipment Co. Ltd., China

Liquan Xie

Department of Hydraulic Engineering, Tongji University, China

CRC Press
Taylor & Francis Group
Boca Raton London New York Leiden

CRC Press is an imprint of the
Taylor & Francis Group, an **informa** business
A BALKEMA BOOK

Published by:
CRC Press/Balkema
P.O. Box 447, 2300 AK Leiden, The Netherlands
e-mail: Pub.NL@taylorandfrancis.com
www.crcpress.com – www.taylorandfrancis.com

ISBN 13: 978-0-8153-7471-8 (hbk)
ISBN 13: 978-1-351-24157-1 (ebk)

Typeset by V Publishing Solutions Pvt Ltd., Chennai, India

Hydraulic Engineering V – Hong, Liu & Xie (Eds)
© 2018 Taylor & Francis Group, London, ISBN 978-0-8153-7471-8

Table of contents

Hydraulic Engineering V – Hong, Liu & Xie (Eds)
© 2018 Taylor & Francis Group, London, ISBN 978-0-8153-7471-8

Preface

Hydraulic research is developing beyond traditional civil engineering to satisfy increasing demands in natural hazards, structural safety assessment and also environmental research. In such conditions, this book embraces a variety of research studies presented at the 5th International Technical Conference on Hydraulic Engineering (CHE 2017) being held in Shanghai (China) during December 15–17, 2017. The conference served as a major forum to promote technological progress and activities, technical transfer and cooperation, and opportunities for engineers and researchers to maintain and improve scientific and technical competence in the field of hydraulic engineering, environment and safety engineering, and other related fields.

40 technical papers were selected for publication in the proceedings. Each of the papers has been peer reviewed by recognized specialists and revised prior to acceptance for publication. The papers embody a mix of theory and practice, planning and reflection participation, and observation to provide the rich diversity of perspectives represented in the conference. This book review recent advances in scientific theories and modelling technologies that are important for understanding the challenging issues in hydraulic engineering and environment engineering. Some excellent papers were submitted to this book, and some highlights include:

- hydraulic structures, e.g. gate dam, enclosure dam, sloping breakwater, seawall and armor.
- stability and risk analysis of hydraulic structures, e.g. reliability of the seawall armor structure, social influence evaluation index of dam-failure risk, risk of the rainfall and tide based on maximum entropy.
- hydraulic engineering technology, e.g. prestressed grid anchorage, regulation and dredging technology.
- regulation engineering technology, e.g. ecological revetment in Yangtze River waterway regulation, geomembrane enveloped concrete structure, flexible mattress with ecological cavities, reinforced concrete piles in river wandering control.
- dredging technology and equipment, e.g. erosion prediction of dredging pump impeller, power of dredger cutter head, hard clay cutting by moving jets, pipeline transport of coarse sand, cutting resistance of rake tooth, bow-spray of TSHD.
- estuary engineering and shoreline restoration, e.g. coastal wetland protection at the Yangtze Estuary, spur-dike wetlands, response of salinity intrusion to winds, land forming and its effect, shoreline restoration after Holocene sea level change in Hong Kong.
- geotechnical aspects, e.g. safety uncertainties of retaining wall, effects of PVD and EVD on vacuum preloading.
- water resources and water treatment, e.g. nitrate leaching related to irrigation and precipitation, effect of sulfate, alkalinity and chloramine on iron release of iron pipes.
- measurement technology, e.g. diffusion measurement under wave action based on tracer method, EIT in sediment concentration measurement.
- numerical software and applications, e.g. flow characteristics around hydrofoil structure, rock-breaking under high pressure waterjet, flow past a submerged vertical cylinder with porous structure, hydrodynamic response of regulation engineering in the Deepwater Navigation Channel of Yangtze Estuary.

– fluid-structure interactions, e.g. simulation of trailing suction hopper dredger under wave action.
– structure safety, e.g. temperature stress of bi-modulus beams, ship bulkheads subjected to combined loading of high-speed fragment and shockwave, wind-induced vibration of steel bridge members in Mountainous City.

At the very time we would like to express our deep gratitude to all authors, reviewers for their excellent work, and Léon Bijnsdorp, Lukas Goosen and other editors from Taylor & Francis Group for their wonderful work.

Hydraulic Engineering V – Hong, Liu & Xie (Eds)
© 2018 Taylor & Francis Group, London, ISBN 978-0-8153-7471-8

Committees

CHAIRMAN

Xiaoming Hou, *Professor, CCCC Shanghai Dredging Co. Ltd., China*
Tsung-chow (Joe) Su, *Professor, Florida Atlantic University, USA*

TECHNICAL COMMITTEE

Qingbo Zhang, *Professor, CCCC National Engineering Reseach Center of Dredging Technology and Equipment Co. Ltd., China*
Guojun Hong, *Dr., CCCC National Engineering Reseach Center of Dredging Technology and Equipment Co. Ltd., China*
Xiping Dou, *Professor, Nanjing Hydraulic Research Institute, China*
Yunfeng Xia, *Professor, Nanjing Hydraulic Research Institute, China*
Hailong Huang, *Professor, Nanjing Hydraulic Research Institute, China*
Wenrui Huang, *Professor, Department of Civil and Environmental Engineering, Florida State University, FL, USA*
Shuguang Liu, *Professor, Tongji University, China*
Nianqing Zhou, *Professor, Head of hydraulic Department, Tongji University, China*
Cuiping Kuang, *Professor, Tongji University, China*
Wenjuan Yao, *Professor, Shanghai University, China*
Bang-Fuh Chen, *Professor, Department of Marine Environment and Engineering, National Sun Yat-sen University, Taiwan, China*
Shengfa Yang, *Professor, Chongqing Jiaotong University, China*
Jisheng Zhang, *Professor, Associate Dean, Hohai University, China*
Lijun Deng, *Dr., Department of Civil and Environ Engineering, University of Alberta, Canada*
Xiuhan Chen, *Dr., Delft University of Technology, The Netherlands*
Hongsheng Zhang, *Professor, Shanghai Maritime University, China*
Xiangzhou Xu, *Professor, Dalian University of Technology, China*
Rongjian Li, *Professor, Xi'an University of Technology, China*
Gang Wang, *Professor, Chongqing University, China*
Liquan Xie, *Associate Professor, Tongji University, China*

ORGANIZING COMMITTEE

Shuai Jiang, *CCCC National Engineering Research Center of Dredging Technology and Equipment Co. Ltd., China*
Gongxun Liu, *Dr., CCCC National Engineering Research Center of Dredging Technology and Equipment Co. Ltd., China*
Jinwen Wang, *CCCC National Engineering Research Center of Dredging Technology and Equipment Co. Ltd., China*
Yiying Sun, *Dr., CCCC National Engineering Research Center of Dredging Technology and Equipment Co. Ltd., China*

Feixin Wang, *Dr., CCCC National Engineering Research Center of Dredging Technology and Equipment Co. Ltd., China*
Shengjun Qian, *CCCC National Engineering Research Center of Dredging Technology and Equipment Co. Ltd., China*
Liquan Xie, *Associate Professor, Tongji University, China*

Hydraulic Engineering V – Hong, Liu & Xie (Eds)
© 2018 Taylor & Francis Group, London, ISBN 978-0-8153-7471-8

Analyzing the influence of parameter value uncertainties on the reliability of the retaining wall

Jian-xu Chen, Wen-wu Song & Jia-ying Yu
School of Energy and Power Engineering, Xihua University, Chengdu, China

ABSTRACT: Using the first-order second-moment center theory for calculating anti-sliding and anti-overturning reliability, this study aimed at analyzing the influence of uncertainty parameters on the retaining wall, such as the internal friction of the filler (φ), the friction angle between the soil and the retaining wall structure (δ), the friction coefficient between the soil and the retaining wall base (f), the unit weight of the soil behind the wall (γ), and means and standard deviations of four random variables. The results indicate the eight major parameters that influence the anti-sliding and anti-overturning reliability of the retaining wall in the order of their importance: μ_φ, μ_f, μ_γ, σ_φ, μ_δ, σ_δ, σ_f, σ_γ and μ_φ, σ_δ, μ_γ, μ_δ, σ_φ, σ_γ, μ_f, σ_f, respectively. Finally, the study analyzes the influence of mixed parameter uncertainty on two types of reliability of the retaining wall.

1 INTRODUCTION

Stability of retaining walls plays a vital role in the field of civil engineering. For the retaining wall design to be reliable, it is necessary to ensure its stability. The traditional method is to use the valuation method for the design of the retaining wall; however, it ignores the influence of the randomness of each parameter and the uncertainty of their means and variances for the stability of the retaining wall. At present, many scholars worldwide have carried out relevant studies on the reliability of the retaining wall. Enrique (2004) analyzed the stability of the retaining wall by combining the randomness of the parameters that affect the retaining wall with the fixed value method, overcoming the deficiency of the fixed value method. Zhang (1997) calculated the anti-skid and anti-overturning stability of a retaining wall model by using the second-order moment center point method, obtaining the degree of influence of the change in each parameter on its stability.

For analyzing the stability of the retaining wall in actual engineering, the means and variances of the parameters are derived from a large number of random sampling. However, because of the limitations of the samples, the means and standard deviations of the parameters often do not fully reflect the overall situation. Zhang Jianren analyzed the degree of variability of four parameters (φ, δ, f, and γ) that has not been analyzed so far; however, it has not been concluded whether the means or standard deviations of the parameters play a major role in reliability. Therefore, in this study, the coefficient of variation was divided into means and standard deviations to analyze the relationship between primary and secondary reliability variables, as well as the influence of mixed uncertainty on two types of reliability.

2 FIRST-ORDER SECOND-MOMENT CENTER THEORY

First, the nonlinear functional function is derived for Taylor series expansion at the mean (center point) of the random variable and then retained to the first term; then, the mean and standard deviation of the functional function are calculated approximately.

Let X_1, X_2......X_n be n independent variables in the structure, μ_{X_i} be the average value, and σ_{X_i} be the standard deviation. Its function is

$$Z = g(X_1, X_2 \ldots\ldots X_n) \tag{1}$$

The function Z is extended at the mean value $P^*(X_1, X_2 \ldots\ldots X_n)$ and is retained to the term

$$Z \approx g(X_1, X_2 \ldots\ldots X_n) + \sum_{i=1}^{n} \frac{\partial g}{\partial X_i}\Big|_{\mu} \left(X_i - \mu_{X_i}\right) \tag{2}$$

Then, the mean value μ_{Z_L} and the standard deviation μ_{Z_L} of the functional function can be calculated.

The reliable index of the structure is

$$\beta = \frac{\mu_{Z_L}}{\mu_{Z_L}} = \frac{g\left(\mu_{X_1}, \mu_{X_2}, \ldots\ldots \mu_{X_n}\right)}{\sqrt{\sum_{i=1}^{n} \left(\frac{\partial g}{\partial X_i}\Big|_{\mu}\right)^2 \sigma_{X_i}^{\ 2}}} \tag{3}$$

The failure probability of the structure is

$$P_f = \Phi(-\beta) \tag{4}$$

The value of Φ $(-\beta)$ can be found by the standard normal distribution table.

3 MODEL ESTABLISHMENT

The most common destabilized damages to the retaining wall are sliding and overturning instability. A number of engineering experiments have proved that the four main random variables that affect the reliability of retaining walls are φ, δ, f, and γ. Therefore, other parameters are considered as constant in this study. The model is established as shown in Figure 1.

3.1 *Establishment of the anti-slip limit functional equation*

Let us consider the following condition:

$$F_R \geq F_S \tag{5}$$

where F_R denotes the retaining wall of the anti-skid and F_S denotes the retaining wall of the sliding force. The available earth pressure of the model is obtained from the Coulomb theory as follows:

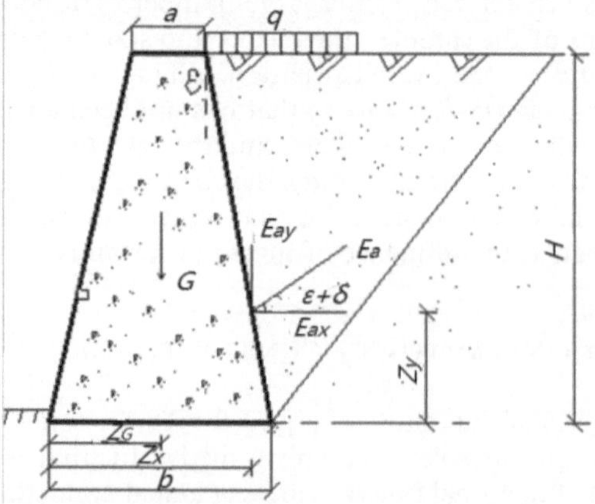

Figure 1. Calculation model of the retaining wall.

$$E_a = \left(\frac{1}{2}\gamma H^2 + qH\right) K_a \tag{6}$$

$$K_a = \frac{\cos^{2(\varphi-\varepsilon)}}{\cos^2\varepsilon\cos(\varepsilon+\delta)\left[1+\sqrt{\frac{\sin(\varphi+\delta)\sin\varphi}{\cos(\varepsilon+\delta)\cos\varepsilon}}\right]^2} \tag{7}$$

The limiting function of anti-sliding is

$$Z_1 = F_R - F_S = W_1(\varphi,\delta,f,\gamma) = f[\gamma_0 V + E_a \sin(\varepsilon+\delta)] - E_a\cos(\varepsilon+\delta) \tag{8}$$

where γ_0 is the unit weight of the soil behind the wall; V is the volume of the retaining wall; q is the overload of the filling behind the wall; H is the height of the filling; φ is the internal friction angle of the filling; δ is the friction angle between the soil and the retaining wall structure; and f is the friction coefficient between the soil and the retaining wall base.

3.2 *Establishment of the anti-overturning limit functional equation*

Let us consider the following condition:

$$M_R - M_S \geq 0 \tag{9}$$

where M_R denotes the resistance of the retaining wall around the toe rotation torque and M_S denotes the retaining wall around the toe rotation torque. The anti-overturning limit function is

$$Z_1 = M_R - M_S = W_2(\varphi, \delta, f, \gamma) = E_a \sin(\varepsilon+\delta) Z_x + \gamma_0 V Z_G - E_a \cos(\varepsilon+\delta) Z_y \tag{10}$$

where Z_x represents the horizontal distance from the point of action of the active earth pressure to the wall; Z_y represents the vertical force of the active earth pressure point to the wall toe; and Z_G represents the horizontal distance from the gravity point of the wall to the toe.

4 RELIABLE INDEX CALCULATION

When calculating two reliable indices, the four random variables φ, δ, f, and γ are denoted as X_1, X_2, X_3, and X_4, respectively, which are usually subject to the normal distribution (Du, 2008). The partial derivative of the two functional functions for each variable can be calculated as follows:

$$\frac{\partial W_1}{\partial X_1} = [f\sin(\varepsilon+X_2) - \cos(\varepsilon+X_2)]\frac{\partial K_a}{\partial X_1} \tag{11}$$

$$\frac{\partial W_1}{\partial X_2} = \left[\cos(\varepsilon+X_2) + \sin(\varepsilon+X_2)\right] + E_a[f\sin(\varepsilon+X_2) - \cos(\varepsilon+X_2)]\cdot\frac{1}{2}\gamma H^2\frac{\partial K_a}{\partial X_1} \tag{12}$$

$$\frac{\partial W_1}{\partial X_3} = \gamma_0 + E_a\sin(\varepsilon+X_2) \tag{13}$$

$$\frac{\partial W_1}{\partial X_4} = \frac{1}{2}H^2K_a\left[f\sin(\varepsilon+X_2) - \cos(\varepsilon+X_2)\right] \tag{14}$$

$$\frac{\partial W_2}{\partial X_1} = \left[\sin(\varepsilon+X_2)Z_x - \cos(\varepsilon+X_2)Z_y\right]\frac{\partial K_a}{\partial X_2} \tag{15}$$

$$\frac{\partial W_2}{\partial X_2} = \frac{1}{2}H^2K_a[f\sin(\varepsilon+X_2)Z_x - \cos(\varepsilon+X_2)Z_y] + [\sin(\varepsilon+X_2)\cos(\varepsilon+X_2)Z_y]\frac{\partial K_a}{\partial X_2} \tag{16}$$

Table 1. Statistical parameters of model random variables.

Variable	Mean	Standard deviation
φ(°)	45	3.6
δ(°)	15	4.2
f	0.5	0.1
$\gamma(KN/m^3)$	18	0.9

$$\frac{\partial W_2}{\partial X_3} = 0 \tag{17}$$

$$\frac{\partial W_2}{\partial X_4} = \frac{1}{2} H^2 K_a \left[\sin(\varepsilon + X_2) Z_x - \cos(\varepsilon + X_2) Z_y \right] \tag{18}$$

$$\frac{\partial K_a}{\partial X_1} = K_a \left[2\tan(\varepsilon - X_1) - \frac{\sin(2X_1 + X_2)}{(\sqrt{M} + M)\cos(\varepsilon + X_2)\cos\varepsilon} \right] \tag{19}$$

$$\frac{\partial K_a}{\partial X_2} = K_a [\tan(\varepsilon + X_2) - \frac{N}{\sqrt{M} + M} \tag{20}$$

$$M = \frac{\sin x_1 \sin(X_1 + X_2)}{\cos\varepsilon \cos(\varepsilon + X_2)}, N = \frac{\sin x_1 (\varepsilon + X_1)}{\cos\varepsilon \cos^2(\varepsilon + X_2)} \tag{21}$$

The other parameters are a = 1 m, b = 3 m, H = 5 m, γ_0 = 24 kN/m^3, Z_G = 1.549 m, Z_x = 2.412 m, Z_y = 1.667 m, ε = 10° and $q = 5$ *KPa*.

5 INFLUENCE OF UNCERTAINTY ON RELIABILITY

The two types of reliability of the retaining wall and the probability of two types of instability are calculated before the parameter is completely determined (uncertainty is 0).

5.1 *Analyzing the influence of uncertainty on slip resistance reliability*

The values of μ_φ, μ_δ, μ_f, μ_γ, σ_φ, σ_δ, σ_f, and σ_γ are calculated by equations (19) and (20) according to the uncertainties of 4%, 8%, 12%, 16%, and 20%. The maximum (β_{1max}) and minimum (β_{1min}) values of the slip reliability β_1 are calculated by the interval endpoint value, and the obtained image of β_{1max}, β_{1min} with anti-skid reliability β_1 and uncertainty U_1 is shown in Figure 2.

It is evident from Figure 2 that the uncertainty of the parameters will lead to fluctuation of the anti-skid reliability and that the upper and lower limits of the fluctuation interval are approximately symmetrical to β_1 = 1.508. The order of the degrees of β_1 is $\mu_\varphi > \mu_f > \mu_\gamma > \sigma_\varphi > \mu_\delta > \sigma_\delta$, The magnitude of the fluctuation amplitude is related to the parameters and the uncertainty. Apparently, β_1 = 1.508, so that the degree of influence on β_1 is $\sigma_f \approx \sigma_\gamma \approx 0$. The influence of variation coefficient and variance coefficient on β_1 is similar to $V_\delta > V_f > V_\varphi$ and $V_\gamma \approx 0$ in the literature [3], with parameters δ, f, and φ and standard deviation of φ.

5.2 *Analyzing the influence of uncertainty on overturning reliability*

Being the same as 4.1, drawn $\beta_{2\,max}$, $\beta_{2\,min}$, the parameters of different uncertainties are shown in Figure 3. It is evident from the figure that the uncertainty of the parameters will lead to the

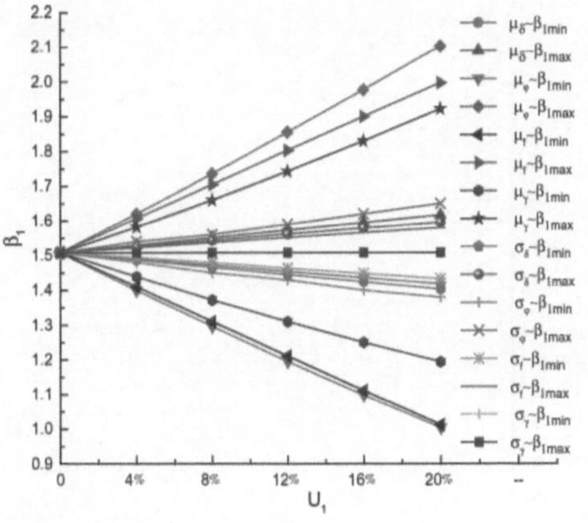

Figure 2. Diagram of the relationship between U_1 and β_1.

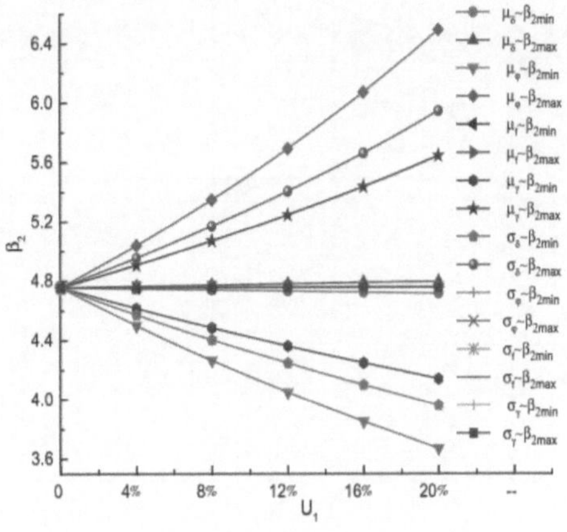

Figure 3. Diagram of the relationship between U_1 and β_2.

fluctuation of the anti-overturning reliability. The upper and lower limits of the fluctuation interval are approximately symmetrical to $\beta_2 = 4.756$. The relationship between the effects of β_2 size is $\mu_\varphi > \sigma_\delta > \mu_\gamma > \mu_\delta > \sigma_\varphi$, and the upper and lower limits of the fluctuation range of μ_φ, σ_δ, μ_γ, μ_δ, σ_φ, σ_γ, μ_f, σ_f are almost coincident with each other, and with the change in uncertainty, there is no effect on β_2. The degree of influence is $\sigma_\varphi \approx \sigma_\gamma \approx \mu_f \approx \sigma_f \approx 0$. The relationship between the coefficients of variation of β_2 is similar to that of $V_\delta > V\varphi$ and $V_f \approx V\gamma \approx 0$ in the literature [3], which also shows that the means and standard deviations of δ and φ have a major impact on β_2.

Figures 2 and 3 show that the uncertainty of μ_φ has the most obvious effect on the two types of reliability. In the range of 20% of influence, the maximum value of β_1 is 2.103 under the influence of the uncertainty of μ_φ. The corresponding failure probability is 0.0178, the minimum value of β_2 is 3.670, and the corresponding failure rate is 1.21E-04. The failure probability is only about 1% of the failure probability corresponding to β_1.

5.3 *Analyzing the reliability of mixed uncertainty*

It can be seen that by using fixed value, reliability, anti-skid safety factor $K_{cs} \geq 1.3$, and anti-overturning safety factor $K_{co} \geq 1.6$ can be calculated simultaneously, and the structure is safe. Figures 2 and 3 show the analysis of the influence of a single uncertainty on the reliability of the project, determine the uncertainty of all parameters simultaneously, and consider the most dangerous state of the retaining wall. In the case of different mixed uncertainties, we show the change of $\beta_{1\,min}$ and $\beta_{2\,min}$ with uncertainty of U_2. Previous analysis shows that the influence of mixed uncertainty on β_1 is neglected and that the two parameters, σ_f and σ_γ, which have no effect on it, are neglected. The influence of mixing uncertainty on β_2 is analyzed, and no influence of σ_φ, σ_γ, μ_f, and σ_f is found. The results of the analysis are shown in Figures 4 and 5.

Figure 4 shows that, when $U_2 > 2.5\%$, the use of fixed value calculation $K_{cs} < 1.3$ retaining wall is not safe. Figure 5 show that, when $U_2 > 40\%$, the use of fixed value method $K_{co} < 1.6$ retaining wall is unsafe. It can be seen that $2.5\% < U_2 < 40\%$; because of the existence of uncertainty, the retaining wall has slipping instability, but no overturning instability.

The common points of Figures 4 and 5 are that the influence of the mixing uncertainty of the parameters on the retaining wall is much higher than that of the reliability of the single parameter and that the minimum values of the two are much higher. The difference is that the effect of mixing uncertainty U_2 on β_1 is greater than that on β_2.

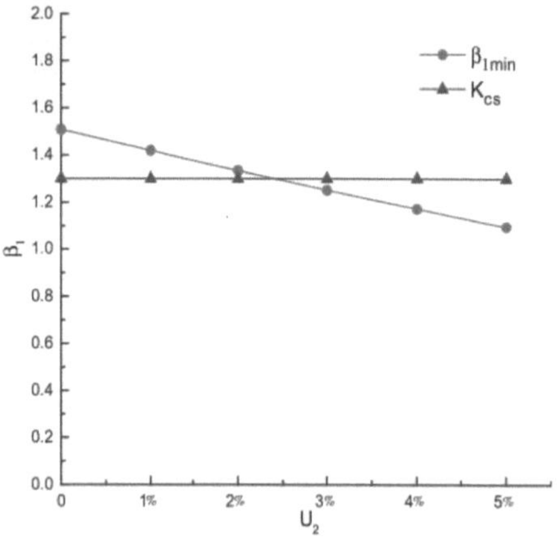

Figure 4. Relationship between U_2 and $\beta_{1\,min}$.

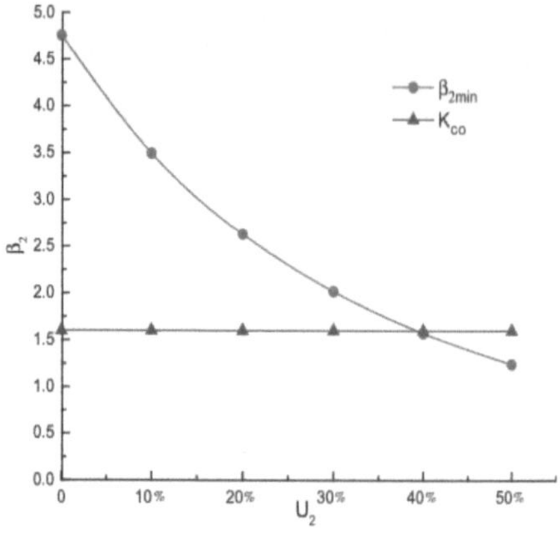

Figure 5. Relationship between U_2 and $\beta_{2\,min}$.

6 CONCLUSION

1. By calculating the two types of reliability for each parameter under different uncertainties, the relationships between the primary and secondary reliability variables are found: $\mu_\varphi > \mu_f > \mu_\gamma > \sigma_\varphi > \mu_\delta > \sigma_\delta > \sigma_f \approx \sigma_\gamma$ (anti-sliding) and $\mu_\varphi > \sigma_\delta > \mu_\gamma > \mu_\delta > \sigma_\varphi \approx \sigma_\gamma \approx \mu_f \approx \sigma_f$ (anti-overturning).
2. When calculating the safety factor of the retaining wall, such as the anti-skid reliability β_1 and the anti-overturning reliability β_2, using the fixed value method, it is necessary to strictly control the mean value μ and the variance σ of the parameter. The degree of determination should be within the normal range.
3. Concerning the uncertainty of same parameters, the anti-overturning failure probability of the retaining wall is much less than its sliding failure probability. Therefore, when calculating reliability, more attention should be paid to the influence of parameter uncertainty on sliding reliability.

REFERENCES

Enrique. 2004. Design and sensitivity analysis using the probability-safety-factor method. An application to retaining walls. *Structural Safety*, 26(2): 159–179.

Jianren Zhang. 1997. Reliability Analysis of Stability for Retaining Wall Structure. *China Journal of Highway and Transport*, 1997, 10(03): 56–61.

Jingxi Shu. 2016. Introduction of the reliability analysis of gravity retaining wall. *Sichuan building materials*, 42(5): 53–55.

Xinhua Song 2016. Analysis of stability of masonry slope based on cusp catastrophe theory. *Rock and Soil Mechanics*, 37(12): 3499–3505.

Yongfeng Du. 2008. Analysis of reliability of structural systems for stability of gravity retaining walls. *Chinese Journal of Geotechnical Engineering*, 30(3): 349–353.

Hydraulic Engineering V – Hong, Liu & Xie (Eds)
© 2018 Taylor & Francis Group, London, ISBN 978-0-8153-7471-8

A method for calculating the trajectory and extension of bow-spray for the TSHD

Feixin Wang, Jifu Yin & Qingbo Zhang
CCCC National Engineering Research Center of Dredging Technology and Equipment, Shanghai, China

Lihua Wang
CCCC Shanghai Dredging Co. Ltd., China

Zheng Han
CHEC Dredging Co. Ltd., China

ABSTRACT: The currently used formulas for calculating the trajectory of bow-spray for the TSHD are based on the projectile theory, but they do not consider the mixing and expending effects due to jet characteristics. Therefore, they cannot be used for calculating the interface between slurry falling and the land/water surface. On the basis of the spray distance calculation model accounting for air resistance, in this study, a new method was established for calculating the trajectory and extension range of bow-spray by introducing the basic characteristics of jet flow. First, the practicability and accuracy of the new model were calibrated and validated by field tests, numerical simulations, and the air resistance-based calculation model of the spray distance. Then, the effects of the jet propagation coefficient, the injection angle, and the nozzle diameter on the bow-spray distance and falling boundary were analyzed and are discussed. The results indicate that by decreasing the jet propagation coefficient through the optimization of the jet nozzle structure, it is possible to extend the optional operating position and reduce sediment loss. For the injection angle of 40–45°, the maximum spray distance and furthest inner boundary can be reached. However, for the injection angle of 30–35°, the falling angle is smaller, which is conducive to sediment loss control. There can be certain effects on the spray distance and falling boundary of the nozzle diameter. Therefore, determining an appropriate injection angle and a nozzle diameter based on specific operation conditions and further analyses is important.

1 INTRODUCTION

Bow-spray is one of the fundamental filling construction technologies for the TSHD, for filling a large amount of sand to the shallows or shores by means of direct discharge of slurry from the bow nozzle. It is commonly used for beach filling, and can also be used for island construction in open water. Bow-spray plays an important role in the initial stage of land reclamation projects when the conditions for water pipeline installation are not met and direct dumping with TSHD is restricted by water depth. This technology is used to facilitate rapid formation of shoals, to create the conditions for subsequent construction machinery landing on the breach. Bow-spray is widely used in reclamation projects, such as the Palm Islands and the World Islands in Dubai, Yangshan Deepwater Port Construction, and other projects in China (Wang et al., 2017; Yin et al., 2016).

Spray distance is one of the most important limiting factors of bow-spray. It is not only the critical factor for the feasibility of the bow-spray operation, but also a key element of operation position determination. At present, the commonly used formula for spray distance is based on the free projectile motion theory. The result is larger than the actual distance due to the neglecting of air resistance. Wang et al. (2017) proposed a calculation method for the trajectory of bow-spray by including air resistance effects used in the projectile and fire gun trajectory calculation, hereinafter referred to as the AR (Air Resistance) model.

The calculation accuracy is significantly improved. Yin et al. (2016) adopted the CFD method to simulate trajectory curve and flow-field characteristics of bow-spray and obtained the spray distance and flow-field characteristics under different conditions.

Most of the existing bow-spray and fire gun motion trajectory models and methods, proposed by Wang et al. (2017), Sun (2009), Min et al. (2011), and Chen & Yang (2016), select micro fluid group as the analysis objective on the basis of the projectile theory; however, the effects of continuity of the jet fluid and entrainment and mixing with the surrounding air are ignored. In addition, the interface between slurry falling and land/water surface, especially the inner boundary, plays an important role in operating position determination and sediment loss control for bow-spray. Therefore, it is necessary to carry out a more thorough discussion on analysis and calculation methods for the bow-spray trajectory and the falling boundary. In this study, a new model for calculating the trajectory and extension range of bow-spray is suggested on the basis of basic characteristics of jet flow and the AR model. Through theoretical calculation, numerical simulation, and experimental verification, we aim to provide a theoretical basis for the analysis and optimization of bow-spray technology.

2 METHOD FOR CALCULATING THE TRAJECTORY AND EXTENSION OF BOW-SPRAY

According to the definition of jet flow (Yu, 1992), bow-spray of the TSHD is a typical kind of jet flow, belonging to the category of circular incompressible non-submerged jet. The motive force is the outflow momentum and gravity. Simultaneously, air resistance has an important impact on the trajectory of bow-spray at a high flow velocity. Bow-spray follows the theory and basic rules of jet flow. Properties of jet flow, for instance, entrainment and mixing of the jet flow, self-preservation of the longitudinal velocity distribution, linear extension of the mixing layer, and turbulent characteristics of the jet flow, can be applied in the analysis and measurement of bow-spray performance. It is necessary to indicate that, unlike that in environment engineering and other industries, for bow-spray and fire gun, the jet fluid density is much larger than that for the surrounding air, and more attention should be paid to the spray distance and extension range, especially contract extension, than to motion characteristics near the jet zone.

On the basis of the above-mentioned considerations, a new model for calculating the trajectory and extension range of bow-spray can be established, as shown in Figure 1, hereinafter referred to as JR (Jet-Resistance) model. First, it is assumed that bow-spray fluid circularly distributed normal to the flow direction and linear extended along the jet direction, due to jet flow properties. Second, the air resistance model is modified by taking acting surface and air density into account.

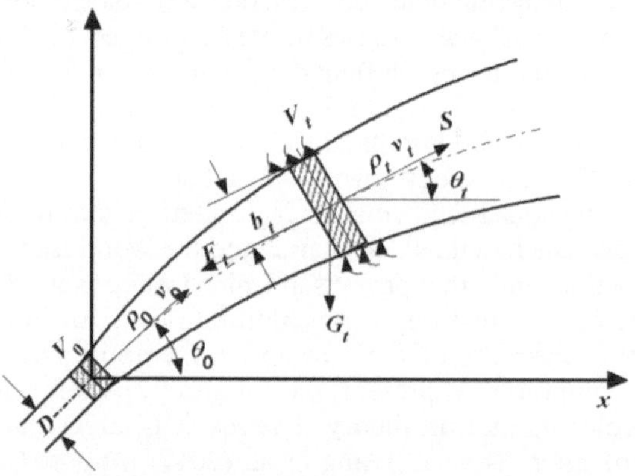

Figure 1. Schematic of the method for calculating the trajectory and extension of bow-spray.

The fluid mass sprayed from nozzle is taken as the analysis objective. At time $t = 0$, the micro fluid mass properties, including density ρ_0 (kg/m^3), velocity v_0 (m/s), and velocity angle θ_0, are determined by the initial conditions of bow-spray and the radius of the micro fluid mass b_0(m), which is 1/2 of the nozzle diameter D(m).

Without considering the mass exchange and interaction force between the micro mass with those before and after, at time t (s), radius of the micro mass, b_t (m), is expended because of the entrainment and mixing with the surrounding air as:

$$b_t = b_0 + \varepsilon \cdot S_t \tag{1}$$

where ε, the jet propagation coefficient, is a constant characterization of jet lateral extension degree, mainly depending on the jet outlet structure, and fluid and environmental fluid entrainment and mixing characteristics, generally estimated by test data. S_t is the motion distance of the fluid micro fluid at time t (m), which can be obtained by the time integral of velocity v_t (m/s):

$$S_t = \int_0^t v_t \, \mathrm{d}t \tag{2}$$

Meanwhile, the micro fluid volume $\mathrm{d}V_t$ (m^3) is extended as follows:

$$S_t = \pi b_t^2 v_t \, \mathrm{d}t \tag{3}$$

The density ρ_t (kg/m^3) reduces to

$$\rho_t = \frac{\rho_0 \mathrm{d}V_0 + \rho_a(\mathrm{d}V_t - \mathrm{d}V_0)}{\mathrm{d}V_t} \tag{4}$$

where ρ_a is the density of air (1.205 kg/m^3).

Gravity G_t (N) and air resistance F_t (N) are the forces acting on the micro fluid at time t. The quadratic relationship between air resistance and velocity is taken into account, according to related literature studies (See Chen & Yang, 2016; Min et al., 2011; Warburton et al., 2010; Sun, 2009; Wang et al. 2017) and previous research results. In addition, the contact surface of fluid and air is selected as the acting surface of air resistance, as the continuity of bow-spray jet flow is taken into account, different from the cross section for the general projectile theory, while Wang et al. (2017) did not consider the effect of acting surface. Thus, the formula of air resistance is as follows, borrowing the concept of drag force and lift force:

$$\mathrm{d}F_t = C_F \cdot \frac{1}{2} \cdot \rho_a A_L v_t^2 = \pi C_F b_t v_t^3 \, \mathrm{d}t \tag{5}$$

where A_L is the lateral area of the micro fluid (m^2) and C_F is the dimensionless air resistance coefficient. Then, the acceleration of the micro fluid at horizontal and vertical directions (a_{tx} and a_{tz}, respectively) can be calculated as follows:

$$a_{tx} = \frac{\mathrm{d}F_t}{\rho_t \mathrm{d}V_t}\cos(\theta_t) = \frac{C_F \rho_a v_t^2}{\rho_t b_t}\cos(\theta_t) \tag{6}$$

$$a_{tz} = \frac{\mathrm{d}F_t}{\rho_t \mathrm{d}V_t}\sin(\theta_t) - g = \frac{C_F \rho_a v_t^2}{\rho_t b_t}\sin(\theta_t) - g \tag{7}$$

Combining the above formulas, we can calculate the trajectory and extension range of bow-spray, and the jet extension constant (ε) and the air resistance coefficient (C_F) remain undetermined. It is recommended to solve these equations by numerical method, due to the complex structure of equations and difficulties in analytic solution directly.

3 MODEL VERIFICATION AND ANALYSES

3.1 *Model calibration and verification with field tests*

Field test of the "Xin Hai Long" TSHD at the Yangtze Estuary is selected for model calibration and validation. The main parameters are as follows: the height from the nozzle to the water surface is 13.5–18.0 m; the mean value is 15.7 m; the angle of the bow nozzle to the ground is 45°; the diameter of the vessel pipe is 1.0 m; and the nozzle diameter is 0.45 m. The measured flow velocities are 5.0 and 6.6 m/s under the low–low and low–high mode conditions for water transport, respectively, whereas for slurry cases (fine sand, with average density of 1250 kg/m^3), the corresponding values are 4.7 and 6.3 m/s. For other parameters, the reader is referred to Wang et al. (2017). The spray distances are measured by total station instrument, whereas the extension range (inner and outer boundaries) of bow-spray is taken from the scene image analysis of the tests and the data obtained by the total station instrument.

After several trial calculations, the values of the two undetermined constants $\varepsilon = 0.06$ and $C_F = 1.41 \times 10^{-5}$ are recommended, when the results of the JR model agree well with the measured data, as shown in Table 1 and Figure 2. The results of the AR model are also shown in the charts for comparison.

According to the above-presented comparison data (Figure 2 and Table 1), the calculated results of the JR model are close to the measured results when the undetermined constants select the above values. The calculated center line, outer boundary, and inner boundary are in good agreement with the measured values for both water and slurry transport under the low–high pump mode condition. The difference between the calculated and measured outer/inner contract boundary is less than 2.5 m, and error in the calculated spray distance is within 1.5%; the difference between the calculated and measured values for slurry transport under the low–low mode condition is 5.7 m, with error 6.73%. For other three cases, differences are within 0.5 m and errors are less than 0.5%.

Table 1. Comparison of the bow-spray distance and falling boundary between the JR model and the measured data.

Experimental conditions		Measured data (m)			JR model (m)			AR model (m)
Fluid	Pump mode	Distance	Inner*	Outer*	Distance	Inner*	Outer*	Distance
Water	Low–low	155.5	142.9	168.9	155.3	141.7	169.8	152.7
Water	Low–high	99.2	–	–	99.3	90.5	108.6	97.1
Slurry	Low–low	146.6	133.7	157.7	146.1	133.3	159.8	144.7
Slurry	Low–high	84.7	–	–	90.4	82.4	98.9	89.4

*Inner boundary and outer boundary, respectively.

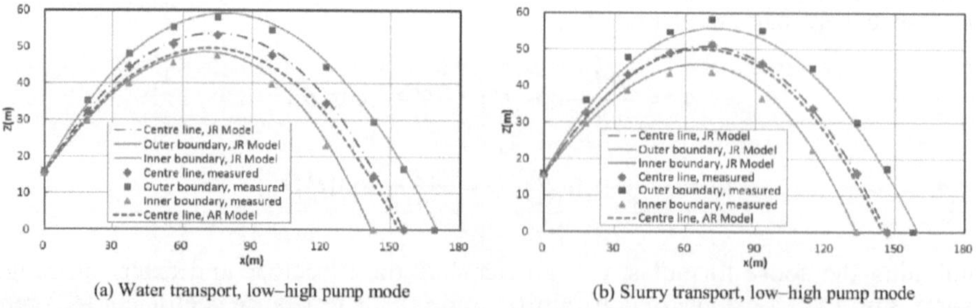

(a) Water transport, low–high pump mode

(b) Slurry transport, low–high pump mode

Figure 2. Comparison between the JR model and the measured data with respect to the trajectory and extension range of bow-spray.

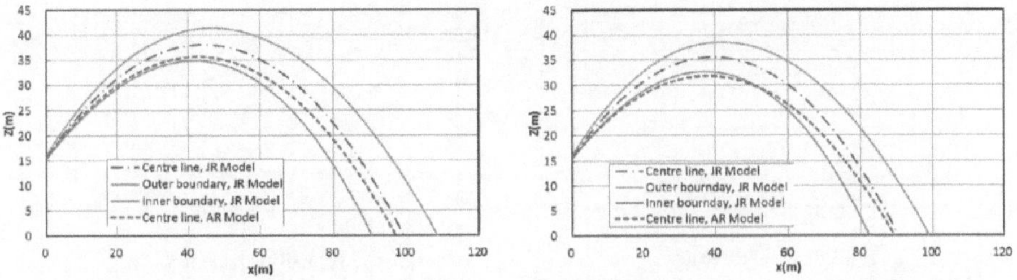

(a) Water transport, low–low pump mode (b) Slurry transport, low–low pump mode

Figure 3. Comparison of the calculated bow-spray trajectory between the JR model and the AR model.

3.2 *Comparison with the AR model*

Comparison of the calculated bow-spray trajectory between the JR model and the AR model is also performed, as shown in Table 1 and Figures 2 and 3. The spray distance calculated by the JR model is slightly larger than that calculated by the AR model in the four cases, with a difference of 1–3.6 m and relative error of 1–2.3%. This is mainly due to the determination of the model parameter values, so that the calculation accuracy of the spray distance is the same for the two models. In the case of slurry transport under the low–low mode condition, the spray distance calculated by the AR model also has a certain deviation from the measured value, with a difference of 4.7 m and error of 5.55%. In the case of the bow-spray center line, the maximum height calculated by the JR model is slightly larger than that calculated by the AR model, except for the case of slurry transport under the low–high mode condition (the results of which are consistent). It seems that the center line calculated by the JR model is more close to the measured value than that calculated by the AR model in the case of clean water transport under the low–high mode condition.

The comparison and analysis results show that the JR model has more advantages than the AR model, for a more comprehensive consideration. The main improvements are as follows: the trajectory and extension range of bow-spray can be calculated by inducing the jet theory, and the contract boundary calculation is more suitable for practical engineering than the single drop point; air resistance is refined and improved by considering the acting surface and air density and by non-dimensionalizing the air resistance coefficient, in order to make the model more reasonable and thus more applicable.

3.3 *Comparison with numerical simulations*

The results of numerical simulation are used to compare and verify the results obtained by the JR model, as the measured bow-spray contract boundary is absent due to the test condition limitation. A 3D numerical simulation method carried out by CFD is used, which is improved on the basis of the 2D numerical analysis by Yin et al. (2016). The effect of air resistance is neglected, so only the rationality of the jet propagation coefficient is verified. Therefore, the coefficient values are $\varepsilon = 0.06$ and $C_F = 0$. A comparison of the two cases, water transport under the low–low mode condition and slurry transport under the low–high mode condition, is performed, as shown in Figures 4 and 5.

Figures 4 and 5 show that the results of the JR model are consistent with the numerical calculation results, without considering the effects of air resistance. The center line of the former agrees well with that of the latter, whereas there is a slight deviation between the outer and inner boundaries. From the results of the contract boundary calculation, both the numerical simulation and the JR model are elliptical; however, the numerical simulation results are smaller than those of the JR model. It is necessary to indicate that specific velocity is selected as the contract boundary discriminant index for numerical simulation due to data treatment problems, which may lead to certain differences between the results. Overall, the jet propagation coefficient $\varepsilon = 0.06$ is reasonable for the JR model, according to comparison and verification with the numerical simulations.

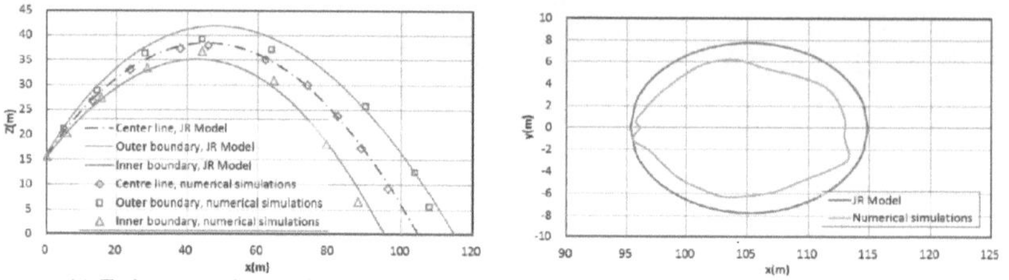

(a) Trajectory and extension range (b) Falling boundary

Figure 4. Comparison of the results between the JR model and numerical simulations in the case of water transport under the low–low mode condition.

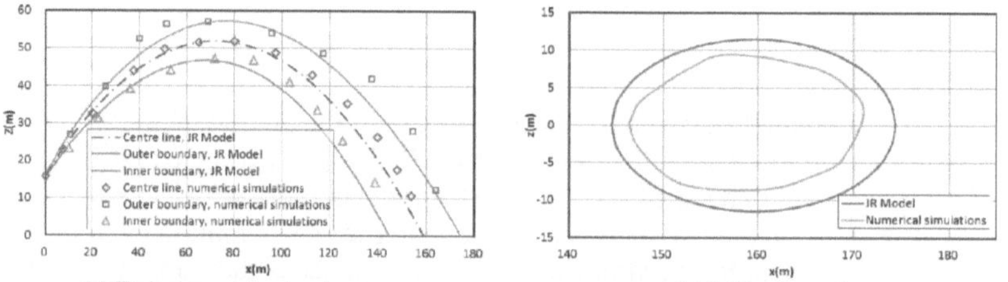

(a) Trajectory and extension range (b) Falling boundary

Figure 5. Comparison of the results between the JR model and numerical simulations in the case of slurry transport under the low–high mode condition.

4 ANALYSES OF INFLUENCING FACTORS

4.1 *Jet propagation coefficient*

Jet propagation coefficient has an important effect on the extension range of jet flow. The trajectory and extension range of bow-spray in the case of slurry transport under the low–high mode condition is calculated and compared under different values of the jet propagation coefficient. The results are shown in Figure 6.

Figure 6 shows that the center line of bow-spray raises slightly with decreasing jet propagation coefficient ε, and the spray distance is also slightly increased. The spray distance increases about 4.0%, that is, 5.9 m, for $\varepsilon = 0.02$, to compare with $\varepsilon = 0.06$. The extension range narrows visibly as the jet propagation coefficient decreases with slight shrink of the outer boundary, evident aggrandizement of the inner boundary, and opposite movement of the contract boundary to the dredger. The inner boundary extends 13.9 m with the ratio 10.4% for $\varepsilon = 0.02$, compared to that for $\varepsilon = 0.06$. Thus, improving the nozzle structure of bow-spray to reduce the jet propagation coefficient can effectively reduce the extension range of bow-spray and obviously expand the inner boundary, thereby increasing the optional area for selecting the dredger operation position and effectively reducing sediment loss during the construction process. Therefore, this is an effective way to improve the research and process of bow-spray.

4.2 *Injection angle*

Injection angle is also an important factor in determining the spray distance and trajectory. The trajectory and contract boundary of bow-spray in the case of slurry transport under the low–high mode condition are calculated and compared under different injection angles, as shown in Figure 6.

Figure 7 shows that the highest spray height and falling angle drop evenly with decreasing injection angle, while the spray distance decrease and flattening of the falling boundary are accelerated. The difference of spray distance and falling boundary between the cases of 40° and 45° is insignificant; however, for the 35° case, the spray distance and inner boundary decrease 4.0 and 4.3 m, respectively; for the 30° case, the spray distance and inner boundary decrease

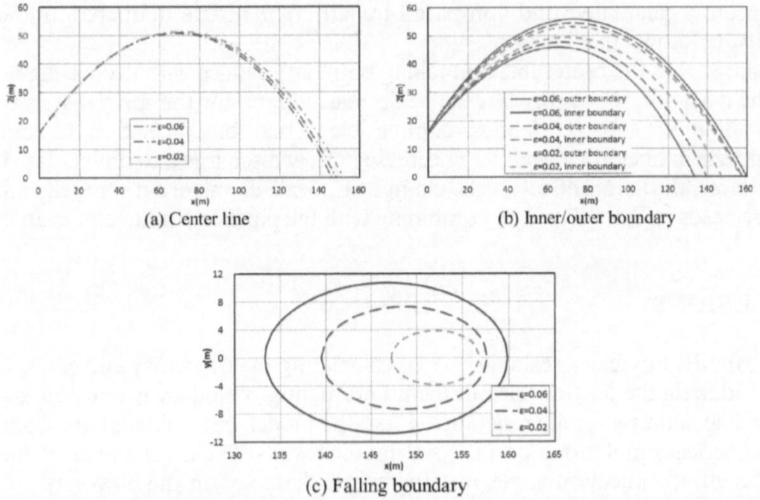

(a) Center line (b) Inner/outer boundary

(c) Falling boundary

Figure 6. Effect of the jet propagation coefficient on the trajectory and extension range of bow-spray.

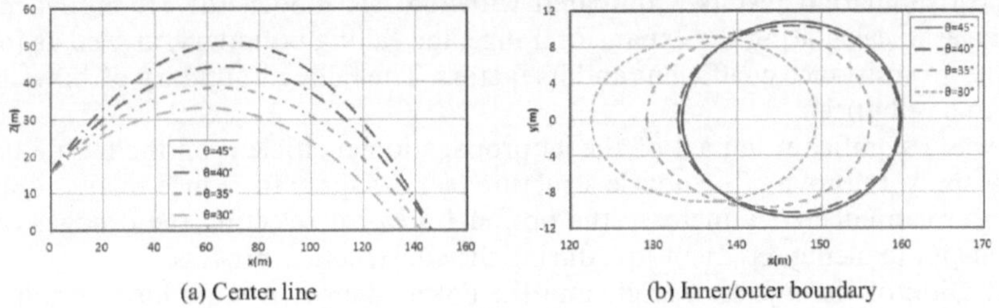

(a) Center line (b) Inner/outer boundary

Figure 7. Effect of the injection angle on the trajectory and extension range of bow-spray.

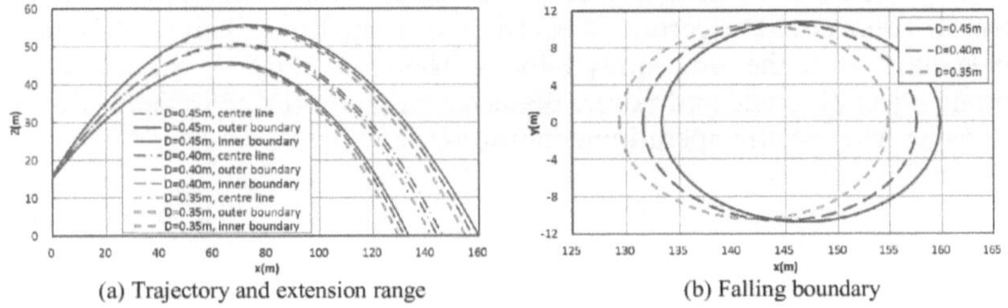

(a) Trajectory and extension range (b) Falling boundary

Figure 8. Effect of the nozzle diameter on the trajectory and extension range of bow-spray.

10.7 and 10.9 m, respectively, with corresponding shrinkage ratios of 7.30% and 8.17%. The falling angles of the slurry are 54.1°, 50.2°, 46.2°, and 42.1° at four different injection angles.

According to the above analyses and calculation results, the maximum spray distance and inner boundary distance can be obtained when the injection angle is within 40–45°, which is beneficial to the optional operating position. For the case of 30–35° injection angle, the spray distance and inner boundary decrease obviously; therefore, it is better to control back-silting and sediment loss than in the case of 40–45°, from the viewpoint of falling angle. The proper injection angle should be carefully selected according to the specific working conditions during actual projects.

4.3 *Nozzle diameter*

Nozzle diameter may have certain effect on spray distance and falling boundary. The trajectory and contract boundary of bow-spray in the case of slurry transport under the low–high mode condition is calculated and compared for different nozzle diameters under the same spray velocity, as shown in Figure 8.

Figure 8 shows that the center line and falling boundary reduce with decreasing nozzle diameter, while the deviation is not very large. The decreasing ratio for the spray distance and falling boundary is about 3.5% for $D = 0.35$ m, comparable to that for $D = 0.45$ m. In actual projects, decrease in nozzle diameter will lead to a decrease in flow discharge and an increase in the spray velocity; therefore, the actual effect of the change of nozzle diameter on spray distance and falling boundary needs to be analyzed by combining with the pipeline characteristic analysis results.

5 CONCLUSIONS

In this study, the JR model was established for calculating the trajectory and extension of bow-spray by introducing the jet flow entrainment and mixing boundary expansion law, based on the AR model accounting for air resistance. First, the model was validated and compared with field tests, numerical simulations, and the AR model results. Then, the effects of the jet propagation coefficient, the injection angle, and the nozzle diameter on the bow-spray distance and falling boundary were analyzed and discussed. The main conclusions are as follows:

1. The practicability and accuracy of the JR model were verified by the measured data and numerical simulation results. Compared with the AR model, the JR model had more advantages in calculating the expansion range and falling boundary, as well as in considering the air resistance coefficient and its factors. The falling boundary of bow-spray was found to be elliptical.
2. There was a significant impact of the jet propagation coefficient on the falling boundary of bow-spray. Improving the nozzle structure of bow-spray to reduce the jet propagation coefficient can effectively increase the optional area for selecting the dredger operation position and reducing sediment loss during the construction process.
3. The selection of the injection angle and the nozzle diameter could have certain impacts on the spray distance and the falling boundary. For the injection angle of 40–45°, the maximum spray distance and the furthest inner boundary was reached. However, for the injection angle of 30–35°, the falling angle was smaller, which was conducive to sediment loss control. There could be certain effects on the spray distance and falling boundary of the nozzle diameter at the same spray velocity. Therefore, it is important to determine an appropriate injection angle and nozzle diameter based on the analysis results of pipeline characteristics and specific operation conditions.

REFERENCES

Chen Xuejun, Yang Yongming. Jet trajectory model and positioning compensation method for fire water cannon [J]. Chinese Journal of Engineering Design, 2016, 23(6).

Min Yonglin, Chen Xiaoyang, Chen Chi, Hu Chuanping. Pitching angle-based theoretical model for the track simulation of water jet out from water fire monitors [J]. Journal of Mechanical Engineering, 2011, 47(11): 134–138.

Sun Jian. Study on the water jet track of water fire monitors [D]. Shanghai: Shanghai Jiaotong University, 2009.

Wang Lihua, Han Zheng, Wang Feixin. Analysis on rainbowing distance calculation and mixture moving rail of TSHD [J]. Port & Waterway Engineering, 2017, (4): 15–19, 27.

Warburton, R.D.H., J. Wang, J. Burgdöfer. Analytic approximations of projectile motion with quadratic air resistance [J]. J. Service Sience & Management, 2010, 3: 98–105.

Yin Jifu, Wang Feixin, Wang Lihua, et al. Numerical simulation on the path curve and flow characteristic of bow-spray for Trailing Suction Hopper Dredger [J]. Port & Waterway Engineering, 2016, (11): 29–33.

Yu Chanzhao. Introduction to environment fluid mechanics [M]. Beijing: Tsinghua University Press, 1992.

Hydraulic Engineering V – Hong, Liu & Xie (Eds)
© 2018 Taylor & Francis Group, London, ISBN 978-0-8153-7471-8

Temperature stress analysis for a bimodulus beam placed on a double-parameter foundation

W.J. Yao & Y. Yang
Shanghai Institute of Applied Mathematics and Mechanics, Shanghai University, China
Department of Civil Engineering, Shanghai University, China

ABSTRACT: In this study, the mechanical response of a double-parameter foundation with different moduli under the temperature effect was investigated. In addition, the position of the neutral axis and strain differential equations were established. Then, an expression for the stress, deflection, and bending moment of the beam placed on the foundation was derived. On the basis of the calculation results from different-modulus theories, we compare the diversity with the classical modulus solution. For the concrete material widely used in road and bridge engineering, it is worth mentioning that relative to the classical same-modulus theory, the maximum inaccuracy of tension stress increased by 26%. However, cracks and damage on the pavement are common in practical engineering, which may be due to the neglecting of the bimodulus characteristic of the foundation beam. Therefore, we propose several reasonable optimal design suggestions for practical engineering.

1 INTRODUCTION

According to classical mechanics engineering materials have the same elastic moduli in tension and compression. However, a myriad of experimental tests show that there are also materials such as concrete, metal alloy, biological materials (Barak M.M. *et al.* 2009), rubber, rock, and nuclear graphite with different tensile elastic and compressive elastic moduli (G. Tsoukleri *et al.* 2009), which are identified as bimodulus materials. Furthermore, this kind of materials contributes significantly to the mechanical performance of the structure. However, the difference between tension moduli and compression moduli is often neglected in practical engineering. In fact, this simplification is not in accordance with the veritable constitutive relation, which will trigger certain safety problems due to the large error.

At present, the component and structure of different-modulus materials have provided great results in both application and analysis. Yao et al. (2006) provided analytical solutions for statically indeterminate structure with different moduli under complex stress. He et al. (2015) provided analytical solutions for bending curved beams with different moduli in tension and compression. However, a few studies have been conducted to study the mechanical response of the bimodulus structures placed in the temperature field. In the early 21st century, Yao et al. (2005, 2014) studied the temperature stress field in the bimodulus single frame and bending beam with different moduli under different constraints and found that the concrete bimodulus property deduced by solar radiation might have a significant effect on the mechanical response of the pavement and bridge. Gao et al. (2017) studied the mechanical response of a bi-modulus Winkler foundation beam under the temperature and developed a finite-element calculation procedure for a bimodulus structure under the temperature effect. In the field of engineering, the influence of temperature stress cannot be ignored, for example, the long exposure of highway roadbed to solar radiation. In addition, the concrete material is typical of the different modulus. Therefore, it can be seen that considering the different modulus of material is necessary in the study of temperature stress field in the foundation of highway.

In view of this, the mechanical response of the bimodulus beam on the double-parameter foundation is studied to identify the calculation error between the bimodulus theory and

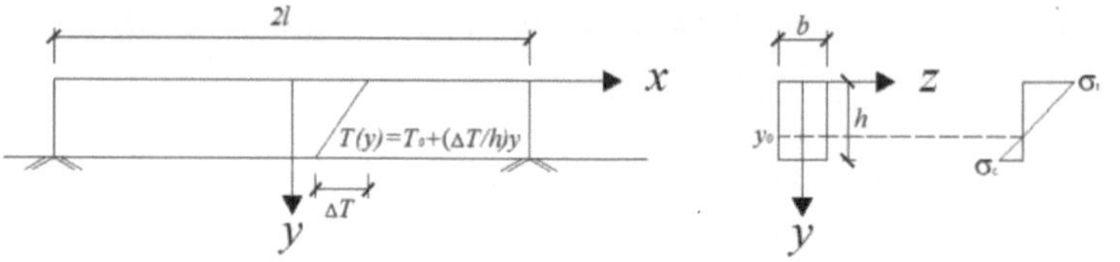

Figure 1. Structural model.

the classical same-modulus theory. Finally, several reasonable optimal design suggestions for practical engineering are proposed.

2 STRUCTURAL MODEL

2.1 *Basic assumption*

The bimodulus foundation beam investigated in this study is assumed to be continuous, homogeneous, and isotropic. Materials produce small elastic deformation under any load and obey the ordinary laws of the equilibrium equations, geometric equations, and deformation continuity equations. The diversity reflects well on the constitutive equations compared with the classic same elastic moduli. In addition, the temperature of the beam section is assumed to be linear due to its height and has no relationship with the length of the beam. Simultaneously, we supposed to use the double-parameter foundation, which adopts k_w and k_g to represent the resistant coefficient in the foundation and the shear transfer parameter between each foundation.

2.2 *Structural model*

As shown in Figure 1, a bimodulus beam ($2l \times b \times h$) is placed on the double-parameter foundation. The temperature distribution function can be described as follows:

$$T(y) = T_0 + (\Delta T / h)y = T_0 + G \cdot y \tag{1}$$

where T_0 is the temperature at the bottom surface of the bimodulus foundation beam, ΔT is the temperature difference between the top and bottom surfaces, and G is the temperature gradient.

3 THEORETICAL ANALYSIS AND ANALYTICAL DERIVATION

3.1 *Establishment of governing equations*

According to the assumption of plane cross section, the normal strain of the random cross section in the bimodulus beam can be obtained as follows:

$$\varepsilon_x = \alpha(A(x) + B(x) \cdot y) \tag{2}$$

where α is the thermal expansion coefficient of the material and $A(x)$ and $B(x)$ are undetermined constants. Therefore, the constrained strain can be obtained as follows:

$$\varepsilon_x = \alpha(A(x) + B(x) \cdot y - T(y) = \alpha(A(x) + B(x) \cdot y - G \cdot y) \tag{3}$$

We obtain the position of the neutral axis y_0, where the strain is zero, as follows:

$$y_0 = \frac{T_0 - A(x)}{B(x) - G} \tag{4}$$

As shown in Figure 1, because of the restriction, the upper zone tends to be the tension region, whereas the bottom zone tends to be the compression region. The stress can be written as follows:

$$\sigma_x = \begin{cases} \alpha E_c(A(x)+B(x)\cdot y - T_0 - G\cdot y) & 0 \le y \le y_0 \\ \alpha E_t(A(x)+B(x)\cdot y - T_0 - G\cdot y) & y_0 < y \le h \end{cases} \tag{5}$$

According to the double-parameter foundation beam model, we can get

$$N = \alpha E_c \int_0^{y_0} (A(x)+B(x)\cdot y - T_0 - Gy) b dy + \alpha E_t \int_{y_0}^{h} (A(x)+B(x)\cdot y - T_0 - Gy) b dy = 0 \tag{6}$$

$$M = \alpha E_c \int_0^{y_0} (A(x)+B(x)\cdot y - T_0 - Gy) b y dy + \alpha E_t \int_{y_0}^{h} (A(x)+B(x)\cdot y - T_0 - Gy) b y dy = 0 \tag{7}$$

Substituting Eq. (4) into Eq. (6) and simplifying, we get

$$A(x) = T_0 + (G - B(x))h \frac{\sqrt{E_t}}{\sqrt{E_t}+\sqrt{E_c}} \tag{8}$$

Then, substituting Eq. (8) into Eq. (7) and simplifying, we get

$$M = -\frac{\alpha b h^3}{3}(G - B(x)) \frac{E_t E_c}{(\sqrt{E_t}+\sqrt{E_c})^2} \tag{9}$$

According to the equilibrium condition, we have

$$\frac{d^2 M}{dx^2} = p(x) \tag{10}$$

Adopting the double-parameter foundation assumption, we have

$$p(x) = b\left(-k_w v + k_g \frac{d^2 v}{dx^2}\right) \tag{11}$$

where v is deemed to be positive when it points down. The geometric equations can get

$$\begin{cases} \varepsilon_x = \dfrac{\partial u}{\partial x} = \alpha(A(x)+B(x)\cdot y) \\ \dfrac{dv}{dx} + \dfrac{\partial u}{\partial y} = 0 \end{cases} \Rightarrow \frac{dv}{dx} = -\alpha \int_0^x B(x) dx \tag{12}$$

Substituting Eqs. (11) and (12) into Eq. (10), we have

$$\frac{d^3 M}{dx^3} = \alpha b k_w \int_0^x B(x) dx - \alpha b k_g \frac{dB(x)}{dx} \tag{13}$$

Computing the derivative of both sides of Eq. (13), we have

$$\frac{d^4 M}{dx^4} = \alpha b k_w B(x) - \alpha b k_g \frac{d^2 B(x)}{dx^2} \tag{14}$$

Substituting Eq. (9) into Eq. (14) and simplifying, the differential equation of the undetermined parameter $B(x)$ is easily obtained:

$$B^{(4)}(x)+k_g\lambda^4 B^{(2)}(x)-k_w\lambda^4 B(x)=0 \tag{15}$$

where $\lambda^4=3(\sqrt{E_t}+\sqrt{E_c})^2/(h^3.E_t\cdot E_c)$, which is expressed in m²/N. When $k_g=0$, the double-parameter foundation model can be regarded as the Winkler foundation model. Particularly, when $k_g=0$ and $E_t=E_c=E$, it can be concluded that $\lambda^4=12k_w/h^3$. Solving Eq. (15), we have

$$B(x)=C_1e^{qx}+C_2e^{-qx}+C_3\cos px+C_4\sin px \tag{16}$$

where C_1, C_2, C_3, and C_4 are undetermined constants.

3.2 *Loading features and boundary condition*

1. When $x=0$, Eq. (13) can be rewritten as

$$\frac{d^3M}{dx^3}=-\alpha bk_g\frac{dB(x)}{dx}\Rightarrow\frac{1}{\lambda^4}\frac{d^3B}{dx^3}+k_g\frac{dB(x)}{dx}=0 \tag{17}$$

2. Considering that both sides of the beam are free ends, where $x=l$, we have

$$M(x=l)=0\Rightarrow B(x=l)=G \tag{18}$$

$$\frac{d^2M}{dx^2}=0\Rightarrow\frac{d^2B}{dx^2}=0 \tag{19}$$

$$Q(x=l)=-k_gb^*av(l) \tag{20}$$

where $b^*=b+\left(k_g/k_w\right)^{\frac{1}{2}}, a=\left(k_w/k_g\right)^{\frac{1}{2}}$. The relationship between the generalized shear Q and the deflection v (Selvadurai, 1979) in any section can be described as follows:

$$Q=\frac{\partial M}{\partial x}+k_gb^*\frac{\partial v}{\partial x} \tag{21}$$

When $x=l$, substituting Eq. (21) into Eq. (20), we have

$$\frac{\partial M}{\partial x}=-k_gb^*av-k_gb^*\frac{\partial v}{\partial x}\Rightarrow\frac{b}{\lambda^4}\frac{dB}{dx}-k_gb^*\int_o^l B(x)dx-\frac{k_g^2b^*a}{k_w}B=0 \tag{22}$$

Combining Eqs. (17), (18), (19), and (22), we get

$$\begin{bmatrix} \frac{1}{\lambda^4}q^3+k_gq & e^{ql} & q^2e^{ql} & (\frac{bq}{\lambda^4}-\frac{k_gb^*}{q}-\frac{k_g^2b^*a}{k_w})e^{ql}+\frac{k_gb^*}{q} \\ -(\frac{1}{\lambda^4}q^3+k_gq) & e^{-ql} & q^2e^{-ql} & (-\frac{bq}{\lambda^4}+\frac{k_gb^*}{q}-\frac{k_g^2b^*a}{k_w})e^{-ql}-\frac{k_gb^*}{q} \\ 0 & \cos pl & -p^2\cos pl & -(\frac{bp}{\lambda^4}+\frac{k_gb^*}{p})\sin pl-\frac{k_g^2b^*a}{k_w}\cos pl \\ -(\frac{1}{\lambda^4}p^3-k_gp) & \sin pl & -p^2\sin pl & (\frac{bp}{\lambda^4}+\frac{k_gb^*}{p})\cos pl-\frac{k_g^2b^*a}{k_w}\sin pl-\frac{k_gb^*}{p} \end{bmatrix}^T \begin{bmatrix} C_1 \\ C_2 \\ C_3 \\ C_3 \end{bmatrix} = \begin{bmatrix} 0 \\ G \\ 0 \\ 0 \end{bmatrix} \tag{23}$$

We solve Eq. (27) to get the undetermined constants C_1, C_2, C_3, and C_4. Then, we substitute C_1, C_2, C_3, and C_4 in sequence into Eq. (16) and Eq. (8), to deduce $A(x)$ and $B(x)$.

3.3 *Derivation of normal stress, bending moment, and displacement*

Substituting $A(x)$ and $B(x)$ into Eqs. (5) and (9), we get

$$\sigma_x = \begin{cases} \alpha E_c(y - h \cdot \frac{\sqrt{E_t}}{\sqrt{E_t}+\sqrt{E_c}}) \cdot (C_1 e^{qx} + C_2 e^{-qx} + C_3 \cos px + C_4 \sin px - G) & 0 \le y \le y_0 \\ \alpha E_t(y - h \cdot \frac{\sqrt{E_t}}{\sqrt{E_t}+\sqrt{E_c}}) \cdot (C_1 e^{qx} + C_2 e^{-qx} + C_3 \cos px + C_4 \sin px - G) & y_0 \le y \le h \end{cases} \quad (24)$$

$$M = \frac{\alpha b h^3}{3} \frac{E_t E_c}{(\sqrt{E_t}+\sqrt{E_c})^2} (C_1 e^{qx} + C_2 e^{-qx} + C_3 \cos px + C_4 \sin px - G) \quad (25)$$

By substituting Eq. (25) into Eqs. (10) and (11) and then taking the derivative of the expression in Eq. (12) and substituting the deformed Eq. (12) into a new equation, the displacement is obtained:

$$v = -\frac{\alpha}{k_w}\left[\left(\frac{q^2}{\lambda^4}+k\right)\cdot C_1 \cdot e^{qx} + \left(\frac{q^2}{\lambda^4}+k_g\right)\cdot C_2 \cdot e^{-qx} - \left(\frac{p^2}{\lambda^4}-k_g\right)\cdot C_3 \cdot \cos px - \left(\frac{p^2}{\lambda^4}-k_g\right)\cdot C_4 \cdot \sin px\right] \quad (26)$$

4 EXAMPLES AND DISCUSSION

4.1 *Calculation model and working condition*

The detailed dimensions of the beam are $2l \times b \times h = 6\ \text{m} \times 0.3\ \text{m} \times 0.6\ \text{m}$, $\alpha = 8 \times 10^{-6}/°\text{C}$, $u = 0.18$, $T_0 = 60°\text{C}$, $\Delta T = 30°\text{C}$ (Kang et al. 2007), $k_w = 2 \times 10^7\ \text{N/m}^3$, and $k_g = 2.4 \times 10^6\ \text{N/m}$ (Wang et al. 2006).

A different elastic modulus is assumed for two cases. Case I: $E_c = 2.4 \times 10^7\ \text{kN/m}^2$, $m_c = 1\text{–}5$; case II: $\bar{E} = (E_c+E_t)/2 = 2.4 \times 10^7\ \text{kN/m}^2$, m_t (m_c) = 1/5–5, where $m_t = E_t/E_c$, $m_c = E_c/E_t$.

As shown in Table 1, the bimodulus analytical solution derived in this study can completely recover the classical same modulus solution when $E_t = E_c$. Simultaneously, the computation error between the analytical solution and the simulative solution is less than 5%. Therefore, the computing methods adopted in this study are valid and rational.

4.2 *Difference between the bimodulus theory and the same-modulus theory*

1. Nomal stress distribution along the height

As shown in Figure 2, after considering the bi-modulus property of the foundation beam, the maximum tensile and compressive stress continues to occur at the bottom and top of the beam. In the bimodulus theory, the normal stress distribution behaves unsymmetrically.

Table 1. Results for the neutral axis position and maximum stress of the bimodulus beam under the temperature effect.

	Analytical solution (case I)			Numerical solution by FEM software (case I)		
E_c/E_t	1	3	5	1	3	5
E_c ($\times 10^7$ kN/m^2)	2.4	2.4	2.4	2.4	2.4	2.4
E_t ($\times 10^7$ kN/m^2)	2.4	0.8	0.48	2.4	0.8	0.48
x (m)	0	0	0	0	0	0
y (m)	0.3	0.38038	0.41459	0.3	0.38038	0.41459
σ^c_{max} (kN/m^2)	−1382.220	−1620.954	−1705.292	−1402.000	−1666.700	−1771.600
σ^t_{max} (kN/m^2)	1382.220	935.858	762.630	1402.000	962.250	792.290

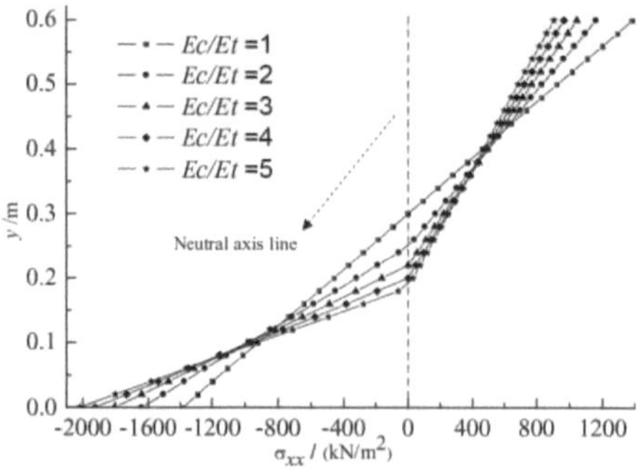

Figure 2. Normal stress of case III.

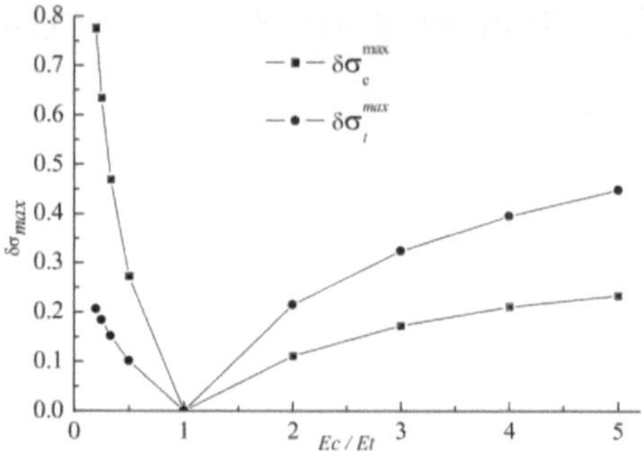

Figure 3. Relationships among E_c/E_t, σ_t^{max}, and σ_c^{max}.

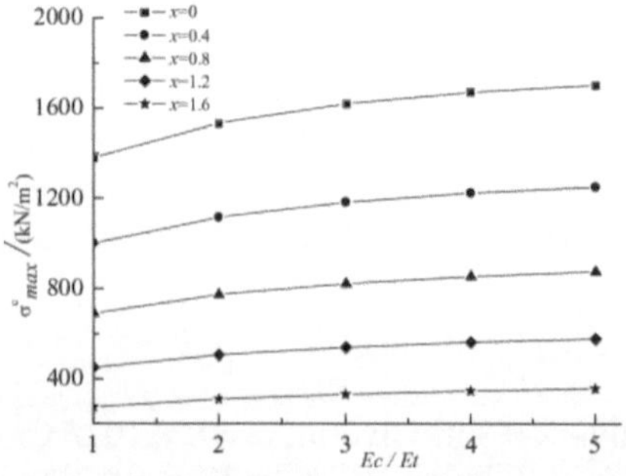

Figure 4. Variations of σ_c^{max} with E_c/E_t of case I.

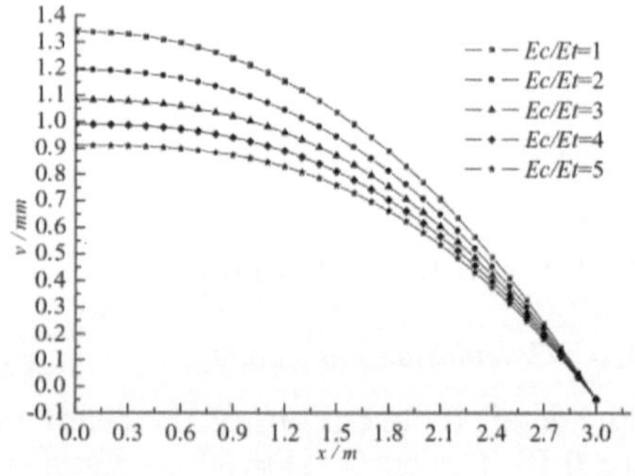

Figure 5. Response diagram for vertical displacements of case I.

The stress values are not the same at the corresponding points in the tensile and compressive regions.

As shown in Figure 3, σ_t^{max} is inversely proportional to E_c/E_t, which indicates that increasing E_c will lower the tensile stress. Analogously, increasing E_t will deduce the compressive stress.

As shown in Figure 4, σ_t^{max} calculated by the bimodulus theory is 26% larger than that calculated by the classical same-modulus theory. Therefore, excessive deformation may appear, and more and more oversize cracks may generate correspondingly.

2. Displacement of the beam

As shown in Figure 5, considering the constant property of total section stiffness, it can be concluded that the inhomogeneity of section stiffness definitely magnifies the displacement after adopting the bimodulus theory compared to the same-modulus theory, which will lead to weakening of the resistance ability of the foundation beam.

5 CONCLUSIONS

In this study, on the basis of the bimodulus theory and the double-parameter foundation assumption, formulas to calculate normal stress, displacement, and bending moment of the bimodulus beam were derived. The conclusions can be summarized as follows:

1. Considering the bimodulus property of the foundation beam, the maximum tensile and compressive stress continues to occur at the bottom and top of the beam. However, the

normal stress distribution behaves unsymmetrically, and the stress values are not the same at the corresponding points in the tensile and compressive regions.

2. For the concrete material widely used in road and bridge engineering, it is worth mentioning that relative to the classical same-modulus theory, the maximum inaccuracy of tension stress increased by 26%. However, cracks and damage on the pavement are common in practical engineering, which may be due to the neglecting of the bimodulus characteristic of the foundation beam. Therefore, it is important to adopt the bimodulus theory to develop the structure design to correct the errors by simply considering the classical same-modulus theory.
3. Enlarging the compressive modulus is an effective way to decrease the tensile stress in the foundation beam. Thus, the rational selection of materials with some difference between the tensile modulus and the compressive modulus in practical engineering will be helpful in the stress state of structures.
4. With regard to the classical same-modulus theory, the structure stiffness discreteness increases and the resistance to the force and deformation decreases accordingly after considering the bimodulus characteristic of the foundation beam.

REFERENCES

Barak M.M., Currey J.D., Weiner S., Shahar R. 2009. Are tensile and compressive Young's moduli of compact bone different. *Journal of the Mechanical Behavior of Biomedical Materials*, 2(1):51–60.

Gao J., Yao W.J., Liu J. 2017. Temperature stress analysis for bi-modulus beam placed on Winkler foundation. *Applied Mathematics & Mechanics*, 1–14.

Georgia T., John P., Konstantinos P., et al. 2009. Subjecting a graphene monolayer to tension and compression. *Small*, 5(21):2397–2402.

He X.T., Xu P., Sun J.Y., et al. 2015. Analytical Solutions for Bending Curved Beams with Different Moduli in Tension and Compression. *Mechanics of Advanced Materials & Structures*, 22(5):325–337.

Kang H., Zheng Y., Cai Y., 2007. Regression analysis of actual measurement of temperature field distribution rules of asphalt pavement. *China Journal of Highway and Transport,* 20(6):13–18.

Selvadurai A.P.S., Gladwell G.M.L. 1979. Elastic Analysis of Soil-Foundation Interaction [M]. Elsevier Scientific Pub. Co.

Wang G., Shi J., 2006. Calculation of the beam on elastic foundation under the parabolic load, *Journal of hefei university of technology (natural science)*, 29(7):905–907.

Yao W.J., Liu J., 2014. Temperature stress for bending beam with different moduli under different constraints. *Multidiscipline Modeling in Materials and Structures*, 10(2):163–175.

Yao W.J., Ye Z.M., 2005. Analytical method and the numerical model of temperature stress for the structure with different modulus. *Journal of mechanical strength*, 27(6):808–814.

Yao W.J., Ye Z.M., 2006. Internal forces for statically indeterminate structures having different moduli, *Journal of Engineering Mechanics*, 132(7): 739–746.

Hydraulic Engineering V – Hong, Liu & Xie (Eds)
© 2018 Taylor & Francis Group, London, ISBN 978-0-8153-7471-8

Research on the law for developing cutting resistance of rake tooth during the dredging of the saturated dense silt soil

Gongxun Liu, Guojun Hong, Qingbo Zhang & Shu Wang
Key Laboratory of Dredging Technology of CCCC, CCCC National Engineering Research Center of Dredging Technology and Equipment Co. Ltd., Shanghai, China

Hongjian Xu
Qingdao First Municipal Engineering Co. Ltd., Qingdao, China

Peipei Zhang
Department of Hydraulic Engineering, Tongji University, Shanghai, China

ABSTRACT: Through a series of cutting tests conducted on saturated dense silt soil in the Huanghua port area with prototype rake tooth, the effects of cutting parameters (including cutting angle, moving velocity of rake tooth, cutting thickness, and the type of rake tooth) on cutting resistance have been fully studied. The relationships between the cutting parameters and cutting resistance of rake tooth were established, respectively. The test results show that during dredging, cutting resistance increases first, then decreases gradually after reaching the maximum value, and finally achieves a certain stable value. With the increase in cutting angle, the maximum cutting resistance also increases, which has a linear relationship with the sine of the cutting angle. With the increase in cutting speed, the maximum cutting resistance increases linearly. With the increase in cutting thickness, the growth rate of the maximum cutting resistance becomes increasingly larger, and the maximum cutting resistance and the cutting thickness shows a quadric relationship. In the soil cutting process, the cutting resistance of three types of rake tooth, chisel rake tooth, plough rake tooth, and semi-plough rake tooth, increases first, then decreases, and finally becomes stable. However, the difference between them is that the cutting resistance of the semi-plough rake tooth is much smaller than that of the other two types at the same cutting angle. The fitting curves of the cutting angle and the maximum cutting resistance show different relationships. The results of this study can be used as an important reference for selecting the type of rake tooth and for confirming construction parameters during the dredging of the saturated dense silt soil.

1 INTRODUCTION

Encountering dense silt, "iron sand", and other similar soils is very common in the dredging process. Such type of soil poses difficulties in cutting and raking, mainly due to their remarkable characteristic of shearing dilatancy. In the cutting process, the dilatancy zone of certain range occurs within the soil, and the pore water pressure in this region is negative, which leads to the increase of soil effective stress and a significant improvement of cutting resistance (Miedema 2010, Zhao & Miedema 2001).

Studies on the cutting mechanism and cutting resistance of rake tooth for dredging soil conducted previously by non-Chinese scholars have shown substantial positive results (Joanknecht 1974, Meijer 1985, Vandycke et al. 2005, Vercruijsse et al. 2005), especially professor Miedema (1984, 1985, 2004) of Delft University of Technology, whose research involved physical model test, mathematical model and numerical analysis method, and results analysis of finite element calculation. In China, the research on soil cutting mechanism and cutting resistance mainly focuses on agricultural equipment (Xie 1983, Zhai 2007, Zhou 2009, Cheng 2011). However, the number of studies conducted on the soil cutting mechanism and cutting resistance of dredging machines and tools is relatively less (Hong 2008, Liu 2015). In order to accelerate the Chinese transformation of dredging in scale to dredging power, it is necessary to make significant efforts to carry out research on the soil cutting mechanism and cutting resistance of dredging machines and tools, especially for special soils such as "iron sand" and hard clay.

In this study, through a series of prototype rake tooth cutting tests on saturated dense silt soil from the Huanghua port, the effects of cutting parameters, such as cutting angle, cutting thickness, cutting speed, and tooth shape, on cutting resistance of rake tooth are systematically investigated, as well as the relationship between the cutting parameters and cutting resistance is explored, which can provide a reference in selecting the type of rake tooth and ensuring construction parameters during dredging saturated dense silt soil.

2 TEST CONDITIONS AND TEST CASES

2.1 *Test apparatus and equipment*

The rake traction system supplies power to the rake tooth cutting test, as shown in Figure 1. The speed variation of the dredging test trolley is in the range of 0–2 m/s, and the maximum cutting resistance is 20 kN. The starting and braking of the test rig are controlled by the control system (Figure 2) installed on the vehicle-mounted console.

As shown in Figure 3, the rake tooth attitude control system consists of a test stand, a hydraulic double cylinder, a hydraulic single cylinder, a and rake tooth.

A laser transmitter and receiver is mounted onto the edge of the port dredging sled, and a laser-reflecting plate is fitted on the distal fixation place of dredge test flume in order to measure tooth cutting speed, as shown in Figure 4.

In order to monitor the change in cutting resistance, two force sensors are installed at the bottom of the rake tooth attitude control device. The measurement accuracy of the force sensor is ± 1 N, and the measurement range is 10–20,000 N, as shown in Figure 5.

Figure 1. Traction system for the rake tooth.

Figure 2. Control system for the trolley.

Figure 3. Attitude control system for the rake tooth.

Figure 4. Arrangement of the laser velocimetry sensor.

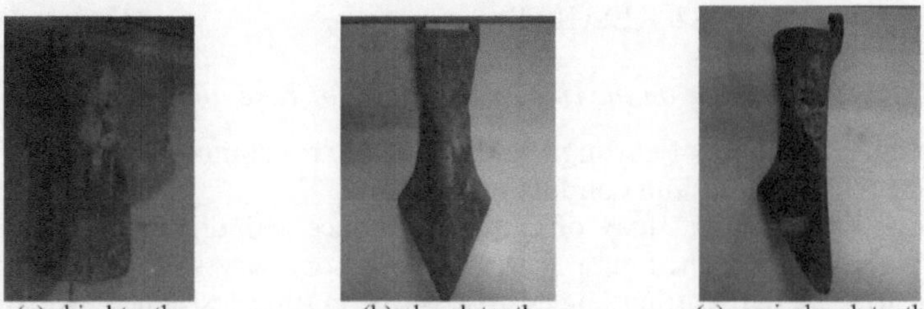

(a) chisel tooth (b) plough tooth (c) semi-plough tooth

Figure 5. Types of tooth.

Table 1. Physical properties of the silty soil for testing.

Median particle diameter, d_{50} (mm)	Relative density, G_s	Plastic limit, w_P (%)	Liquid limit, w_L (%)	Plasticity index, I_P	Permeability coefficient, k (cm/s)
0.038	2.70	16.3	21.6	5.3	1.5×10^{-5}

Table 2. Test cases.

Case	Cutting angle (°)	Cutting speed (m/s)	Cutting thickness (cm)	Remark (analysis factor)
1	60	0.17	5.5	Chisel tooth
2	60	0.29	5.5	Chisel tooth
3	60	0.47	5.5	Chisel tooth
4	60	0.54	5.5	Chisel tooth
5	60	0.30	4.5	Chisel tooth
6	60	0.29	6.5	Chisel tooth
7	60	0.31	7.5	Chisel tooth
8	45	0.31	5.5	Chisel tooth
9	75	0.31	5.5	Chisel tooth
10	90	0.30	5.5	Chisel tooth
11	45	0.17	5.5	Plough rake tooth
12	60	0.18	5.5	Plough rake tooth
13	75	0.17	5.5	Plough rake tooth
14	45	0.17	5.5	Semi-plough rake tooth
15	60	0.17	5.5	Semi-plough rake tooth
16	75	0.18	5.5	Semi-plough rake tooth

2.2 *Test soil*

Silty soil samples are taken from the Huanghua port of Hebei Province, China, whose basic physical properties are listed in Table 1.

2.3 *Test case*

In this study, 16 test cases (see Table 2) were studied. Two sets of parallel tests were conducted under all conditions to study the effects of tooth cutting angle, soil cutting depth, cutting speed, and tooth style on the cutting resistance of rake tooth.

2.4 *Tooth type*

In this study, three types of tooth were used to carry out a series of cutting tests, including the chisel tooth, plough tooth, and semi-plough tooth (Figure 5).

3 ANALYSIS OF THE TEST RESULTS

3.1 *Effect of the cutting angle on the cutting resistance of rake tooth*

In order to examine the effect of cutting angle on cutting resistance, cases 2, 8, 9, and 10 were selected to make comparisons and conduct analysis.

Figure 6 shows the variation curves of cutting resistance with displacement at different cutting angles. It is evident from the figure that the time-history curves of cutting resistance show the same trend for different cutting angles, that is, the cutting resistance increases first, then decreases gradually after reaching the maximum value, and finally achieves a certain stable value.

Figure 6 also shows that the maximum cutting resistance increases with increasing cutting angle. The maximum cutting resistance and the sine of cutting angle show a linear relationship. The fitting curve is shown in Figure 7, whose slope is calculated as:

$$y = -5.007 + 9.719x \tag{1}$$

3.2 *Effect of the cutting speed on the cutting resistance of rake tooth*

In order to examine the effect of cutting speed on cutting resistance, cases 1, 2, 3, and 4 were selected to make comparisons and conduct analysis.

Figure 8 shows the variation of cutting resistance with different cutting speeds. It is evident from the figure that the time-history curves of cutting resistance show the same trend for different cutting speeds, that is, as the rake tooth moves forward, the cutting resistance increases first, then decreases gradually after reaching the maximum value, and finally achieves a certain stable value.

Figure 8 also shows that the maximum cutting resistance increases with increasing cutting speed. The maximum cutting resistance and cutting speed show a linear relationship. The fitting curve is shown in Figure 9, whose slope is calculated as:

$$y = 11.94x - 0.27 \tag{2}$$

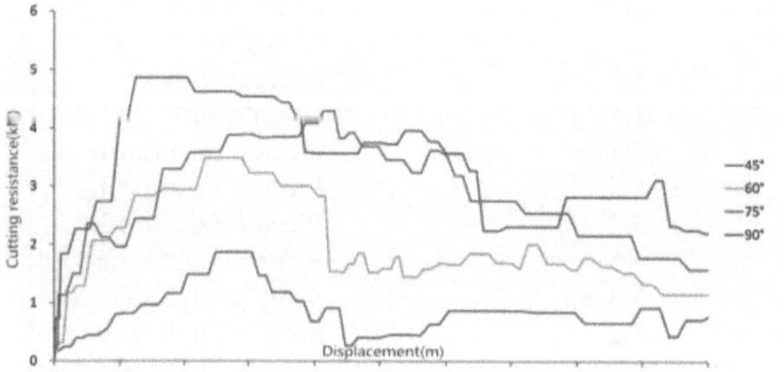

Figure 6. Time-history curves of cutting resistance with different cutting angles.

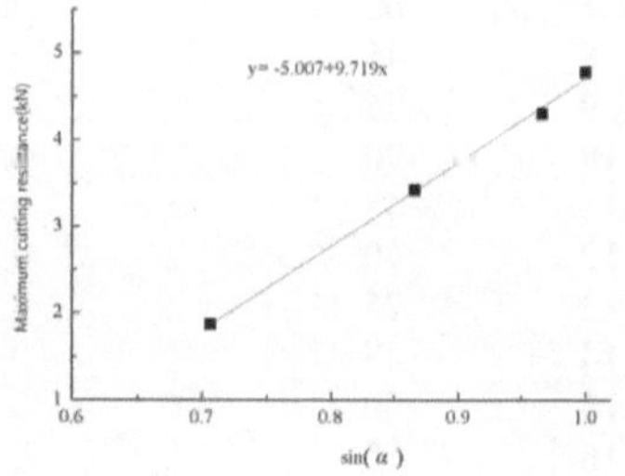

Figure 7. Relationship between maximum cutting resistance and cutting angle.

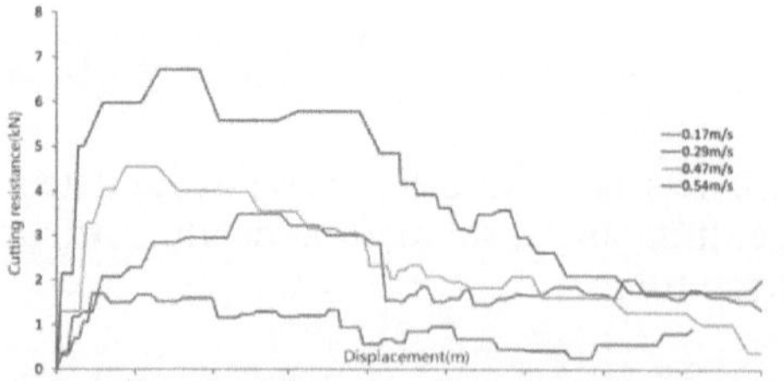

Figure 8. Time-history curves of cutting resistance with different cutting speeds.

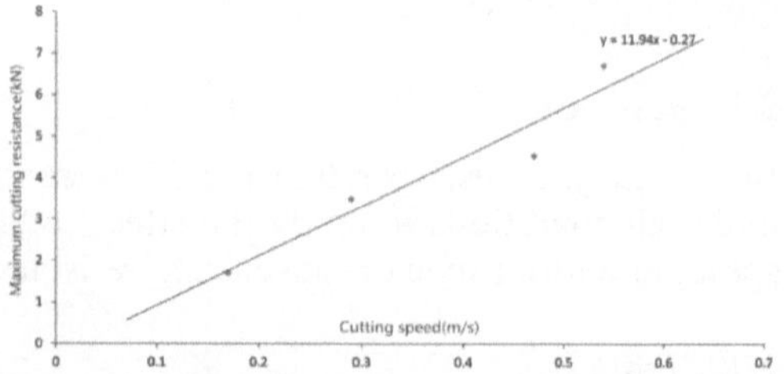

Figure 9. Relationship between maximum cutting resistance and cutting speed.

3.3 *Effect of the cutting thickness on the cutting resistance of rake tooth*

In order to examine the effect of cutting thickness on cutting resistance, cases 2, 5, 6, and 7 were selected to make comparisons and conduct analysis.

Figure 10 shows the variation of cutting resistance with displacement at different cutting thicknesses. It is evident from the figure that the time-history curves of cutting resistance show the same trend for different cutting thicknesses, that is, as the rake tooth moves forward, the cutting resistance increases first, then decreases, and finally becomes stable.

Figure 10 also shows that the maximum cutting resistance increases with increasing cutting depth. However, with the increase of cutting depth, the growth rate of the maximum cutting resistance becomes increasingly larger. The relationship between the maximum cutting resistance and cutting thickness is quadratic. The fitting curve is shown in Figure 11, whose slope is calculated as:

$$y = 0.2825x^2 - 2.011x + 5.8954 \tag{3}$$

3.4 *Effect of the type of rake tooth on the cutting resistance of rake tooth*

Here, three types of rake tooth are analyzed, including chisel rake tooth, plough rake tooth, and semi-plough rake tooth, by monitoring the variation of their cutting resistance during cutting.

The time-history curves of cutting resistance corresponding to different cutting angles of plough and semi-plough rake tooth are shown in Figures 12 and 13, respectively. Comparing Figures 6 and 12 with Figure13, it can be found that the three types of tooth have the following features in common: the cutting resistance increases first, then decreases, and eventually tends to be stable. The differences between the three tooth types are: (a) Under the condition of same angle, the maximum cutting resistance of the semi-plough rake tooth is much smaller than that of the other two types. For example, when the cutting angle is 45°, the maximum cutting resistance of the semi-plough rake tooth is 0.23 kN, whereas the maximum cutting resistances of chisel rake tooth and plough rake tooth are 1.87 and 2.71 kN, respectively.

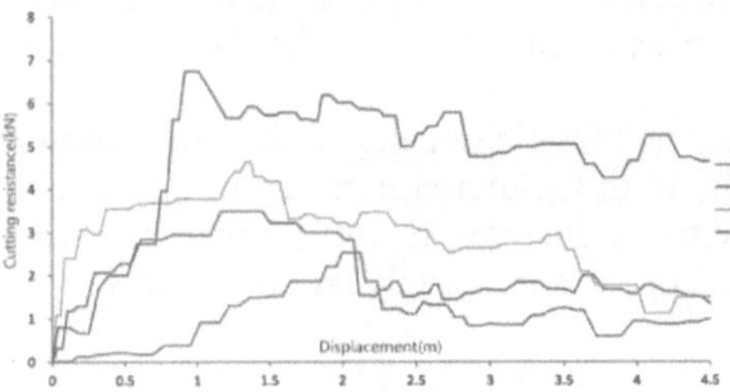

Figure 10. Time-history curves of cutting resistance with different cutting thicknesses.

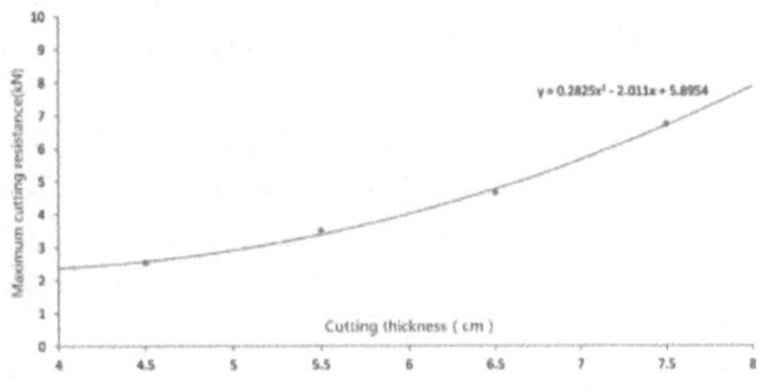

Figure 11. Relationship between maximum cutting resistance and cutting thickness.

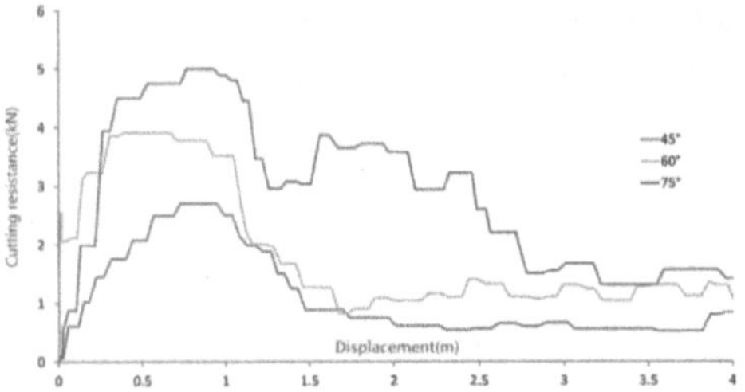

Figure 12. Time-history curves of cutting resistance with different cutting angles for the plough rake tooth.

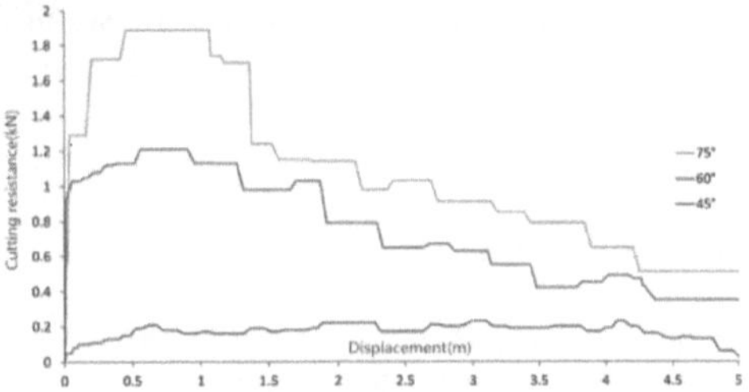

Figure 13. Time-history curves of cutting resistance with different cutting angles for the semi-plough rake tooth.

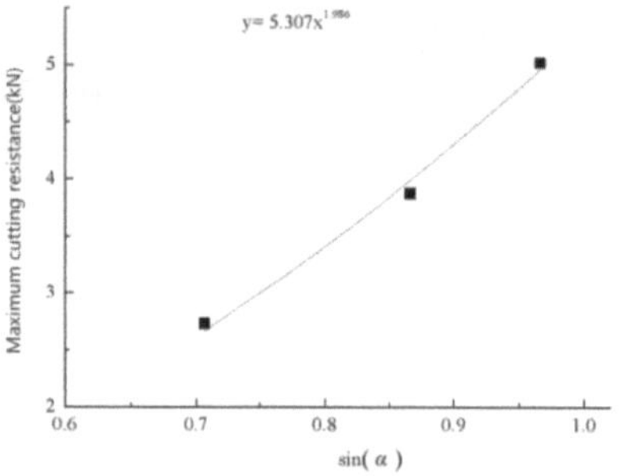

Figure 14. Relationship between maximum cutting resistance and cutting angle for the plough rake tooth.

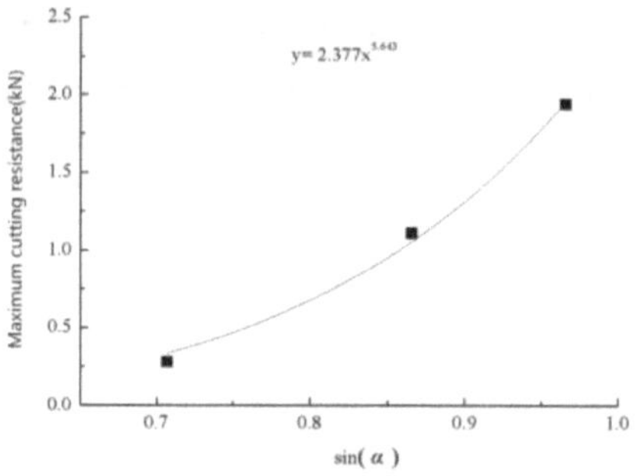

Figure 15. Relationship between maximum cutting resistance and cutting angle for the semi-plough rake tooth.

(b) For the chisel rake tooth, the fitting curve of cutting angle and the maximum cutting resistance is shown in Figure 7, and their relationship conforms to eq. (1). For the plough rake tooth, the fitting curve of cutting angle and the maximum cutting resistance is shown in Figure 14, and their relationship conforms to eq. (4). For the semi-plough rake tooth, the fitting curve of cutting angle and the maximum cutting resistance is shown in Figure 15, and their relationship conforms to eq. (5).

$$y = 5.307x^{1.986} \tag{4}$$

$$y = 2.377x^{5.643} \tag{5}$$

5 CONCLUSION

1. During the cutting process, when rake tooth moves forward, the cutting resistance increases first, then decreases gradually after reaching the maximum value, and finally achieves a certain stable value.
2. The maximum cutting resistance increases with the increase in cutting angle, which shows a linear relationship with the sine of cutting angle. With the increase in cutting speed, the maximum cutting resistance increases linearly. With the increase in cutting thickness, the growth rate of the maximum cutting resistance becomes increasingly larger, which shows a quadratic relationship.
3. The cutting resistance of all the three types of rake tooth increases first, then decreases gradually, and finally achieves a relatively stable value. The only difference between them is that the cutting resistance of the semi-plough rake tooth is much smaller than that of the other two types at the same cutting angle. The fitting curves of the cutting angle and the maximum cutting resistance show different relationships.
4. The results of this study serve as a significant reference for selecting the type of rake tooth and for ensuring the parameters of construction during the dredging of the saturated dense silt soil.

REFERENCES

Cheng, Z., Ding, Q., Ding, W., et al. 2011. Fragmentation-based soil cutting test and its simulation. Experimental Technology and Management. 28(2): 98–100, 181.

Hong, G., Lin, F., and Wang, J. 2008. Experimental study on cutting of saturated hard compacted soil with draghead teeth. Port & Waterway Engineering. (8): 98–104.

Joanknecht, L.W.F. 1974. Cutting forces in submerged soils. The Netherlands: H.Delft.
Liu, X., He, Y., and Zhao, Y. 2015. Simulation of soil cutting using rake based on FEM. Port & Waterway Engineering. (1): 32–36, 41.
Meijer, K.L. 1985. Computation of stresses and strains in saturated soil. Proefschrift T.H. Delft.
Miedema, S.A. 1984. The cutting of densely compacted sand under water. Terra et Aqua. (10): 4–10.
Miedema, S.A. 1985. Mathematical modelling of the cutting of densely compacted sand under water. Dredging & Port Construction. (7): 22–26.
Miedema, S.A. 2004. The cutting mechanisms of water saturated sand at small and large cutting angles. International Conference on Coastal Infrastructure Development-Challenges in the 21st Century. HongKong.
Miedema, S.A. 2010. New developments of cutting theories with respect to offshore applications, the cutting of sand, clay and rock. 20th International Offshore and Polar Engineering Conference. Beijing, China.
Vandycke, S., Van den Broeck, M., and Thomas, K. 2005. The DRAC-ULA—system pushes the limits of hopper dredgers in cemented sand. CEDA Dredging Days. Rotterdam, the Netherlands.
Vercruijsse, P.M., Versteeg, G.J.P., Ooijens, S.C., et al. 2005. Wild Dragon—Developing a draghead for dredging extreme fine hard packed aquatic soils. CEDA Dredging Days. Rotterdam, the Netherlands.
Xie, X. 1983. Nonlinear finite element analysis of two-dimensional cutting problems in agricultural soil. Transactions of the Chinese Society for Agricultural Machinery. (1): 73–82.
Zhai, L. 2007. Study on the effect of plough's working and structure parameters on its resistance under theological soil conditions. Nanjing Agricultural University.
Zhao, Y., and Miedema, S.A. 2001. Finite element calculations to determine the pore pressures when cutting water saturated sand at large cutting angles. CEDA Dredging Days. Amsterdam, The Netherlands.
Zhou, M., Zhang, G., Xu, Q., et al. 2009. Study on Viscous Soil Retangular Cutting FEM Simulation. Journal of Huazhong Agricultural University. 28(4): 491–494.

Hydraulic Engineering V – Hong, Liu & Xie (Eds)
© 2018 Taylor & Francis Group, London, ISBN 978-0-8153-7471-8

A computational model for devising an intelligent densitometer for cutter suction dredgers

C. Wei, F. Ni, L. Xu & X. Chen
Engineering Research Center of Dredging Technology, Ministry of Education, Hohai University, China

ABSTRACT: Slurry density is an important parameter that indicates the performance of dredging operators. Currently, nuclear-based gamma densitometer is the most widely used instrument for measuring slurry density in cutter suction dredgers. The use of nuclear sources may cause problems of operation, safety, and cost. Therefore, this study aims at proposing a computational model for calculating the slurry density in a non-nuclear manner. To this end, real dredging data were collected to train a supervised back-propagation neural network, which can be considered as the calibration stage of a densitometer. In order to improve the computational accuracy of the neural network, we present a revised Gaussian filter to process the original input data for the neural network. From the measurement results, it can be concluded that the proposed model is effective and feasible.

1 INTRODUCTION

Cutter Suction Dredgers (CSDs) are widely used in dredging projects to keep the prescribed scope and depth of ports and waterways, as shown in Figure 1. In such a dredging project, the soil should be removed by a CSD from its original place and then transported to the disposal place through a slurry pipeline (Tang et al. 2008). The economic benefits of constructing a dredging project are mainly based on the performance (i.e., monthly production) of the dredger in the project. In order to improve the performance, the operator is required to carefully manipulate the dredger so as to maintain good production in real time (Tang et al. 2009). Thus, controlling and monitoring the density of slurry flowing through the pipeline are the critical tasks of a CSD operator.

Maintaining good production of a CSD requires the level of solids in the pipeline to be neither too low nor too high, as it can potentially increase the risk of solid setting, thereby leading

Figure 1. Photo of a cutter suction dredger.

to pipeline blockage. Therefore, obtaining the information of slurry density in pipeline is very crucial for the operator to manipulate the dredger (Todd et al. 2015). The typical instrument to measure the slurry density is the nuclear-based gamma-ray densitometer, which may cause potential problems of operation, safety, and cost. Thus, the application of a non-nuclear densitometer has always been of great interest for both dredging industries and operators.

To this end, in this study, we aim to produce a computational model that can be used to calculate the slurry density so as to replace the functionality of the nuclear-based densitometer. The paper is organized as follows. We first introduce the working process of a CSD and the related parameters for building the computational model. The proposed computational model is presented in Section 3. In Section 4, the simulated results of the proposed model are compared with the measurement data of the nuclear-based densitometer. Finally, we conclude this paper in Section 5.

2 DREDGING PROCESS ANALYSIS

2.1 *Dredging process*

The main function of a cutter suction dredger (CSD) is to transfer soil from its original place to disposal place. The key components of a cutter suction dredger include a cutter head, a cutter ladder, a suction pipe, a discharge pipe, several dredger pumps, a working spud pole, and an auxiliary spud pole, as shown in Figures 2 and 3. The cutter head (combined with its drive) and the suction pipe are mounted on the ladder that is suspended by the pontoon of the dredger. The cutter head excavates the soil that is sucked by the dredger pumps through the suction pipe.

During the dredging process, the cutter suction dredger does not use any propulsion to drive it. Instead, it is a stationary dredger, but it moves around the working spud pole. Such a movement is called swing process, which is achieved by pulling and slacking on the four sideline wires of two side anchors, as shown in Figure 3. Thus, the dredging trajectory is an arc around a fixed point (i.e., the working spud pole). After a swing movement, the cutter suction dredger can move forward by means of a hydraulic cylinder.

2.2 *Related parameters*

In order to build a computational model in the next section, we first analyze the parameters that are related to calculating the slurry density. Those parameters are obtained from the data collected from the central control system of a CSD during the dredging process. The original raw data sheets have many variables, and we selected the following independent ones that directly affect the slurry density in pipeline transportation.

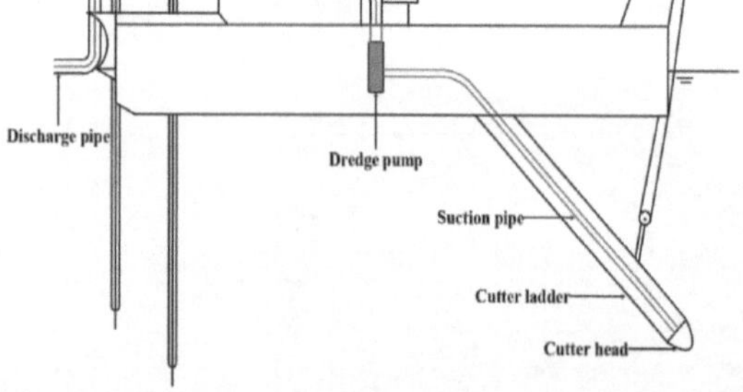

Figure 2. General layout of a CSD.

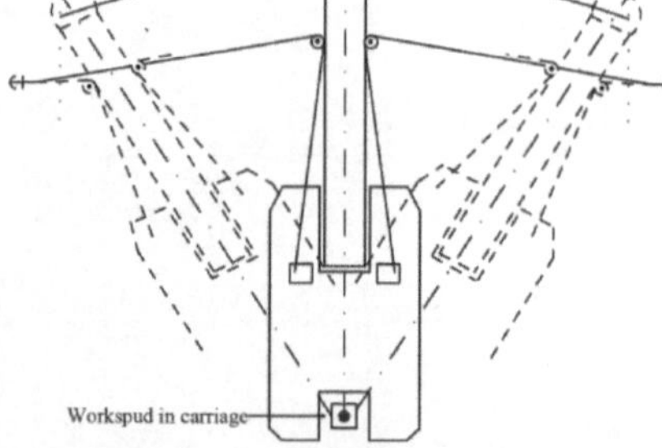

Figure 3. Swing process of a CSD.

- **Swing speed** can reflect the extent to which the dredger is cutting soil. In a swing pattern, the dredger moves around the working spud pole by slacking the cable of the starboard anchor and pulling the cable of the port side anchor, or vice versa. These anchor cables are connected via sheaves close to the cutter head to winches on deck.
- **Motor current of the cutter head** can indicate the cutting forces as well as the type of soil. For the same swing movement, the rotation of the cutter head in the same direction or in the opposite direction results in different reaction forces. In addition, the type of soil to be dredged has a string influence on the cutter head power. For example, the reaction forces of cutting sand and hard rock must be different.
- **Motor current of the underwater pump** can indicate pipeline concentration between the cutter head and the underwater pump. The motor current of the underwater pump will increase with the increase of the slurry density in the pipeline. This is because high density means that the slurry requires more power to be sucked.
- **Suction vacuum** can also indicate the pipeline concentration between the cutter head and the underwater pump, but the principle is this parameter is different from the motor of the underwater pump. The basic principle is that when the slurry density in the pipeline is higher, it will be harder for the underwater pump to suck the slurry, which will lead to the increase of the vacuum (Vaezi and Kumar 2015).
- **Flow rate** can also be considered another parameter to indicate the slurry density to some extent. If the rotating speed of the pumps keeps constant, the increase of the flow rate decreases the slurry density, because it becomes easier for the pump to transport the slurry with lower density.

3 COMPUTATIONAL MODEL FOR DENSITY PREDICTION

3.1 *Back-propagation neural network*

In order to develop a non-nuclear densitometer, in this study, we attempt to establish a back-propagation neural network that can calculate the slurry density in real time. The related parameters (i.e., swing speed, motor current of the cutter head, motor current of the underwater pump, suction vacuum, and flow rate) are selected as the inputs of the network, whereas the slurry density is the output. We define a multilayer neural network, in which there is a single layer of hidden nodes, one output node, and five input nodes. The back-propagation method uses the gradient descent for training a multilayer, feed-forward network (Gelenbe 2014). As usual, we use an error function on the final output of the network as well as adjust each weight in the network so as to minimize the error. To simplify the notation, each layer will have its own index variable: k for output nodes, j and h for hidden nodes, and i for input nodes. In a feed-forward network, the input vector is propagated through a weight layer:

$$y_j(t) = f(net_j(t)) \tag{1}$$

$$net_j(t) = \sum_i^n x_i(t)v_{ji} + \theta_j \tag{2}$$

where n is the number of input values, θ_j is the bias, and f is the output function. The back-propagation method can be used to train the above network if desired output patterns exist and each function that has been used to calculate the actual output patterns is differentiable. We use the most commonly used cost function, the summed squared error, in which each pattern p adds to the cost over all the output nodes k:

$$C = \frac{1}{2}\sum_p^n \sum_k^m (d_{pk} - y_{pk})^2 \tag{3}$$

where d is the desired output, n is the total number of available training samples, and m is the total number of output nodes.

According to the gradient descent, each change of weight in the network should be proportional to the negative gradient of the cost with respect to the specific weight (Rodriguez-Galiano et al. 2015); thus, we can get

$$\Delta\omega = -\eta\frac{\partial C}{\partial \omega} \tag{4}$$

where η is the learning rate. The change of weight is achieved by distinguishing between an error component, $\delta = -\partial C/\partial net$, and $\partial net/\partial \omega$. Thus, the error for output nodes will be

$$\delta_{pk} = -\frac{\partial C}{\partial y_{pk}}\frac{\partial y_{pk}}{\partial net_{pk}} = (d_{pk} - y_{pk})g'(y_{pk}) \tag{5}$$

and that for hidden nodes will be

$$\delta_{pj} = -\left(\sum_{k}^{m}\frac{\partial C}{\partial y_{pk}}\frac{\partial y_{pk}}{\partial net_{pk}}\frac{\partial net_{pk}}{y_{pj}}\right)\frac{\partial y_{pj}}{\partial net_{pj}} = \sum_{k}^{m}\delta_{pk}\omega_{kj}f'(y_{pj}) \tag{6}$$

3.2 *Gaussian filter optimization*

In this subsection, we will propose a Gaussian filter to optimize the input values of the network discussed above. The reason why we need to process the original raw data collected from a cutter suction dredger is that there is a time lag between slurry density and its related parameters. As shown in Figure 4, the peak of the slurry density curve has a 7-min time delay.

Therefore, in order to optimize the performance of the neural network, the input values should be processed. The basic idea is that each expected input value is influenced by several past discrete values:

$$E(f(x)) = \sum_{x \in S} f(x)p(x) \tag{7}$$

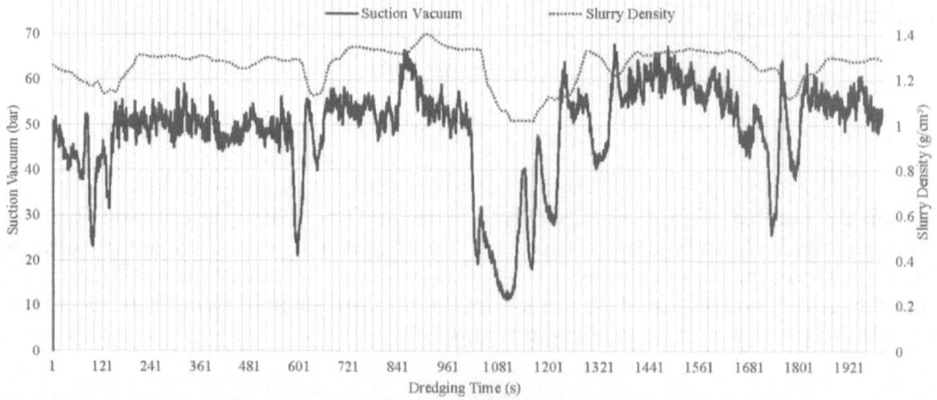

Figure 4. Time lag existing between suction vacuum and slurry density.

where $p(x)$ is the probability factor that indicates the extent to which the past value affects the expected input value. Thus, we can have $\sum_{x \in S} p(x) = 1$, where $p(x)$ is calculated by the probability distribution function

$$p(x) = \frac{1}{\sqrt{2\pi}} \exp\left[-\frac{1}{2} \left(\frac{x - \mu}{\sigma} \right)^2 \right] \tag{8}$$

where μ is the mean and σ^2 is the variable.

The importance of the use of the Gaussian function is underlined by the following fact. The combined effect of the sum of a large number of independent random variables leads to a normal distribution. This means that, for each input of the network, the corresponding past values, including swing speed, motor current of the cutter head, motor current of the underwater pump, suction vacuum, and flow rate, need to be processed again by means of normal distribution. As patterns can be considered as ideal prototypes corrupted by a large number of random processes, the Gaussian is usually a very good model to represent the probability distribution of the patterns. Therefore, we use such a method to develop a Gaussian filter to process the raw captured input values.

4 SIMULATION AND RESULTS

In this study, the dredging data collected from a cutter suction dredger are used to train a neural network and evaluate the proposed approach. To train the neural network, working data of about 4 h are used. In this working period, the operators perform almost all the procedures of dredging. Thus, we can ensure that the training data can delegate the typical situations in dredging projects.

As shown in Figure 5, the measurements of the nuclear densitometer are compared with the calculated values. We demonstrate two calculated values: (1) calculated by the basic back-propagation neural network and (2) optimized by the proposed Gaussian filter.

Figure 5 shows that the back-propagation neural network can be used to calculate the slurry density in a cutter suction dredger, and the trend of the curve is generally consistent with the results of nuclear densitometer. Moreover, the optimized curve performs better than the curve of the basic neural network in two aspects. First, the optimized curve is smoother than the original one, and the trend is almost consistent with the results of nuclear densitometer. Second, the optimized curve eliminates the slippage phenomenon in peaks. Therefore, we can ensure that the proposed computational model is feasible and effective for density prediction.

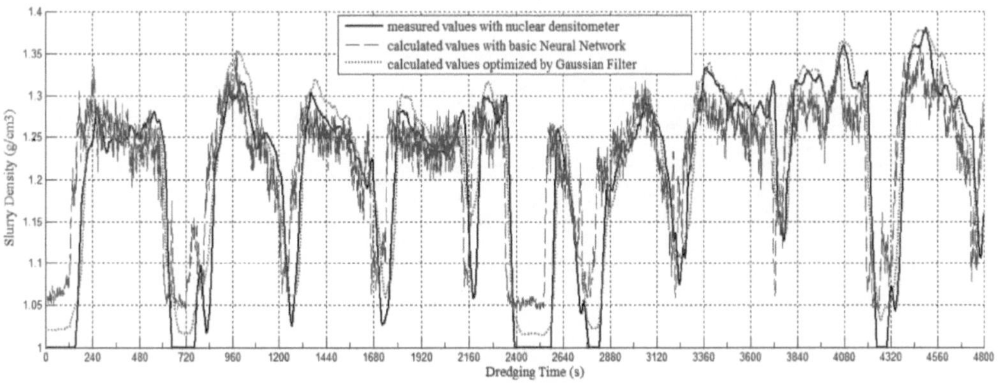

Figure 5. Comparison between the measured and calculated values.

5 CONCLUSIONS

In this study, we proposed a computational model for devising a densitometer to replace the commonly used nuclear-based gamma densitometer. This is because the nuclear densitometer may cause potential problems of operation, safety, and cost. The proposed computational model consists of a back-propagation neural network and a Gaussian filter, which are used to train the sample data and improve the prediction accuracy of the neural network. The simulated results indicate that the proposed approach can be used to calculate the slurry density. Moreover, the predicting curve is in good accordance with the measurements of the nuclear densitometer.

REFERENCES

Gelenbe, E. (2014). Stability of the random neural network model. *Neural Computation*, 2(2), 239–247.

Rodriguez-Galiano, V., Sanchez-Castillo, M., Chica-Olmo, M., & Chica-Rivas, M. (2015). Machine learning predictive models for mineral prospectivity: an evaluation of neural networks, random forest, regression trees and support vector machines. *Ore Geology Reviews*, 71, 804–818.

Tang, J.Z., Wang, Q.F., & Bi, Z.Y. (2008). Expert system for operation optimization and control of cutter suction dredger. *Expert Systems with Applications*, 34(3), 2180–2192.

Tang, J., Wang, Q., & Zhong, T. (2009). Automatic monitoring and control of cutter suction dredger. *Automation in Construction*, 18(2), 194–203.

Todd, Victoria L.G., et al. "A review of impacts of marine dredging activities on marine mammals." Ices *Journal of Marine Science*. 72.2(2015).

Vaezi, M., & Kumar, A. (2015). Pipeline hydraulic transport of biomass materials: A review of experimental programs, empirical correlations, and economic assessments. *Biomass and Bioenergy*, 81, 70–82.

Hydraulic Engineering V – Hong, Liu & Xie (Eds)
ISBN 978-0-8153-7471-8

Risk analysis of rainfall and tidal levels using maximum entropy: The secondary return period method

J.W. Fan, J.L. Huang & Z.Z. Tang
Guangdong Research Institute of Water Resources and Hydropower, Guaugzhou, Guangdong, China
State and Local Joint Engineering Laboratory of Estuary Hydropower Technology, Guangzhou, Guangdong, China
Guangdong Engineering and Technological Research Center of Flood Disaster Prevention, Guaugzhou, Guangdong, China

ABSTRACT: Rainfall and tides are two important factors of urban waterlogging. A risk analysis of rainfall and tidal levels is important to reduce waterlogging disasters and protect people's lives and property. It has been difficult to define and calculate return periods and design values. In this study, by constructing the joint distribution of 1-hour maximum annual rainfall and the corresponding tidal levels, we defined the secondary return period and proposed calculation methods, to point out the differences in the identification of dangerous regions between the secondary return period and traditional return periods. The results of data analysis show that the definition of the secondary return period is more precise. The design value based on the secondary return period is reasonable. It can provide useful information about the risk analysis of waterlogging.

Keywords: Copula function; Secondary return period; risk analysis; rainfall and tide

1 INTRODUCTION

Calculating return period and design value is important in risk analysis. Under univariate conditions, the return period and the corresponding design value are usually unique. However, it is more complicated to determine the return period and design values of variables in a multivariate environment. In recent years, hydrological risk analysis under multivariate conditions has become a central research topic. The definitions of the multivariate return period are commonly used as the "OR" return period and the "AND" return period. However, both of them have computational errors, leading to too large or too small estimated return period and design value.

On the basis of the definition of traditional return periods, Salvadori proposed a new method to define the multivariate return period, secondary return period, which has been applied in flood risk analysis and coastal engineering design. However, the risk analysis of rainfall and tidal levels based on the secondary return period is rarely conduced. Therefore, in order to solve this critical problem, the author applied the secondary return period to establish the rainfall–tide distribution. The maximum entropy principle was used to estimate the parameters of water frequency distribution. We also compared the calculated results with traditional return periods. The purpose of this study was to introduce a new method to define the return period. We can also ensure that the secondary return period is more accurate in estimating the return period and the corresponding design value using the hydrological data of Zhongshan City.

2 RESEARCH METHOD

2.1 *Maximum entropy principle*

The maximum entropy principle was proposed by the scientist ET Jaynes in 1957 and has been widely used in hydrological frequency analysis in recent years. The most objective distribution is when the uncertainty of the elements becomes the largest, that is, the maximum entropy distribution. Through a large number of research experiments, experts have indicated that the error of calculation result was minimal using the maximum entropy principle. Therefore, in the hydrological frequency analysis, the probability distribution generated by the maximum entropy method could reflect the characteristics of hydrological variables more accurately. On the basis of the maximum entropy principle, for any hydrological variable, there are

$$H(X) = -\int_k^l f(x)\ln f(x)dx \tag{1}$$

$$G_j = \int_k^l g_j(x)f(x)dx, j = 0,1,2,\cdots,n \tag{2}$$

where $f(x)$ is the probability density function, $g_j(x)$ is the constraint condition, and n is the number of constraints.

When the parameter values satisfy the maximum value of the objective function under the constraint condition, the method is called parameter estimation based on the maximum entropy principle. At that time, the uncertainty between the random variables is the largest. In this study, the P-III distribution is estimated by the maximum entropy principle. The probability density function of P-III distribution is

$$f(x) = \frac{1}{a\Gamma(b)}\left[\frac{x-c}{a}\right]^{b-1} \exp\left\{-\left[\frac{x-c}{a}\right]\right\} \tag{3}$$

Taking the natural logarithm of equation (3) and multiplying by $-f(x)$ from c to ∞, we obtained the constraints on the basis of the principle of the maximum entropy as follows:

$$\int_c^\infty f(x)dx = 1; \int_c^\infty xf(x)dx = E[x]; \int_c^\infty \ln(x-c)f(x)dx = E[\ln(x-c)] \tag{4}$$

The Lagrangian method could be used to derive the probability distribution corresponding to the maximum entropy, which is expressed by Eq. (5), where λ_0, λ_1, and λ_2 are Lagrangian multipliers:

$$f(x) = \exp[-\lambda_0 - \lambda_1 x - \lambda_2 \ln(x-c)] \tag{5}$$

Substituting formula (5) into formula (4), we obtain the values of λ_0, λ_1, and λ_2 and the probability density function of P-III distribution on the basis of the corresponding maximum entropy.

2.2 *Joint distribution function*

The joint distribution function Copula is closely related to the secondary return period. Let H be a d-dimensional distribution function with a univariate distribution of F_1, F_2, ... F_d. Then, there is a Copula function C, such that, for any $x \in R_n$, there is

$$H(x_1, x_2 \cdots x_d) = C(F_1(x_1), F_2(x_2), \cdots F_d(x_d)) \tag{6}$$

AMH Copula is a type of function commonly used in Copula function cluster, which is suitable for describing the positive correlation between variables. Therefore, AMH Copula function was used to construct the joint distribution of rainfall and tidal levels. The specific expression could be derived as follows:

$$C(U,V)=UV/[1-\theta(1-U)(1-V)];u=F_1(p),v=F_2(z) \tag{7}$$

where θ is the parameter of AMH Copula.

Copula parameter θ is estimated by the correlation index method. The relationship between Kendall rank correlation coefficient τ and θ is established as follows:

$$\tau=\left(1-\frac{2}{3\theta}\right)-\frac{2}{3}\left(1-\frac{1}{\theta}\right)^2\ln(1-\theta),\theta\in[-1,1) \tag{8}$$

$$\tau=\frac{2}{n(n-1)}\sum_{i=1}^{n-1}\sum_{j=i+1}^{n} sign\left[(x_i-x_j)(y_i-y_j)\right] \tag{9}$$

where τ is the Kendall rank correlation coefficient, (x_i, y_i) is the measurement point, *sign*(.) is the symbol function, and *n* is the length of series.

2.3 *Calculation method of return period*

Return period refers to an average interval when hydrological events may occur. "OR" return period and "AND" return period are traditional return periods that are generally recognized in the hydrological community. The calculation of the return period is closely related to the joint distribution function $C(u, v)$. When the value of $C(u, v)$ becomes smaller, the corresponding return period will also decrease.

The "OR" return period is used to describe the design value of P and Z when one of the design values exceed the set threshold (formula (10)). The "AND" return period is used to describe the design value of P and Z when both of them exceed the set threshold (formula (11)):

$$T_{\cup}(p,z)=\frac{1}{P[P>p\cup Z>z]}=\frac{1}{1-C(u,v)} \tag{10}$$

$$T_{\cap}(p,z)=\frac{1}{P[P>p\cap Z>z]}=\frac{1}{1-u-v+C(u,v)} \tag{11}$$

The secondary return period method is a new form of definition of multidimensional return period proposed by Salvadori. It is applied to calculate the return period using the Kendall measure K_c, which is only associated with the joint distribution function Copula that is used to indicate the probability of joint distribution probability less than or equal to t:

$$K_c(t)=P(W\le t)=P(C(u,v)\le t) \tag{12}$$

where $\varphi(t)$ is the generator of Copula function, and K_c can be expressed as follows:

$$K_c(t)=t-\frac{\varphi(t)}{\varphi'(t)} \tag{13}$$

Therefore, the secondary return period can be expressed as follows:

$$T_k=\frac{1}{P(C(u,v)\ge t)}=\frac{1}{1-K_c(t)} \tag{14}$$

According to traditional return periods, the hazard region is directly determined by the given variable value. This definition is not reasonable because different combinations of variables correspond to different hazard regions at the same level of joint distribution probability. Thus, some of the areas defined are contradictory. The advantage of the secondary return

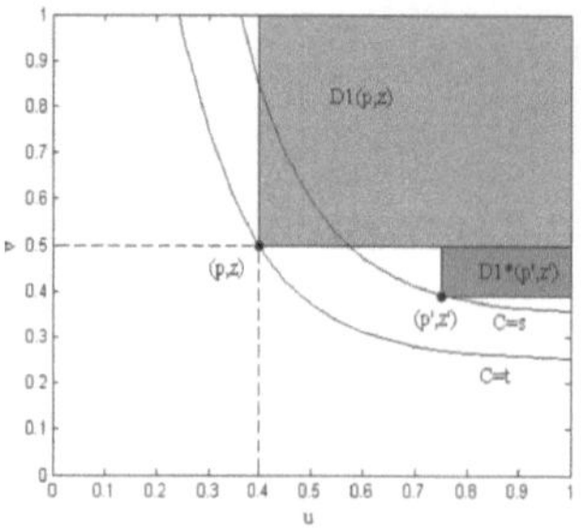

Figure 1. Risk regions for the "AND" return period.

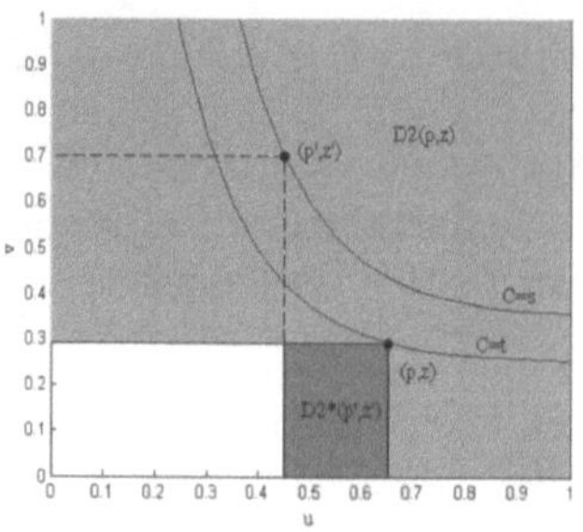

Figure 2. Risk regions for the "OR" return period.

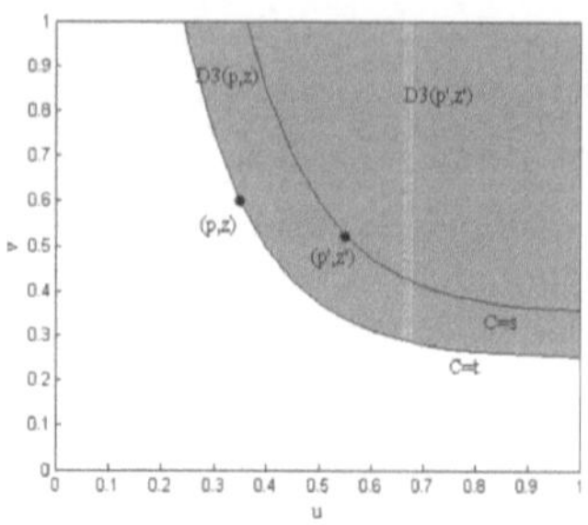

Figure 3. Risk regions for the secondary return period.

period is to determine the same dangerous area at the same probability level. The dangerous areas of three types of return period are shown in Figures 1–3.

Under the "AND" return period condition, selecting the design value on $C=t$, the risk area determined by (p,z) is D_1. Selecting (p',z'), which has a higher return period value, the partial risk area D_1^* determined by (p',z') is located outside D_1. Theoretically, the risk area with lower return period should contain the risk area with higher return period. However, D_1^* is included in the risk area determined by the higher point (p',z'), but it has not been included in the risk area determined by the lower point (p,z). Therefore, it is unreasonable.

Under the "OR" return period condition, D_2 is the risk area identified by (p,z) (light gray area). We can also get the risk area determined by (p',z') in the same way, where D_2^* is in the range determined by (p',z'), but not in D_2 determined by (p,z). That is, the event in D_2^* is considered to be safe by the lower return period (p,z), but dangerous, by the higher return period point (p',z').

The secondary return period method can solve the problem effectively in both cases. Therefore, it is a rational method to define the dangerous events in hydrology compared with the traditional return periods. In the secondary return period method, the risk area determined by $C=t$ is $D_3(p,z)$ – the risk area corresponding to (p,z). The blank area outside the risk area is considered to be safe. $C=t$ can be regarded as the critical line that distinguishes the danger domain and the security domain. Obviously, $D_3(p',z')$ determined by (p',z') is all in D_3. In the secondary return period, the smaller the C, the greater the risk area, and it will definitely cover the risk area determined by the larger C.

3 EXAMPLE ANALYSIS

Zhongshan City is located in the Pearl River Delta and greatly influenced by tropical cyclones. In addition, due to the high incidence of heavy rainfall, floods occur frequently. Besides, the outer river floods and the tide sometimes would cause river water intrusion in low-lying areas, which is also an important factor causing waterlogging in Zhongshan City. The frequent occurrence of waterlogging was not only a threat to the safety of life and property of urban residents, but also seriously affected the development of urban economy. The author collected the 1-hour maximum rainfall and the corresponding tidal levels data from the Hengmen Hydrological Station of Zhongshan City from 1964 to 2013.

The univariate distribution function was estimated by $P-\text{III}$ distribution using the maximum entropy method. The parameters and eigenvalues of the univariate distribution were estimated, and the results are shown in Table 1.

The AMH Copula function parameter θ was estimated by the analytical relationship of τ and θ in equation (8). Thus, we could obtain the τ and θ values of the 1-hour maximum rainfall and the corresponding tidal levels were 0.12 and 0.46, respectively. We used the OLS criterion for further test. The OLS value was small (0.0285), which indicated the better fit of the AMH Copula. The joint distribution model of the 1-hour maximum rainfall and tidal levels is shown in Figure 4.

The author calculated the values of $T_\cup$, $T_\cap$, and T_k at return periods ranging from 2 to 200 years. It is evident from Table 2 that $T_\cup$ and $T_\cap$ were lower and higher than in the univariate return period, respectively. Besides, T_k was between $T_\cup$ and $T_\cap$ and closest to the univariate return period of the sample.

There are numerous combinations of (p,z) to meet the design requirements at the same level of return period. Therefore, it is critical to determine the optimal solution. From the above discussion, it can be seen that the combination of rainfall–tide with the same joint distribution probability can constitute a critical line $C(u,v) = t$ in which the joint probability density of (p_m, z_m) satisfying equations (15) and (16) is the largest, that is, this design value combination is most likely to appear:

$$(u_m, v_m) = arc\max\{F_1(p)*F_2(z)*f(u,v)\} \tag{15}$$

$$p_m = F_1^{-1}(u_m), z_m = F_2^{-1}(v_m) \tag{16}$$

$f(u,v)$ is the probability density function of Copula.

The design value of the rainfall–tidal level could be obtained separately at different return periods using the P-III distribution function. From formulas (15) and (16), we obtained the design values of rainfall–tide combination with the greatest possibility of occurrence at different return periods. The results are shown in Table 3.

Table 1. Estimated parameters of marginal distribution.

	Ex	*Cv*	*Cs*	5*a*	10*a*	20*a*	50*a*	100*a*	200*a*
p (mm)	48.61	0.38	1.14	62.5	73.81	84.35	97.92	107.17	117.06
z (m)	1.43	0.41	1.25	1.88	2.24	2.58	3.04	3.43	3.90

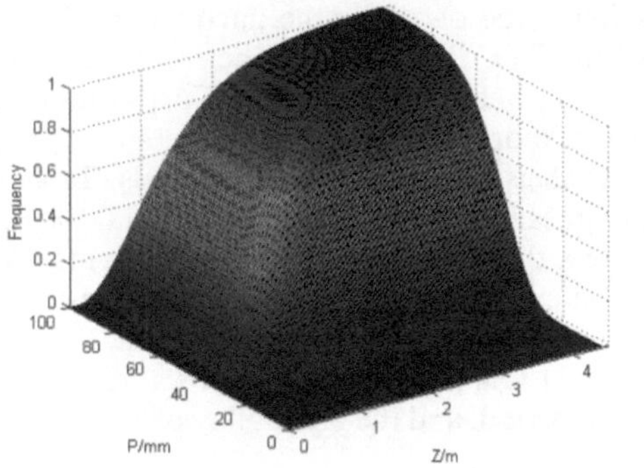

Figure 4. Joint distribution model.

Table 2. Three joint return periods of P and Z for Hengmen Station.

T	$T_\cup(p,z)$	$T_\cap(p,z)$	$T_k(p,z)$
5	4.1	6.2	5.6
10	8.0	12.8	11.6
20	16.1	25.3	22.7
50	41.1	65.39	58.38
100	81.7	130.2	116.3
200	162.9	259.7	232.0

Table 3. Design values of P and Z for different return periods.

	P				Z			
T	$T_{\cup}(p,z)$	$T_{\cap}(p,z)$	$T_{k}(p,z)$	$T(p)$	$T_{\cup}(p,z)$	$T_{\cap}(p,z)$	$T_{k}(p,z)$	$T(z)$
5	65.8	57.6	59.5	62.5	2.02	1.80	1.86	1.88
10	77.9	68.3	71.1	73.8	2.34	2.16	2.20	2.24
20	88.3	77.8	82.4	84.3	2.68	2.44	2.52	2.58
50	100.5	92.0	95.7	97.9	3.18	2.87	2.98	3.04
100	111.6	104.1	105.4	107.1	3.52	3.26	3.34	3.43
200	122.0	110.2	114.6	117.0	4.20	3.73	3.84	3.90

Table 4. Comparison of design values for different return periods.

	P			Z		
	$\frac{T_{\cup}(p,z)-T(p)}{T(p)}$	$\frac{T_{\cap}(p,z)-T(p)}{T(p)}$	$\frac{T_{k}(p,z)-T(p)}{T(p)}$	$\frac{T_{\cup}(p,z)-T(p)}{T(p)}$	$\frac{T_{\cap}(p,z)-T(p)}{T(p)}$	$\frac{T_{k}(p,z)-T(p)}{T(p)}$
5	0.0528	–0.0784	–0.0480	0.0745	–0.0426	–0.0106
10	0.0556	–0.0745	–0.0366	0.0445	–0.0357	–0.0179
20	0.0475	–0.0771	–0.0225	0.0388	–0.0542	–0.0233
50	0.0266	–0.0603	–0.0225	0.0461	–0.0559	–0.0197
100	0.0420	–0.0280	–0.0159	0.0262	–0.0496	–0.0146
200	0.0427	–0.0581	–0.0205	0.0769	–0.0436	–0.0154

In general, the design values based on $T_{\cup}$ are higher than those using $T_{\cap}$ and T_k, and the design values based on T_k are higher than $T_{\cap}$, which is due to the fact that $T_{\cup}$ expands the range of hazardous events. Thus, the design value of $T_{\cup}$ is larger at the same return period level.

The reason for the higher value of $T_{\cap}$ is the narrowing of the range of the risk event; therefore, the design value of $T_{\cap}(p,z)$ is smaller. The secondary return period can overcome the shortcomings of the traditional return periods. Furthermore, it can predict the actual situation of dangerous events more accurately.

We compared the design values in different return periods (Table 4). It can be seen that the maximum error of $T_k(p,z)$ appears at the level of 5 years – 4.80% under the 1-hour maximum annual rainfall conditions, less than the error values of $T_{\cup}(p,z)$ (5.28%) and $T_{\cap}(p,z)$ (7.84%). The design error of $T_k(p,z)$ is also less than the others in the rest of return period level. Similarly, with Z, the maximum error of $T_k(p,z)$ is not more than 2.33%, less than the error values of $T_{\cup}(p,z)$ (3.88%) and $T_{\cap}(p,z)$ (5.42%). The result shows that the secondary return period is closer to the univariate return period, and the probability of risk can be defined more accurately.

4 CONCLUSIONS

The G–H Copula function was used to construct the joint distribution of the 1-hour maximum rainfall and the corresponding tidal levels. In this study, a new method was proposed to define the return period. The analysis shows that the error in the secondary return period is the smallest compared with the traditional return periods. Compared with the design value of the rainfall–tide, the design error of the secondary return period is less than the traditional return periods under the same univariate return period. In fact, the risk regions of the design values based on the traditional return periods are different at the same joint distribution probability level, and some of the regions defined are contradictory. The secondary return period can determine the same risk region in the joint distribution at the same level, which can effectively avoid false identification of flood events. The results of the analysis can serve as a reference for urban flood control.

REFERENCES

Chen Z.S., Huang Q., Liu Z.M. 2015, Analysis on the joint distribution of low flows between Xijiang River and Beijiang River under a changing environment. *Advances in Water Science*, (26)1:20–26.

Corbella S., Stretch D.D. 2012, Multivariate return periods of sea storms for coastal erosion risk assessment. *Natural Hazards and Earth System Sciences*, 12:2699–2708.

Du J., Chen X.Z., 2009, Parameter estimation of Archimedean copula. *Science Technology and Engineering,* (3):637–640.

Fan J.W., Huang J.L. 2017, Combined Frequency Analysis of Peak and Duration Based on Copula Function. *China Rural Water and Hydropower*, 2:204–208.

Genest C., Ghoudi K., Rivest L.P. 1995, A semiparametric estimation procedure of dependence parameters in multivariate families of distributions. *Biometrika,* 82(3):543–552.

Guo S.L., Yan B.W. 2008, Multivariate hydrological analysis and estimation. *Journal of China Hydrology*, 28(3):1–6.

Huang J.L., Fan J.W. 2017, Correlation Study of Rainstorm and Tidal Level Using the Maximum Entropy—Copula Method. *Journal of Catastrophology,* 32(1):65–71.

Jaynes E.T. 1982, On the rationale of maximum entropy methods. *Proceedings of the IEEE,* 70(9):939–952.

Liu Z.M., Chen Z.S. 2011, Study on the standard of waterlog prevention based on two waterlog-inducing factors. *Journal of Hydroelectric Engineering,* 30(3): 39–44. Shiau J.T. 2006, Fitting drought duration and severity with two-dimensional copulas. *Water Resources Management*, 20:795–815.

Salvadori G., De Michele C., Durante F. 2011, On the return period and design in a multivariate framework. *Hydrology and Earth System Sciences*, 15:3293–3305.

Salvadori G., De Michele C. 2010, Multivariate multiparameter extreme value models and return periods: A copula Approach. *Water Resources Research*, 46(10):W10501.

Salvadori G., De Michele C. 2007, On the use of copulas in hydrology theory and practice. *Journal of Hydrologic Engineering*, 12(4):369–380.

Salvadori G. 2004, Bivariate return periods via 2-copulas. *Statistical Methodology,* 1(1):129–144.

Shi L.X., Song S.B. 2015, Calculation of return periods and design values of bivariate floods based on Copula function. *Journal of Hydroelectric Engineering*, 34(10):28–33.

Sklar. M. 1959, Fonctions de répartition à n dimensions et leurs marges. *Université Paris* 8, 8:229–231.

Zhang L., Singh V.P. 2006, Bivariate flood frequency analysis using the copula method. *Journal of Hydrologic Engineering*, 11(2):150–164.

Zhang L. Singh V.P. 2007, Trivariate flood frequency analysis using the Gumbel-Hougaard copula. *Journal of Hydrologic Engineering*, 12(4):431–439.

Hydraulic Engineering V – Hong, Liu & Xie (Eds)
© 2018 Taylor & Francis Group, London, ISBN 978-0-8153-7471-8

Adapting the Afsluitdijk to climate changes

Y. Li
Nanjing Hydraulic Research Institute, Nanjing, China

P. Arecco, C. Miranda Eguez & M.J. Ruiz Fuentes
Delft University of Technology, Delft, The Netherlands

ABSTRACT: After 80 years of successfully protecting central Netherlands from the adverse effects of the North Sea, the Afsluitdijk (Enclosure Dam) is now required to upgrade in order to adapt to the expected climate changes. This study proposes a riprap overlayer cross section to meet the expected critical boundary conditions in the future. It presents the typical parameters of the proposed cross section. The study completes the calculations by checking run-up and overtopping, which is the limiting factor, to define the crest height. It also checks the overall stability of the structure by using Finite Element Models (FEMs). The results of the study show that the proposed cross section can provide the required protection until 2050.

1 INTRODUCTION

The Afsluitdijk (Enclosure Dam) is a major causeway in The Netherlands, which was constructed between 1927 and 1933, running from Den Oever on Wieringen in North Holland Province to the village of Zurich in Friesland Province. The dam is 32 km long and 90 m wide, with an initial height of 7.25 m above sea level and a slope of 1:4 (original value; now 1:3.6) on each side (S. Liu et al., 2012).

The Afsluitdijk was built to protect the coast along the Zuiderzee from the adverse effects of North Sea as well as to make it a freshwater lake. It is evident that the Afsluitdijk protects central Netherlands from the effects of North Sea (S. Liu et al., 2012). After 80 years of remarkable performance, expected changes in boundary conditions and an increase of safety requirements make it necessary to think of an adaptation of the structure (J. C. Aerts and W. J. Botzen, 2013). The higher the temperature, the higher the ocean water level, the higher the storm surges, and the higher the discharge in rivers. Because of the high water level in the river during the storm surge season, there will be an increased risk of failure of the Afsluitdijk. Therefore, urgent planning is necessary to overcome the adverse effects of climate change (X. Zhu et al., 2010).

In this paper, we propose an upgrade of the existing cross section of the Afsluitdijk, to resist overtopping, which provides the required protection until 2050.

2 ISSUES BEING ADDRESSED

The Afsluitdijk was built with boulder clay, clay, and sand, which were found in the bottom of former Zuiderzee. The dam is erected on top of a smaller closure made of boulder clay, which was emplaced previously to separate the two water masses. On the Lake IJsselmeer side of this dam, a wide body of sand, which is covered with boulder clay and clay, was built. The body of the dam below the waterline is protected with osier wood and

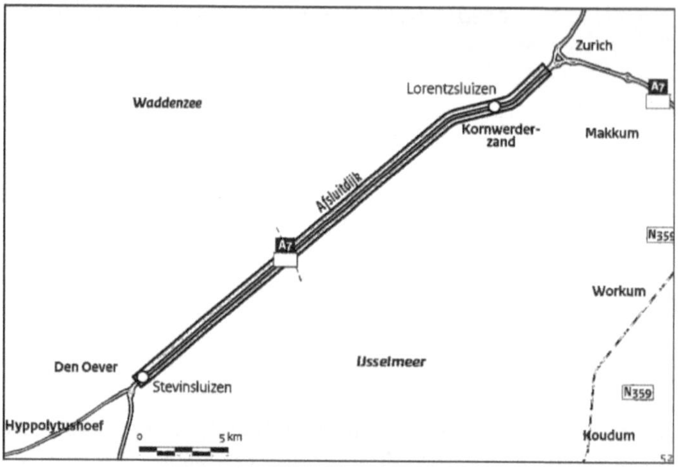

Figure 1. General map of the Afsluitdijk, The Netherlands (Rijkswaterstaat, 2009).

reed mattresses, on which riprap is placed. Above the waterline, the dike is protected with basalt blocks. The crest and inner slope are covered with grass. Furthermore, a motorway stretches out at the top of the dam, enabling road traffic between the provinces of North Holland and Friesland.

For the current situation, the dike reaches a level of +7.8 m NAP with an outside slope of 1:3.6 covered with basalt blocks as a revetment. There is also a permeable clay layer on top of the crest covered with grass with an inside slope of about 1:2.7. The total width of the dike is between 80 and 100 m. Severe storm conditions facing the current structure would mean that a wave overflow of approximately 366 L/s/m will occur, rising the water level of Lake IJsselmeer by 0.5 m over a storm duration of 12 h (S. Liu et al., 2012).

3 DESIGN BASIS CRITERIA

In this section, we summarize the input information and the design criteria that would lay the foundation for the development. The boundary conditions are defined, setting the methodology to be followed.

3.1 *Typical cross section*

As a starting point, an existing cross section from the Afsluitdijk is used as starting point, as shown in Figure 2, and is chosen as the typical cross section, based on which designs are proposed. The mean sea-bottom depth at the Afsluitdijk is about 4–5 m, NAP referred. Therefore, the adopted sea-bottom elevation for this cross section is −5 m NAP. This value is used in the proposed configuration aiming to have a representative design for the whole dike length. However, it is important to note that the actual existing depths at the area where the selected cross section is located are about −10 m NAP. In addition, other existing cross sections are in shallower areas about −2 m NAP.

3.2 *Wave climate*

The definition of wave climate (return period, 10,000 years) is obtained from reference document *"Dijkvakken_ontwerprvw_Waddenzeezijde_T10000"*, which includes details about

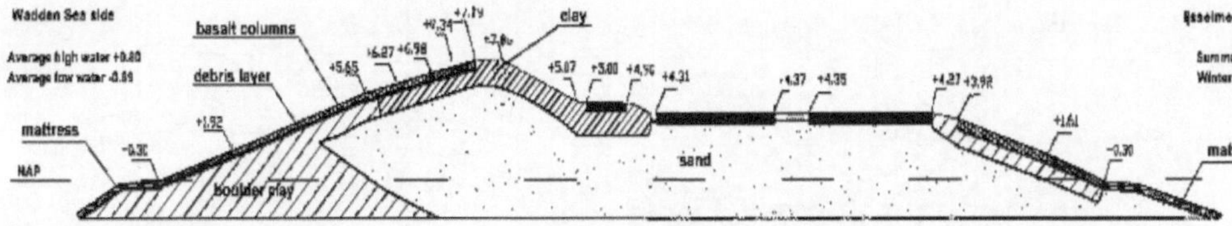

Figure 2. Typical current cross section of the Afsluitdijk (Rijkswaterstaat, 2009).

Table 1. Wave climate data for 2050 (P. Arecco et al., 2013).

	Sea level elevation	Significant wave height	Mean wave period	Peak period
	h (m, NAP)	$H_{0\,m}$	$T_{m1\text{–}0}$ (s)	T_p (s)
Overtopping	5.31	3.83	6.2	7.6
Concrete blocks	5.00	3.72	6.1	7.5
Overall stability	5.52	–	–	–

several cross sections of the Afsluitdijk. The available information includes mean sea level elevation, significant wave height, mean wave period, and peak period. For design purposes, the end of the structure's service life is considered as 2050; therefore, the wave climate data used will be the one corresponding to this year. The used wave climate data are presented in Table 1.

3.3 *Currents*

Currents of low magnitude are omitted from this study.

4 PROPOSED UPGRADED CROSS SECTION

A design based on a layer of riprap over the existing dike has been studied as a possible solution. This type of cross section is considered of interest because it requires minimal intervention (no demolition or excavations and fewer materials). For this design, the existing basalt columns would not be removed. Therefore, a filter layer is not necessary to keep the core material in place, as this condition is already met by the existing armor layer.

The size of the armor layer was calculated using the Van der Meer formula, with a d_n50 of 3000 kg. The base of the armor layer is composed of a toe rock of 750 kg, which is calculated using the Gerding and Van Leeuwen formula (S. Baart et al., 2011). Calculations are completed by checking run-up and overtopping, which is the limiting factor, to define the crest height. Figure 3 shows the proposed upgraded cross section. For the current cross section configuration and the hydraulic boundary conditions of 2050, the overtopping resistant alternative has to cope with an overtopping of about 360 L/m/s. Meanwhile, for the same situation, the proposed riprap over-layer alternative reduces overtopping up to 10 L/m/s.

Despite the severe boundary conditions used for calculating the upgraded cross section, the crest elevation obtained is very similar to the existing one. This is a consequence of riprap's higher capacity for dissipating wave energy compared with basalt columns, which results in lower run-up values and thus smaller overtopping rates.

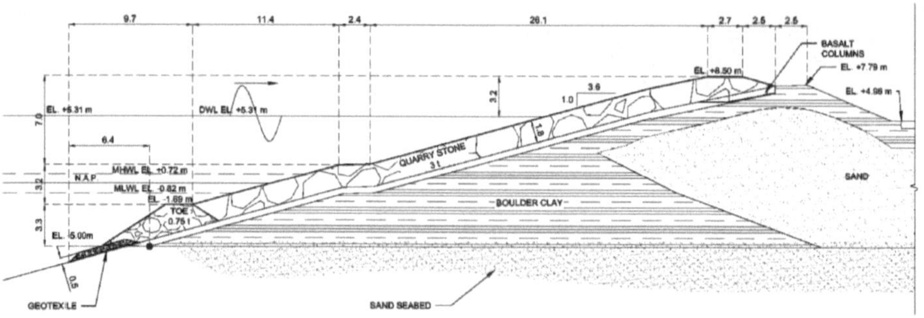

Figure 3. Typical parameters of the upgraded cross section.

5 OVERALL STABILITY CHECK

To check the overall stability, the current Afsluitdijk cross section is first modeled and then analyzed using finite element models. First, the finite element model (FEM) is considered for the current situation. Then, to check the overall stability of the proposed riprap over-layer cross section, another finite element model is set up in the same way. During the FEM calculation, every construction stage, from the current arrangement to the proposed alternative, is idealized, in order to represent critical situations during the service life and construction of the structure.

5.1 *Analysis*

The dimensions of the FEM are 320 m × 60 m. Model units are meter (m) for length, second (s) for time, kilopascal (kPa) for stress and pressure, and kilo-Newton (kN) for force. The current maximum crest elevation of the dike is set at +7.86 m NAP, whereas the Waddenzee elevation is at −10.00 m NAP and the Lake IJsselmeer elevation is at −4.00 m NAP.

Dead loads considered in this FEM are soil and self-weight of materials. Live loads included in the FEM are the highway and bike path service loads. For simplicity, these live loads are assumed as static loads of 10 and 2 kN/m^2, respectively. The aforementioned information is graphically presented in the following figures. In addition, water levels of both Waddenzee and Lake IJsselmeer are set at +0.00 m NAP under normal conditions. Then, according to different load combinations defined, the water level of Waddenzee is modified up to the maximum elevation given for 2050, which is +5.52 m NAP, in order to obtain the structural and geotechnical safety coefficient.

The proposed alternative has its maximum crest elevation at +8.50 m NAP, and the Waddenzee and IJsselmeer elevations are the same as the existing one. In addition, dead loads and live loads are similar. FEM geometries are shown in Figure 4.

5.2 *Finite element model calculation*

5.2.1 *Normal conditions*

The finite element models (FEMs) used in these cases are 2D nonlinear plain strain models, where 15-node triangular elements are applied, and an example is shown in Figure 5. Soil and rock behave in a highly nonlinear way under loads. This stress–strain behavior is modeled with Mohr–Coulomb (M. D. Bolton and C. Lau, 1993). However, for pavement and concrete, a linear elastic model, which is much more suitable for these types of materials, is used. Active pore pressures for Waddenzee and Lake IJsselmeer under normal conditions (+0.00 m NAP) are shown in Figure 6. Besides, initial effective soil stresses are calculated and shown in Figure 7.

5.2.2 *Storm conditions for 2050*

The active pore pressures that are expected to originate during a storm in Waddenzee (in 2050) are also computed. It is evident that the values of the boulder clay and the deep sand

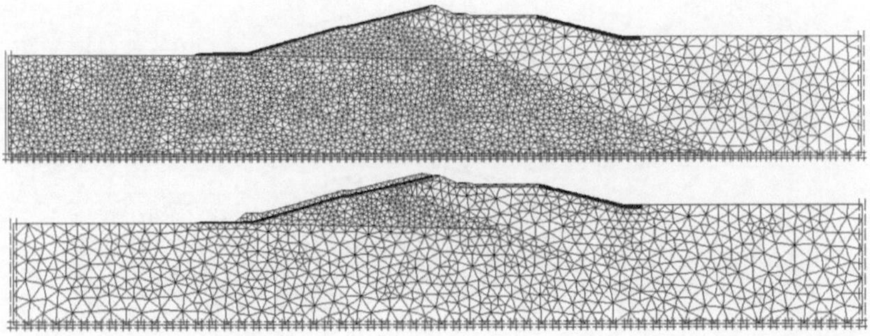

Figure 4. FEM geometry of the proposed cross section (top: existing cross section; bottom: the proposed riprap overlayer cross section).

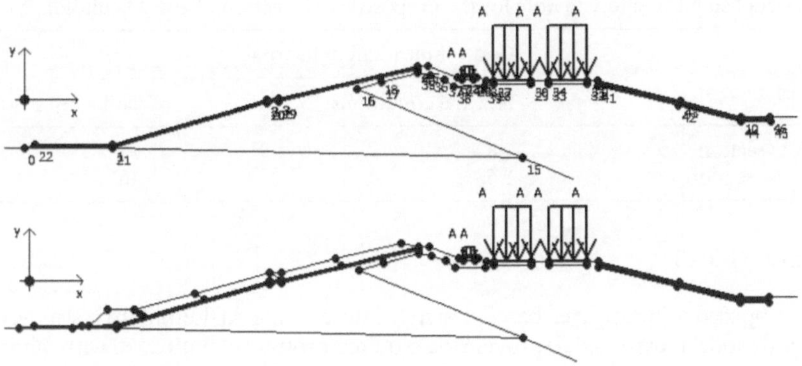

Figure 5. FEM coarse mesh, which is refined in key zones for possible failure mechanisms (top: existing cross section; bottom: the proposed riprap overlayer cross section).

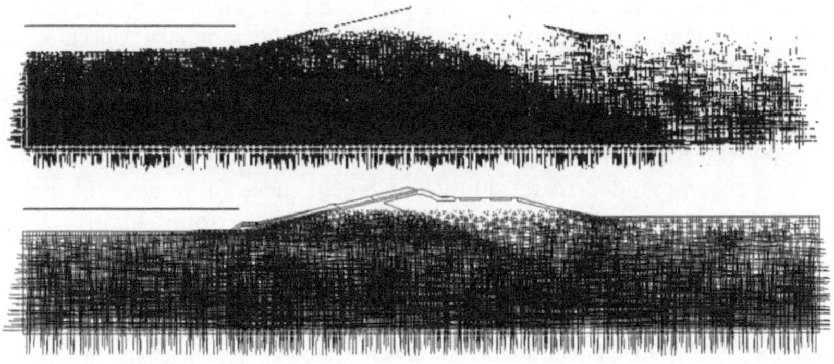

Figure 6. FEM active pore pressure. The extreme active pore pressure at the bottom of the FEM is 499 kN/m^2, which is in accordance with the dimensions of the FEM (top: existing cross section; bottom: the proposed riprap overlayer cross section).

layers below Waddenzee are higher than those in the IJsselmeer influence area. It is assumed that the effect of a Waddenzee storm influences the water table of the Afsluitdijk up to the diagonal geometry line crossing the dike geometry. The previous assumption is based on boulder clay with very low permeability.

As explained earlier, one of the safety factors is obtained for normal conditions and the other safety factor is calculated for future Waddenzee water elevations during a design storm (in 2050). Both the values are presented in Table 2.

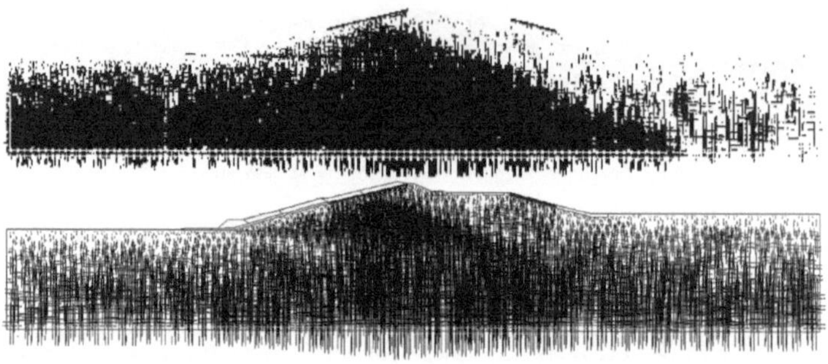

Figure 7. FEM effective stresses. The extreme effective principal stress at the bottom of the model below the dike crest is –632 kN/m² (top: existing cross section; bottom: the proposed riprap overlayer cross section).

Table 2. Overall stability safety factors for the proposed cross section of the Afsluitdijk.

	Geotechnical safety factors	
	Normal conditions	Waddenzee storm in 2050
Existing cross section	1.82	1.73
Proposed cross section	1.53	1.46

6 CONCLUSIONS

This study proposed a riprap layer-based design for the existing Afsluitdijk to resist overtopping. The proposed model is expected to provide the required protection until 2050 with minimal intervention, without demolition or excavations, and with fewer materials. Furthermore, it is expected to make the upgrading of the dike easier and less expensive throughout the structure's service life.

The size and base of the armor layer were calculated to check the main failure mechanisms, overtopping and instability, of the armor layer, as well as run-up. The results showed that the riprap's higher capacity for dissipating wave energy compared with basalt columns results in lower run-up values and thus much smaller overtopping rates for the hydraulic boundary conditions of 2050.

Finite element models were used to check the overall stability of the structure. The proposed cross section was calculated by using FEMs and the results were compared with the existing one, both under the normal and storm conditions. The results indicated that the proposed cross section will enable the Afsluitdijk to confront critical hydraulic conditions until 2050.

However, it is recommended to recalibrate or repeat these finite element model (FEM) calculations with more detailed material information in order to obtain highly reliable results.

REFERENCES

Aerts, J.C. & Botzen, W.J. 2013. Climate adaptation cost for flood risk management in The Netherlands. *Storm surge barriers to protect New York City: Against the deluge.*

Arecco, P., Li, Y., Miranda Eguez, C. & Ruiz Fuentes, M. 2013. Afsluitdijk upgrading, non typical Dutch solutions. *Master project report.*

Baart, S., Ebbens, R., Nammuni-Krohn, J. & Verhagen, H.J. 2011. Toe rock stability for rubble mound breakwaters. *Coastal Engineering Proceedings,* 1, 35.

Bolton, M.D. & Lau, C. 1993. Vertical bearing capacity factors for circular and strip footings on Mohr–Coulomb soil. *Canadian Geotechnical Journal,* 30, 1024–1033.

Liu, S., Huang, J., Rayo, S. & Lim, T.Y. 2012. Afsluitdijk project" The Monument" location. *Master project report.*

Rijkswaterstaat 2009. Legger Afsluitdijk. *In:* RIJKSWATERSTAAT (ed.). The Netherlands: Ministerie van Verkeer en Watestaat.

Zhu, X., Linham, M.M. & Nicholls, R.J. 2010. Technologies for climate change adaptation-Coastal erosion and flooding. Danmarks Tekniske Universitet, Risø Nationallaboratoriet for Bæredygtig Energi.

Hydraulic Engineering V – Hong, Liu & Xie (Eds)
 ISBN 978-0-8153-7471-8

Adjustment of a tridimensional network structure of ecological revetment to the local flow pattern

Ai-xing Ma
State Key Laboratory of Hydrology Water Resources and Hydraulic Engineering Science, Nanjing Hydraulic Research Institute, Nanjing, China

Min-xiong Cao
Nanjing Hydraulic Research Institute, Nanjing, China

Qing-hua Xiao
Changjiang Waterway Planning Design and Research Institute, Wuhan, China

Ying Hu
Nanjing Hydraulic Research Institute, Nanjing, China

Zhong-min Fu
Changjiang Waterway Planning Design and Research Institute, Wuhan, China

ABSTRACT: Revetment structures in the Yangtze River waterway regulation are usually of impermeable or semi-ecological type, which cannot adapt to nutritional soil conservation measures on a slope surface with fluctuating water levels. Recently, a type of tridimensional network structure of ecological revetment was developed based on the concept of energy dissipation of permeable structures. In this study, a laboratory experiment was carried out in a variable-slope flume for accurately measuring the local flow pattern around a tridimensional network structure using a three-dimensional velocity meter (Vectrino). The results indicated that the vertical flow velocity and turbulence intensity near the bottom of the slope decreased by 75% and 20%, respectively, after the flow passed through the tridimensional network structure, thereby creating a less harmful hydrodynamic environment to prevent soil erosion. Meanwhile, because of the presence of holes on the permeable network structure, water can be more easily exchanged between inside and outside of it with good ecological effect. The results of the study can be used as a reference for the design and application of tridimensional network structures.

1 INTRODUCTION

Revetment engineering is common on the banks along the middle and lower reaches of Yangtze River to control river trend and stabilize the channel boundary. Traditional revetment projects often use impermeable structure for the purposes of safety and durability of the revetment works, which hinders the function of the biological habitat of the riparian zone and affects the integrality of the ecosystem structure. In recent years, with the construction and development of the ecological channel in Yangtze River, several types of ecological revetment, such as steel mesh (Li 2016) and interlocking slope protection (Cao 2014), have been proposed. Most of them are semi-ecological revetment structures with good stability and ecological effect; however, they have defects in soil conservation and energy dissipation. Nutrient soil and plant seeds can be easily eroded in areas with fluctuating water levels after a flood, disrupting ecological effects. With regard to the shortcomings of the existing ecological revetment structures in Yangtze River, the waterway engineering and technical personnel have developed a tridimensional network structure of ecological revetment (Xiao 2016), in order to meet the requirements of durability and ecology of the revetment structures in areas with fluctuating water levels along the river.

The tridimensional network structure will lead to the adjustment of the local fluid structure of the internal and surrounding parts of the structure, with varying flow velocity and turbulence characteristics, which will directly affect the protection ability of the slope surface. In this study, the local flow pattern of the internal and the surrounding parts of the tridimensional network structure was measured accurately in a variable-slope flume, and the structure's ability to reduce the near-bottom velocity and its underlying mechanism were analyzed. The results of the study can be used as a technical reference for the design and application of the tridimensional network structure.

2 TRIDIMENSIONAL NETWORK STRUCTURE AND FLUME EXPERIMENT

2.1 *Tridimensional network structure*

The tridimensional network structure of ecological revetment is composed of an outer frame member and a built-in cross-shaped member. The internal and external components are open holes on the panels. After comparison of the size of the members by controlled fluid experiments, the size, position, and arrangement of the holes on the panels and the main dimensions of recommended tridimensional structures are determined as follows: outer frame size, 0.57 m × 0.57 m × 0.25 m (length × width × height); panel thickness, 3.5–7.0 cm; inner cross-member panel thickness, 4.0 cm; net distance between the inner and outer panels, 16.8 m; diameter of the holes on the panels, 7.5 cm; and the distance of the bottom hole from the bottom of the structure, 9.0 cm. The tridimensional network structure is shown in Figure 1.

2.2 *Flume experiment*

1. Experimental flume

Hydrodynamic experiments of the ecological revetment structure were carried out in a variable-slope flume with length 44 m, width 0.8 m, depth 0.8 m, and slope gradient 1%, in Nanjing Hydraulic Science Research Institute. The flume adopts the self-circulation system of water flow, and the discharge of the inlet flow is controlled by an electric valve and an electromagnetic flowmotor with PVC pipeline for energy dissipation. The water level at the outlet of the flume is controlled by a stepping motor by controlling the opening degree of the tailgate.

2. Model scale and test flow conditions

In view of the scale of the tridimensional network structure and the laboratory dynamic conditions, the flume experiment was carried with a geometric scale ratio of 1:3.

The structure was used on the bank of Kouanzhi Waterway located in the lower reach of Yangtze River. The annual mean peak discharge ($Q = 57,500$ m^3/s) was taken as the flow condition

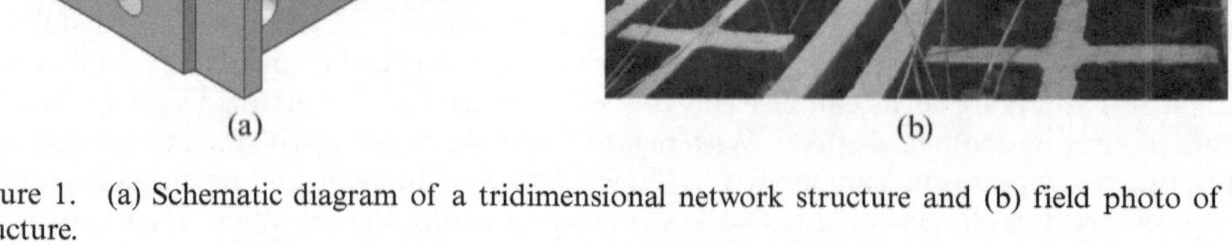

(a) (b)

Figure 1. (a) Schematic diagram of a tridimensional network structure and (b) field photo of the structure.

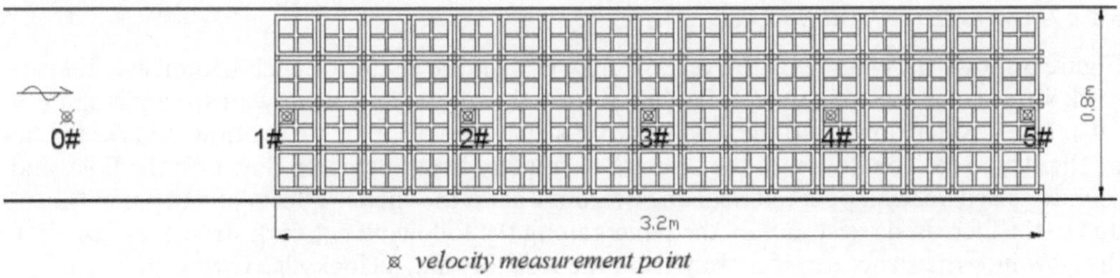

Figure 2. Layout of measuring points.

(Ma 2016) in the laboratory experiment, with corresponding data of 2.81 m water depth and 1.14 m/s depth-averaged velocity. Assuming the logarithmic distribution of the vertical flow velocity, the velocity of the flow on the top of the tridimensional network structure was 0.95 m/s.

As a result of a large-scale model design, the prototype water depth condition could not be directly scaled to obtain the test flow condition. By reducing the water depth to 1.2 m and according to the principle of the same flow velocity on the top of the structure, the model test flow condition was determined, with the flow velocity upstream of the structure group of 0.55 m/s and water depth of 0.4 m.

3. Experimental layout

In the middle of the flume, the tridimensional structures were arranged in 17 rows with a total length of 3.2 m and the corresponding prototype length of 9.6 m (Figure 2). Six observation points were arranged along the test section with one point (0#) at 2 m upstream in front of the structure group and five points in the range of the structure group. Each point was located at the center of the single-network structure, and the vertical velocity would be accurately measured with nine points in the structure and nine points from the top of the structure to the water surface.

4. Velocity measurement instrument

The flow pattern around the tridimensional network structure was measured by the Acoustic Doppler Point Velocity Meter (Vectrino). It can measure the spatial distribution of complex hydraulic elements such as flow velocity and turbulence intensity near the structure by measuring the instantaneous three-dimensional velocity. In the experiments, the velocity-measuring frequency of vertical point was 90 Hz, the measuring time was 40 s, and the number of velocity sample data of each vertical point was 3600.

3 ANALYSIS OF EXPERIMENTAL RESULTS

The flow pattern near the slope is altered due to the layout of the tridimensional network structure group, with the change in local velocity distribution and the turbulence intensity. The geometric mean values of the instantaneous flow velocity in the x direction (longitudinal direction, parallel to the flow direction), y direction (horizontal direction, perpendicular to the flow direction), and the vertical z direction are taken as the time-averaged flow velocity u, v, and w, respectively, and the root mean square values of pulsating flow rate in the corresponding directions are taken as the turbulence intensities σ_x, σ_y, and σ_z, respectively. According to the model similarity law, in an undistorted model, the model flow velocity (u_m) and prototype flow (u_p), and the model turbulence intensity (σ_{xm}) and prototype turbulence intensity (σ_{xp}), respectively, satisfy the following relations:

$$u_p = (\alpha_L)^{0.5} u_m,\ \sigma_{xp} = (\alpha_L)^{0.5} \sigma_{xm} \quad (1)$$

where α_L is the model geometric scale (equal to 3) for the model.

In order to facilitate a comparative analysis, velocity, turbulent intensity, and flow measuring vertical position adopt the values of the prototype.

3.1 *Changes in the local flow field*

Figure 3 shows the change of the vertical flow field at the center of each tridimensional network structure along the flume. It is evident from the figure that the flow upstream is smooth (0#). When water flows into the first network structure (1#), part of the flow is deflected up by the structure. The flow into the structure is mainly from the outer flow into the hole, and a small-scaled vortex appeared near the bottom due to the shelter effects of the panel. Then, due to the increased resistance of the panels along the following network structures (2#–5#), the flow into the structures is mainly from the top, forming a clockwise vortex, and the flow velocity decreased relative to the velocity upstream of the structure group.

Figure 4 shows the flow field structure of a single tridimensional network structure in the central part of the structure group area. It can be seen that the clockwise vortex (sections A1 and A2) is formed in the vertical direction of the structure, and the vortex scale is closely

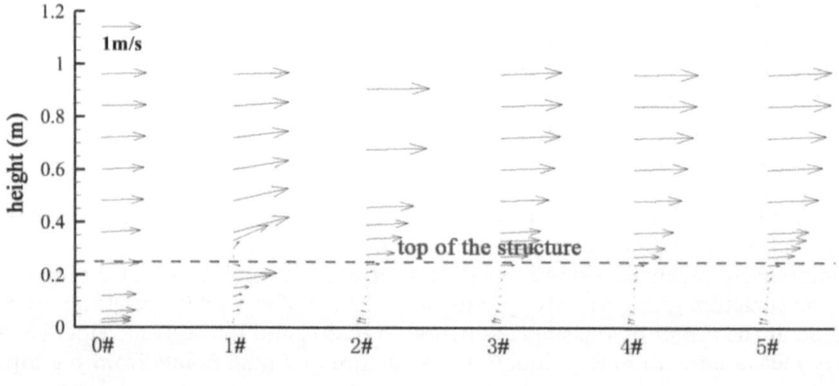

Figure 3. Vertical flow field in the center of each tridimensional network structure.

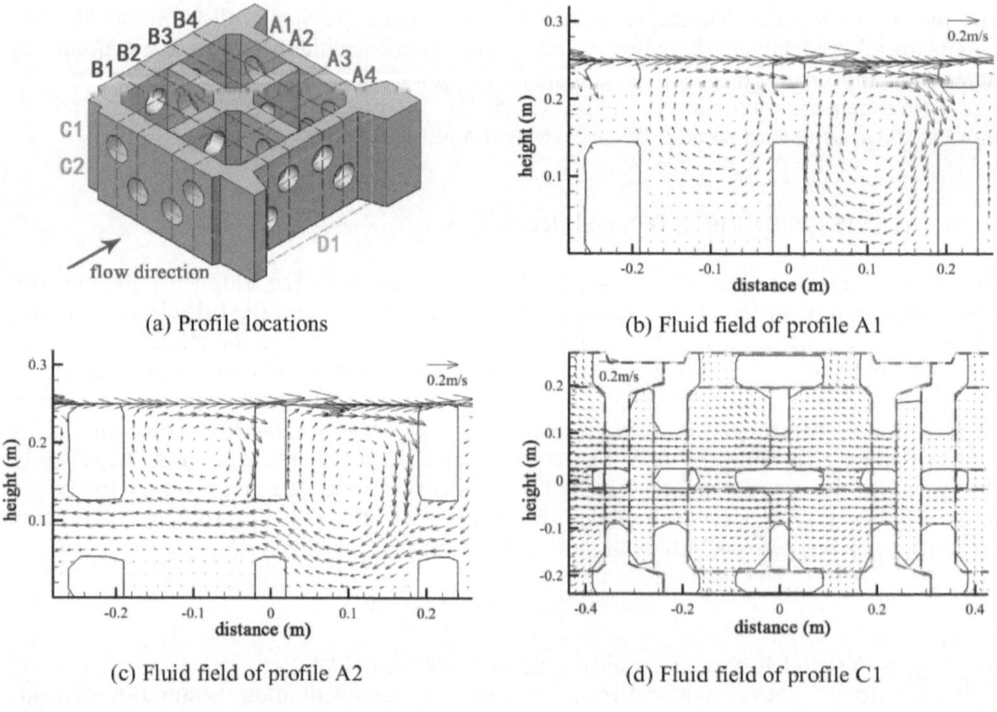

Figure 4. Fluid field at different profiles.

related to the hole position on the panel: When the hole is in the upper part, the complete vortex is formed in the structure height and the center of the vortex is at the same height as the center of the hole. When the hole is in the lower part, the vortex is between the hole and the top of the structure. From the horizontal profile (C1) of the flow field, the main flow is located near the line of the holes between front and rear panels. The internal water flow into the tridimensional network structures has water exchange in vertical and longitudinal directions, and the lateral water exchange is relatively weak.

3.2 *Vertical changes in velocity and turbulent intensity*

Figure 5 shows the flow velocity tendency in *x*, *y*, and *z* directions from each observation point, with the relative depth z/h (z refers to the height from the slope and h refers to the water depth) along the tridimensional network structures. It can be seen that the upstream flow velocity is mainly in the longitudinal direction (x direction) with 0.7–1.0 m/s flow velocity and zero lateral and vertical velocity. After the water flows into the tridimensional network structure, the velocity near the top of the structure increased due to the sudden shrinkage of the flow section caused by the structures, especially the structures in the first row. Furthermore, because of the blocking effect of the panels, the longitudinal velocity in the lower part, especially near the river bed, of the structure apparently decreases to 0–0.4 m/s. In general, the internal flow of the first row of the structure group is still in the flow direction (*x* direction).

The vertical velocity distribution in the longitudinal direction (*x* direction) is close in the observation points 2#–5#, showing that the internal flow velocity of the structures has been adjusted and stabilized after the transition through the front area of the structure group. There also exists clockwise vortex inside each structure, with negative longitudinal flow velocity in the lower part and positive longitudinal flow velocity in the upper part. The longitudinal near-bottom velocity is about –0.2 m/s. The deceleration rate of the near-bottom velocity is about 75%, while the flow direction is reverse.

Because of the presence of holes on the panels of each structure and internal flow turbulence, a certain amount of water is exchanged inside the lateral and vertical directions. In the test section, the lateral velocity (v) in the structure is generally less than 0.05 m/s and the vertical velocity (w) is generally less than 0.04 m/s.

The tridimensional network structure group not only adjusts the vertical velocity distribution but also changes the vertical distributions of flow turbulence intensity. Figure 6 shows the vertical distributions of flow turbulence intensity in *x*, *y*, and *z* directions in each tridimensional network structure (1#–6#). It can be seen that the peak value of flow turbulence intensity occurs near the top of the structure. The maximum turbulence intensities σ_x, σ_y, and σ_z are, respectively, 0.23, 0.13, and 0.23 m/s after the flow has been stabilized after the adjustment of the front transition area of the structure group. The flow turbulence intensities in the *x* and *z* directions are close, and larger than that in the *y* direction. Compared with the flow turbulence

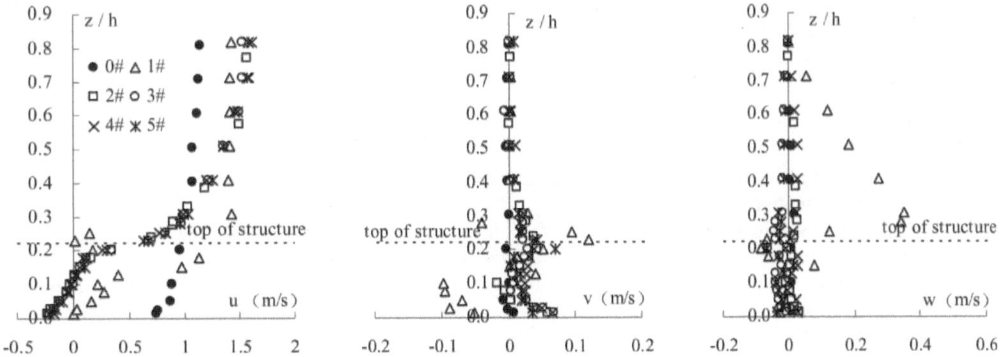

Figure 5. Vertical velocity distribution in the *x*, *y*, and *z* directions.

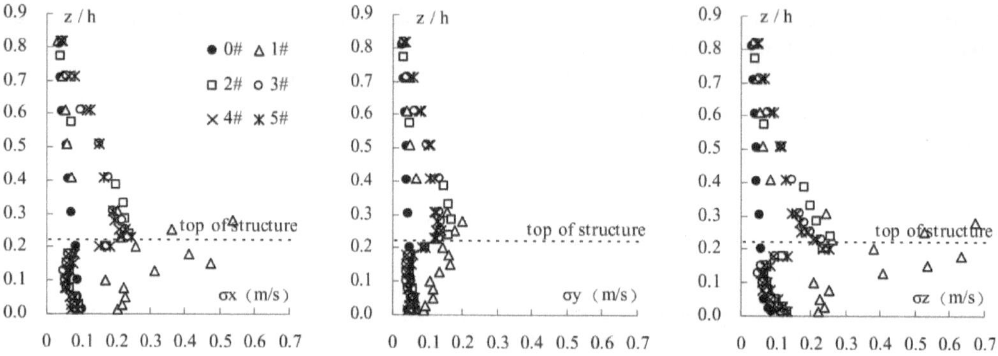

Figure 6. Vertical turbulence intensity distribution in the *x*, *y*, and *z* directions.

intensity upstream of the structure group, the flow turbulence intensity in the *x* direction is smaller with 20% deceleration rate near the bottom, whereas the flow turbulence intensity in the *y* direction is closed. The flow turbulence intensity in the *z* direction of the structures in the settling section is 30% larger than that in the upstream of the structure group.

Precisely, water exchange in the structure mainly occurs in the *x* and *z* directions, while that in the *y* direction is comparatively small.

4 CONCLUSION

A laboratory model experiment with a 1:3 scale was carried out to study the adjustment of a new type of tridimensional network structure to the local flow pattern in a variable-slope flume. The main results of the experiment are as follows. (1) The tridimensional network structure reduces the flow velocity near the bottom of the slope to protect the nutrient soil from erosion, through the shelter effects and the energy dissipation of the structure panels. (2) When the water flows into the tridimensional network structure, the clockwise vortex structure is formed and the water inside the structure exchanges with the surrounding fluid. (3) After the water flow is dissipated by the tridimensional network structure, the deceleration rate of the near-bottom velocity is about 75% in the opposite direction. The decay rate of the longitudinal turbulence intensity is about 20%, whereas the near-bottom vertical turbulence intensity is increased by about 30%.

ACKNOWLEDGMENTS

This study was supported by the National Key Research and Development Program (No. 2016YFC0402100) and the National Natural Science Foundation of China (Grant Nos. 51309160 and 51579216).

REFERENCES

Cao Z.H. 2014. Application of the ecological interlock type structure in revetment project. *Water Resources Planning and Design*, (1): 51–53.

Li F.H. & Yang Y.P. & Sun W.H. 2016. Design of ecological revetment structure of Tiepu-Xiongjiazhou Reach waterway regulation in Jinjiang River. *China Water Transport*, (1): 35–38.

Ma A.X. 2016. Flume experiment study on tridimensional network structure of ecological revetment. Nanjing Hydraulic Research Institute.

Xiao Q.H. & Ma A.X. 2016. Study on ecological revetment structures. Changjiang Waterway Planning Design and Research Institute, Nanjing Hydraulic Research Institute.

Hydraulic Engineering V – Hong, Liu & Xie (Eds)

Experimental study on the application of EIT to the measurement of concentration distribution in the dredge pipeline

Chaozhe Yuan, Jin Xing, Runli Tao, Jifu Yin & Guojun Hong
CCCC National Engineering Research Center of Dredging Technology and Equipment, Shanghai, China

ABSTRACT: Electrical Impedance Tomography (EIT) has been widely used in the medical field. In this study, the EIT technology was used to measure the concentration distribution in the dredge pipeline, which has the advantage of being noninvasive, cheap, safe, non-toxic, simple to operate and rich in information. A set of EIT system was designed and developed independently by the project team, which includes excitation power supply, measuring system, and inversion algorithm based on the Newton–Raphson method. The developed system was tested indoors in a DN300 model tube with a fine sand concentration of 21%, salinity of 3.5%, and temperature of 20°. The results of the study indicate that the system can invert the concentration distribution in the DN300 model tube, with the inversion precision of the system being about 5%. Therefore, the EIT system has a very high application value.

1 INTRODUCTION

EIT (Electrical Impedance Tomography) technology has been widely used in the medical field, which is a noninvasive type of medical imaging used to acquire human biomedical data from its electric properties and the changes of tissues and organs. The EIT system consists of three main parts, namely excitation power supply, measuring system, and inversion algorithm. In the process of inversion of the conductance distribution, the excitation power supply is used to add an electric field of the test body part, and the current injection mode and the measured voltage data are used to image the internal conductance distribution, according to a certain image reconstruction algorithm. Electrical impedance tomography is an important means for the detection of human information in medicine. This technology is characterized by noninvasive property, low cost, safety, nontoxic side effects, simple operation, and rich information. Furthermore, it is widely accepted by doctors and patients (S.F. Rao, 2016).

In the field of dredging, the main instruments used for the measurement of the concentration of slurry in the pipeline include γ-ray concentration meter and acoustic impedance concentration meter. These two instruments realize linear measurement and point measurement, respectively. However, they cannot obtain the distribution of slurry concentration in the dredge pipeline (G.X. Xu, 2004).

Because of its unique characteristics, EIT has been developed rapidly in the fields of medicine, geological prospecting, industrial process, and multiphase flow detection. However, there are a few applications in the monitoring of slurry transportation in dredging field. In this study, the EIT technology is applied to measure the conductance distribution of slurry in the dredging pipeline, and a method to obtain the concentration distribution in the dredge pipeline is established. It can not only guide the construction but also promote the study of the critical velocity in dredge pipeline.

2 METHODS OF EIT

2.1 *Numerical calculation method of EIT technology*

The development history of EIT is not long, and more than 30 scholars and organizations are engaged in its basic research work. At this stage, the generalized pulse spectrum method, perturbation method, back-projection method, double-limit method, Newton–Raphson method, layer stripping method, D-Bar method, and sensitivity matrix methods were presented internationally. The electrical impedance imaging algorithm can be divided into two categories: linear approximation method and nonlinear reconstruction method. The linear approximation method mainly includes: Calderon approximation method, back projection method, one step Newton method and sensitivity matrix method. In the nonlinear method, the D-Bar method and the layer stripping method belong to the direct optimization algorithm. Most of the nonlinear methods are based on the least square method, such as the classical Newton-Raphson method, the damped least squares method and the generalized inverse method.

According to the imaging method, electrical impedance tomography can be divided into two categories, namely dynamic imaging (or differential imaging) and static imaging. Dynamic imaging uses two different time measurement data through the image reconstruction algorithm to obtain the difference between the two electrical impedance distributions, thus reconstructing a differential image. The advantage of the dynamic algorithm is that the noise of the measurement system can be eliminated at the time of subtraction, so its image reconstruction algorithm is not too high for the measurement system. Therefore, its operation and calculation are easy. The disadvantage of the dynamic algorithm is that the application range is narrow. When the electrical impedance distribution does not change at two moments of data acquisition, it cannot be imaged. Although the development of static imaging reconstruction algorithm is relatively late, it drew wide attention because of its extensive application and relatively good imaging results. At present, the Newton–Raphson algorithm is the most popular algorithm. It is developed to minimize the difference of the sum of the squares of the measured values and calculated values.

EIT consists of two problems, namely positive problem and inverse problem. Solving the boundary electric field distribution by the known internal conductivity distribution is the positive problem, and the inverse is known as the inverse problem. The inverse problem is solved by using the solution of the positive problem to correct or iterate the calculated conductivity to obtain the true value.

2.2 *Positive problem*

The positive problem in the electrical impedance imaging is to calculate the potential distribution under the known conditions of the conductivity distribution, the current injection position, and the current value in the imaging field. In general, the current field is regarded as a quasi-static field, and the potential distribution function (Φ) and the conductivity distribution function (σ) of the electric field region satisfy the Laplace equation:

$$\nabla \cdot \sigma \nabla \phi = 0 \tag{1}$$

The Dirichlet boundary condition is

$$\phi\big|_{\partial\Omega_1} = \phi_0 \tag{2}$$

The Neumann boundary condition is

$$\sigma \frac{\partial \phi}{\partial n}\bigg|_{\partial\Omega_2} = J \tag{3}$$

where $\partial\Omega$ denotes domain boundaries, with boundary conditions involving Dirichlet and Neumann conditions; Φ_0 is the known boundary potential, J is the known boundary current density, and n is the outward normal direction of boundary. It is a positive problem to calculate the potential distribution in the field based on the known conductivity σ distribution and boundary conditions.

2.3 *Inverse problem*

Kohn and Vogelius theoretically proved that the isotropic conductivity distribution in the measured area can be uniquely determined by obtaining infinite high-accuracy voltage (or current) measurements of all the measured regions (Kohn R.V. *et al.*, 1989). However, the number of electrodes that can be placed outside the measured area is limited in practical applications, and the accuracy of measured data is limited by the influence of the acquisition system, so that the reconstructed image is often the approximation of the true distribution image.

The inverse problem of electrical impedance imaging is a nonlinear problem. In general, the iterative method is used to solve the nonlinear problem to obtain satisfactory results. The realization of the reconstruction algorithm in static electrical impedance imaging is usually to convert the electrical impedance imaging problem into a nonlinear system, and then the iterative method is used to solve the nonlinear problem. The absolute conductivity distribution is obtained, and the static image reconstruction is realized. In the static EIT algorithm, the most widely used method is the Newton–Raphson (and modified Newton–Raphson) method based on the least-squares technique.

The Newton–Raphson iterative algorithm needs to constantly solve the positive problem, until the calculation accuracy meets the requirements.

The fundamental processes of the algorithm are as follows (Y. Ma *et al.,* 2002):

1. Set the σ initial value; σ_0
2. Calculate the control equation to obtain the voltage ϕ_b of the measuring boundary element;
3. Compare the calculated ϕ_b and the measured ϕ_b^m; if the condition $\|\phi_b - \phi_b^m\| > \varepsilon$ is satisfied, then the calculation is completed, otherwise move to step 4;
4. Update the conductivity value σ, that is, $\sigma^{k+1} = \sigma^k + \Delta\sigma^k$;
5. Repeat steps 2, 3, and 4 until the calculation accuracy meets the requirement.

3 PROCESS AND RESULTS

3.1 *Testing process*

The framework of the electrical impedance tomography system is shown in Figure 1. The excitation power supply and measuring system provided the input boundary conditions for the EIT algorithm. Using the EIT algorithm, the conductivity distribution in the test pipeline can be obtained. Finally, the true mud concentration can be obtained on the basis of the conductivity and concentration of slurry corresponding database.

During the indoor test, the DN300 plexiglass pipe was used (Figure 2) and 18 electrodes (numbered 1, 2, 3 … 16, 17, and 18) were arranged on a specified section of the DN300 pipe. In the measurement process, any two electrodes can be selected as the excitation electrode, and the other 16 electrodes as the electric potential measurement electrodes.

In theory, there are many ways to arrange the electrodes. Practically, there are two ways of electrode arrangement, namely adjacent electrodes energization and opposite electrodes energization. Because of the complex measurement environment, low measurement accuracy, and other reasons, the measurement error is inevitable, thereby affecting the accuracy of the image reconstruction result. The error of the adjacent electrodes energization mode is more sensitive to the other mode. Therefore, it is more reasonable to choose the opposite electrodes energization mode in complex environment when the measurement accuracy cannot be guaranteed (J.X. Yang, 2014, T. Tong, 2011).

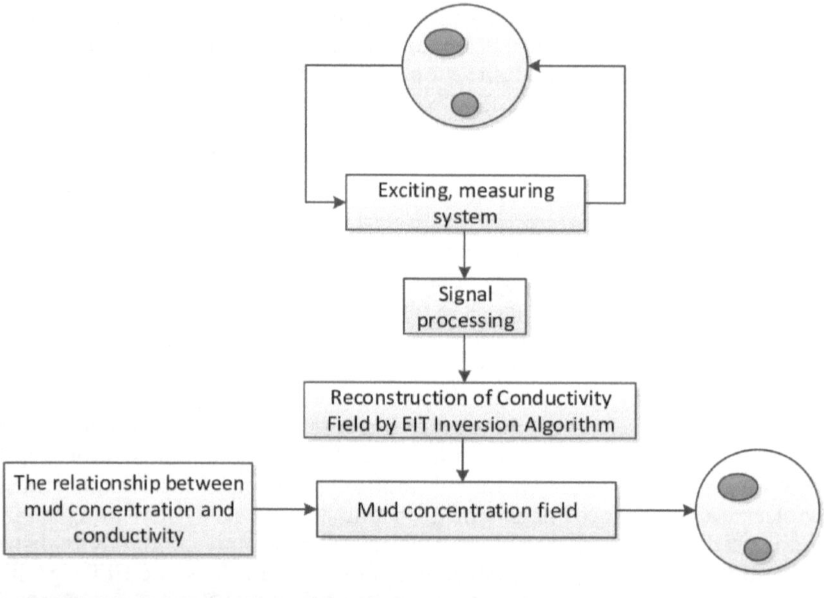

Figure 1. Framework of the electrical impedance tomography system.

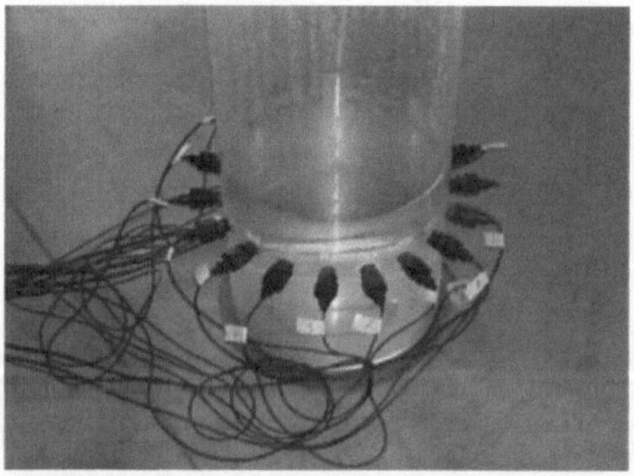

Figure 2. DN300 plexiglass pipe.

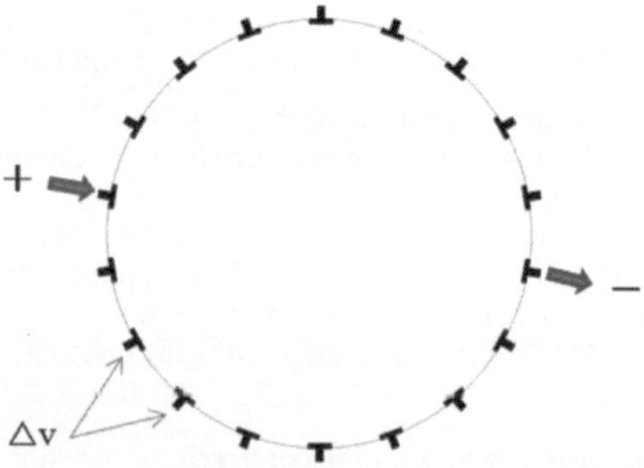

Figure 3. Schematic diagram of opposite electrodes' energization mode.

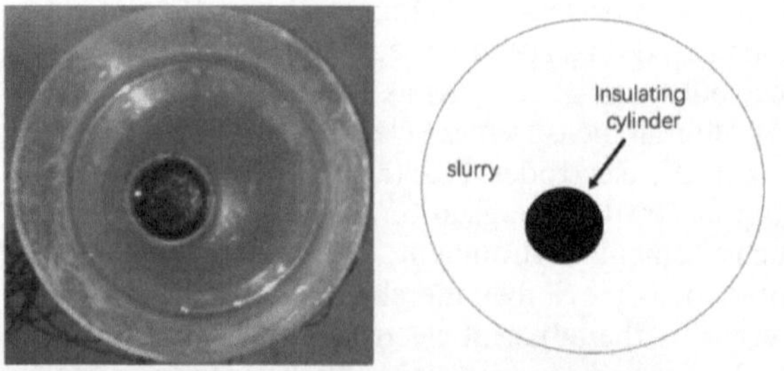

Figure 4. DN300 test pipe slurry and distribution diagram of the insulating cylinder.

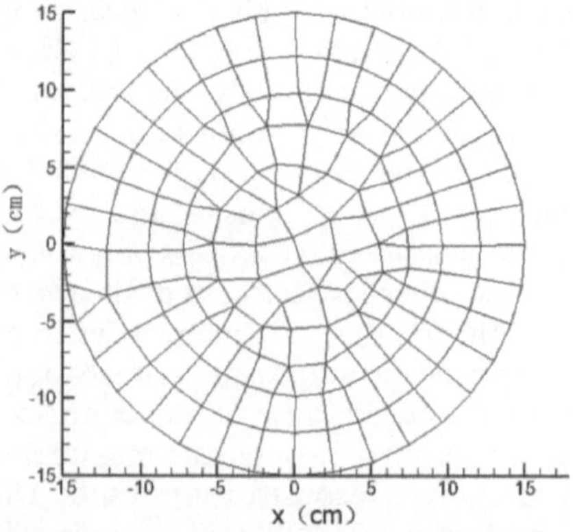

Figure 5. Unstructured grid for conductivity reconstruction.

In this study, the slurry in the plexiglass pipe is a mixture of fine sand and 3.5% saline. The median diameter of the fine sand is about 0.16 mm, and the volume concentration of the slurry is about 21% (F.X. Wang *et al.*, 2012). Because of the complexity of the test environment and unpredictable factors, the mode of the indoor test is the energization mode of opposite electrodes (Figure 3). The excitation electrodes are electrode 1# and electrode 10#, and by measuring the potential difference between the other electrodes, we obtain a set of 14 measurements. Similarly, with electrodes 2# and 11# energized, another 14 measurements were obtained. Continuing in this manner, 126 measurements can be obtained. These 126 measurements form a complete set of the corresponding conductivity field under the corresponding excitation.

In order to verify the accuracy of the reconstruction algorithm, a graduated cylinder was placed in the homogeneous slurry, as shown in Figure 4. The cylinder was used to simulate the large insulating blocks in the slurry. In the process of the conductivity reconstruction, the inversion algorithm is carried out using the unstructured grid, as shown in Figure 5.

3.2 *Inversion results*

In the indoor test, the measured electrical potential difference exhibits asymmetric parabolic forms, as shown in the left- and right-hand sides of Figure 6. The reconstruction result of the conductivity distribution of the actual situation (shown in Figure 4) in the cylinder is shown in Figure 7. The measured value of the slurry conductivity is 3.9 mS/cm.

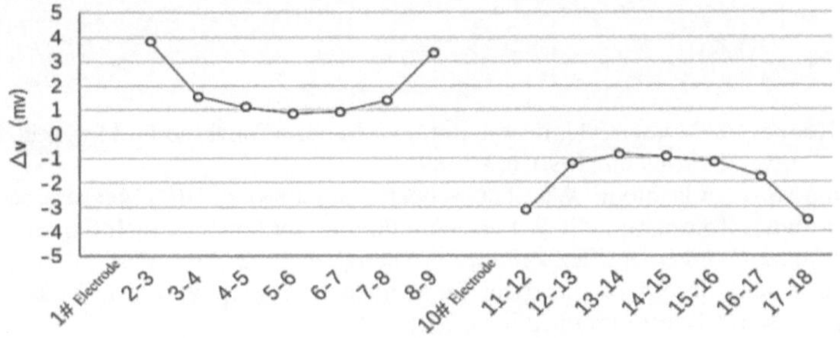

Figure 6. Potential difference between measuring electrodes under the energization mode of opposite electrodes (1# and 10#).

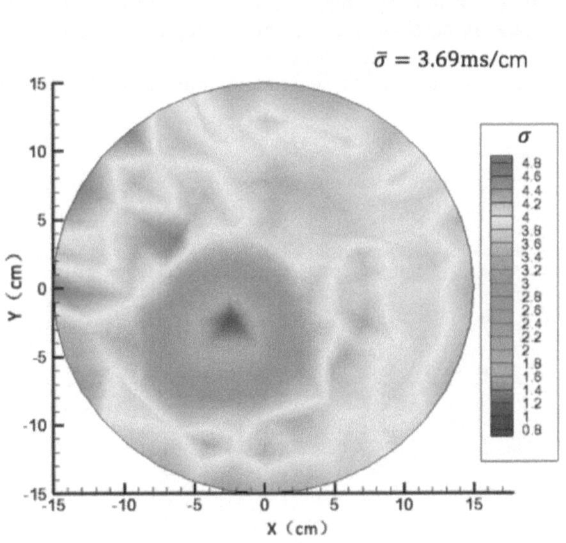

Figure 7. Schematic diagram of the image reconstruction of conductivity distribution.

Comparing Figures 4 and 7, we can easily obtain the distribution of the slurry concentration in the cylinder by the image reconstruction algorithm. The value of the blue area is very low, which corresponds to the location of the insulated glass cylinder. And according to the reconstruction of the conductivity distribution, the average conductivity in the pipeline is 3.69 mS/cm, which is about 5% smaller than the measured conductivity of 3.9 mS/cm of the slurry.

4 CONCLUSION

In this study, the EIT technology was used to measure the conductivity distribution of slurry in the dredging pipeline. The developed EIT system was tested indoor in a DN300 model tube with a fine sand concentration of 21%, salinity of 3.5%, and temperature of 20°. Several conclusions can be drawn from the test:

1. The energization mode of opposite electrodes meets the accuracy requirements of the image reconstruction algorithm.
2. The EIT system (using the Newton–Raphson algorithm method and opposite electrodes' energization mode) can invert the concentration distribution of slurry as well as clearly distinguish the insulated cylinder from the slurry in the DN300 test pipe.
3. The inversion accuracy of the EIT system can reach about 5%, which meets the general requirements of accuracy in the dredging field.
4. Most importantly, the EIT system has a high application value in the dredging field.

REFERENCES

Kohn, R.V., McKenney, A. Numerical implementation of a variational method for electrical impedance tomography [J]. Inverse Problems, 1990, 6(3): 389.

Ma, Y., Miedema, S.A., Vlasblom, W.J. Theoretical Simulation of the Measurements Process of Electrical Impedance Tomography [C]//Asian Simulation Conference/5th International Conference on System Simulation and Scientific Computing, Shanghai. 2002: 261–265.

Rao, S.F. Research and design of electrical impedance tomography testing system [D]. Nanjing University of Posts and Telecommunications, 2016.

Tong, T. Designing exciting circuit for electrical impedance tomography [D]. Shanghai Jiao Tong University, 2011.

Wang, F.X., G.J. Hong, C. Zhang, R.M. Luo. Primary study on conductivity characteristics of dredging soil-water mixture [J]. Port & Waterway Engineering, 2012, (12): 111–114.

Xu, G.X. Theoretic and applied research of electrical impedance tomography [D]. Chongqing University, 2004.

Yang, J.X. Electrical impedance tomography research and application [D]. Northwest A&F University, 2014.

Hydraulic Engineering V – Hong, Liu & Xie (Eds)
© 2018 Taylor & Francis Group, London, ISBN 978-0-8153-7471-8

The flow characteristics around a hydrofoil and the stall mechanism influenced by the electromagnetic force

Jinfu Yin & Guojun Hong
National Engineering Research Center of Dredging Technology and Equipment, Shanghai, China

Ke Chen, Yunxiang You & Tianqun Hu
State Key Laboratory of Ocean Engineering, Shanghai Jiaotong University, Shanghai, China

ABSTRACT: This paper numerically researches the flow characteristics around a hydrofoil and stall mechanism influenced by the electromagnetic force at a chord length Reynolds number $Re_L = 1.63 \times 10^5$ with the method of Detached Eddy Simulation (DES). The regulation and mechanism of the three dimensional flow characteristics at the normal and stall attack angle were investigated. The results show that the electromagnetic force can completely restrain the laminar separation at the normal attack angles. The large scale vortex structures were changed to the laminar ones at stall attack angles. Meanwhile, the electromagnetic force can increase the lift, reduce the drag, decrease the fluctuations of lift/drag, and raise the lift-drag ratio. Moreover, the electromagnetic force can significantly delay the stall problem and increase the stall angle, improving the working performance.

1 INTRODUCTION

The hydrofoil structure is widely used in ship engineering. In the process of navigating with the ship, the flow separation of the hydrofoil surface will decrease its lift and increase the resistance at a certain attack angle, even resulting in stall that directly affects the propulsion efficiency and control of the ship performance. Stall is a state in which the lift is suddenly reduced when the angle of attack of the hydrofoil increases to a certain critical value. This is a phenomenon that is not expected during the voyage of the ship (Ekaterinaris, 1998). Therefore, how to effectively suppress and delay the stall of hydrofoil structure is of great engineering value and academic significance in ship engineering.

Flow separation is one of the essential causes of hydrofoil stall phenomenon, so effective control of surface flow separation is important to suppress and delay the phenomenon. Numerous flow control methods can be found, including passive and active ones. The primary advantage of the passive control approaches is the simplicity but the change of the body shape is required. In contrast, active flow control can be performed by blowing, suction, synthetic jet and electromagnetic forces in some cases (Henoch, 1995).

In recent years, using the electromagnetic force to control the flow separation becomes an international research hotspot. The results show that the electromagnetic force can increase the velocity of the near the hydrofoil wall and improve the maximum stall angle of the hydrofoil. It can also effectively suppress and retard the occurrence of the hydrofoil stall and improve the bucket efficiency of the hydrofoil (Weier, 2004; Cierpka, 2010; Chen, 2008; Chen, 2009; Liu, 2011; Albrecht, 2010; Mutschke, 2006). The researchers also studied the mechanism of the Lorentz force-controlled hydrofoil around the flow field and discussed the control efficiency (Yin, 2014). At present, the two-dimensional approximation formula is used to study the electromagnetic force control problem.

However, the three-dimensional characteristic and stall mechanism of the electromagnetic force control hydrofoil structure are still not very clear yet. Thus we present this paper in which the Maxwell equation is used to directly analyze the three-dimensional electromagnetic force of the hydrofoil surface. The DES numerical simulation method is used to simulate the flow characteristics and the stall mechanism at the sub-critical Reynolds number $Re_L = 1.63 \times 10^5$. The regulation and mechanism of the three dimensional flow characteristics at the normal and stall attack angle are investigated.

2 NUMERICAL METHODS

A hydrofoil covered by electrodes and magnets is placed in a weakly conductive fluid. The model of the hydrofoil is shown in Figure 1. The electrode and the magnetic pole are arranged on the upper surface of the hydrofoil. The magnets are permanent magnets, so the size and direction of magnetic field are constant all the time. The electric field is varying with the voltage loaded on the electrodes. When the hydrofoil is in the weakly conductive fluid to form a complete circuit, the characteristic of flow field around will be changed by the Lorentz force.

The electromagnetic force is added to the momentum equation, and the incompressible N-S equation is:

$$\left.\begin{aligned} &\frac{\partial u_i}{\partial t} + u_j \frac{\partial u_i}{\partial x_j} = -\frac{1}{\rho}\frac{\partial p}{\partial x_i} + \frac{\partial}{\partial x_j}\left(v \frac{\partial u_i}{\partial x_j}\right) + f_i \\ &\frac{\partial u_i}{\partial x_i} = 0 \end{aligned}\right\} \tag{1}$$

where t is time variable, ρ the density of the fluid, v the dynamic coefficient of viscosity, $u_i (i = 1,2,3)$ the instantaneous velocity components, p the pressure and f_i the Lorentz force, x_i and x_j are the spatial vectors.

The Lorentz force in the weakly conductive fluid is given by

$$\mathbf{F} = \mathbf{J} \times \mathbf{B} \tag{2}$$

$$\mathbf{J} = \sigma(\mathbf{E} + \mathbf{u} \times \mathbf{B}) \tag{3}$$

where **J** is the current density vector, **B** and **E** are the magnetic and electric field vectors respectively, and σ is the electric conductivity. In weakly conductive fluids ($\sigma \sim 10\, S \cdot m^{-1}$) like seawater or electrolytes, the electrical currents originating from $\sigma \mathbf{u} \times \mathbf{B}$ are usually too small to produce any noticeable effect on the flow. The Lorentz force in the weakly conductive fluid is changed to be

$$\mathbf{F} = \sigma(\mathbf{E} \times \mathbf{B}) \tag{4}$$

The DES method is often based on the one-equation Spalart-Allmaras, realizable $k - \varepsilon$ and SST $k - \omega$. Among the turbulent models, the transition SST model is based on the coupling of the SST transport equations with two other transport equations. The SST $k - \omega$ model

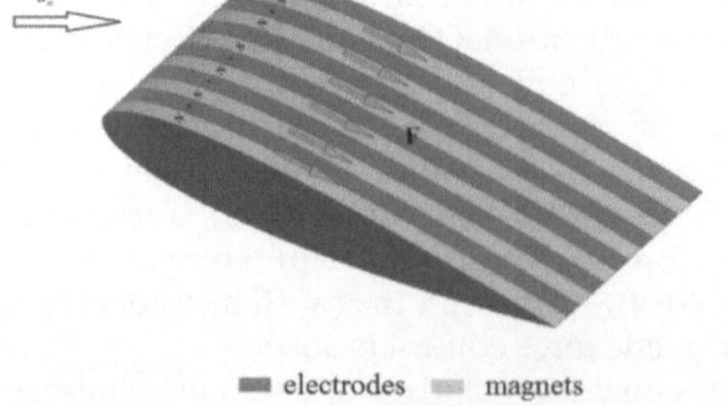

Figure 1. Configuration of the hydrofoil covered with electrodes and permanent magnets.

gives a highly accurate prediction of the flow separation under adverse pressure gradients by the inclusion of transport effects into the formulation of the eddy-viscosity.

To obtain the $k-\omega$ formulation, the turbulent length scale $l_{k-\omega}$ in the k equation dissipation term is defined as:

$$l_{k-\omega} = k^{1/2}/\beta_k \omega \tag{5}$$

The model constant β_k has a value of 0.09.

In the DES method, the RANS and LES grid scale L_{DES} is modified by:

$$l_{DES} = \min\left(l_{k-\omega}, C_{DES}\Delta\right) \tag{6}$$

where C_{DES} is the calibration constant, and the Δ is the maximum local grid spacing. If open the subscale stress model prematurely, it would result in the sharp decline of Reynolds stress in the boundary layer which leads to the grid induce separation phenomenon, so C_{DES} has a value of 0.61. In the case that $l_{k-\omega} \le C_{DES}\Delta, l_{DES} = l_{k-\omega}$, the model will operate in SST mode in the boundary layer. In the other case $l_{k-\omega} > C_{DES}\Delta, l_{DES} = C_{DES}\Delta$, the model will operate in LES mode and develop a flow instability past the step with the correct balance of the stresses.

As the vertical distance from the hydrofoil wall in the middle section y_d, the electrode pole width $a = 0.01\ m$. The Maxwell equation is used to solve the electromagnetic force distribution of the airfoil surface formed by electrodes and magnets and is discretized by the finite element method. Fig. 2(a) gives the normal electromagnetic force density and (b) gives the distribution of equipotential lines at a cross section, where x is the wing chord, y is the airfoil normal, and $|F|$ is the electromagnetic force. The results show that the electromagnetic force is exponentially decaying along the normal direction and becomes zero at about $y_d/c = 0.015$, i.e. The periodic variation characteristic appears along the orientation and the maximum is located at the junction of the electromagnetic pole. The variation is similar to that of the electromagnetic force distribution on the cylindrical surface (Yin 2015).

Taking the NACA0018 as the research object, the chord length c is 0.3 m and the angles of attack are between 5°~30°. The length of the chord has a significant effect on the characteristics of the hydrofoil during the numerical simulation. In order to simulate the complete three-dimensional characteristics, the length of the chord is selected about 5c. The origin of the coordinate system is located at the front of the hydrofoil. The computational domains in the front of and behind the hydrofoil are set to 12.5c and 21c, respectively. Uniform velocity profile is applied to the velocity inlet boundary and the pressure outlet to the outlet. A symmetry boundary condition is employed to the lateral surface and the no slip wall is used to the hydrofoil surface. In the present simulation, the Finite Volume Method (FVM) is employed to calculate the 3-D unsteady incompressible Navier-Stokers equation. A QUICK format scheme is used for the convective terms and a central differencing scheme is used for momentum discretization. The PISO scheme is used to deal with the pressure-velocity coupling algorithm and the initial time step is about 0.001 s.

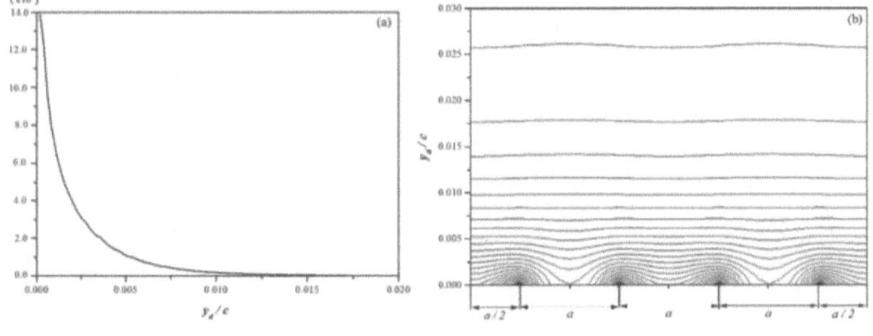

Figure 2. Three dimensional distribution of the Lorentz force around the hydrofoil (a) Normal direction (b) Spanwise direction.

3 RESULTS AND DISCUSSION

The total resistance coefficient is defined as C_d, the total lift coefficient C_l and the electromagnetic force parameters N during the simulations.

$$C_d = \frac{2F_d}{\rho u_\infty^2 S}, C_l = \frac{2F_l}{\rho u_\infty^2 S}, N = \frac{j_0 B_0 c}{\rho u_\infty^2} \tag{7}$$

where u_∞ is the velocity, B_0 the main magnetic field intensity, E_0 the main electric field intensity, and $j_0 = \sigma E_0$. The interaction parameter N is the ratio of the Lorentz force to the inertial force in the fluid flow. F_l and F_d denote the total lift and drag forces, respectively. S is the projected area of the hydrofoil to the direction of the free-stream.

3.1 *No control flow*

In the numerical simulation, the chord length Reynolds number Re_L is 1.63×10^5, weak electrolyte density ρ is 1025 kg/m^3, fluid conductivity σ is 10 $S.m^{-1}$, and kinematic viscosity v is 10^{-6} m^2/s. The flow around the hydrofoil at different attack angles is simulated without electromagnetic force. Fig. 3 shows the numerical results for the vorticity distribution of the flow around a hydrofoil when $Re_L = 1.63 \times 10^5$ and $t = 50$ s. The vortex field of the hydrofoil around the flow field shows laminar flow separation and a long vortex structure in the rear of the hydrofoil before the stall at $\alpha = 12°$. The hydrofoil is already in the stall state and the large-scale flow separation phenomenon appears at $\alpha = 18°$. The trailing edge vortex interacts with the separation vortex show a more complicated flow phenomenon.

Fig. 4 shows the comparisons of the time-averaged value of lift/drag coefficients with the experimental values (Jacobs, 1937) when $Re_L = 1.63 \times 10^5$ and $t = 50$ s. It can be seen from Fig. 4 that the coefficient of lift and drag agree with the experimental results. The stall angle of the NACA0018 hydrofoil is about $\alpha = 14°$ at the Reynolds number $Re_L = 1.63 \times 10^5$ in the subcritical region. When the angle of attack is greater than the stall angle, the lift coefficient decreases sharply and the drag coefficient increases.

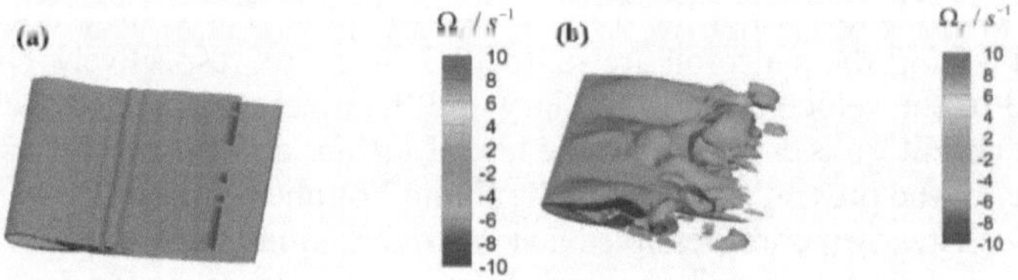

Figure 3. Numerical results for the vorticity distribution of the flow around a hydrofoil when $Re_L = 1.63 \times 10^5$ and $t = 50$ s, (a) $\alpha = 12°$, (b) $\alpha = 18°$.

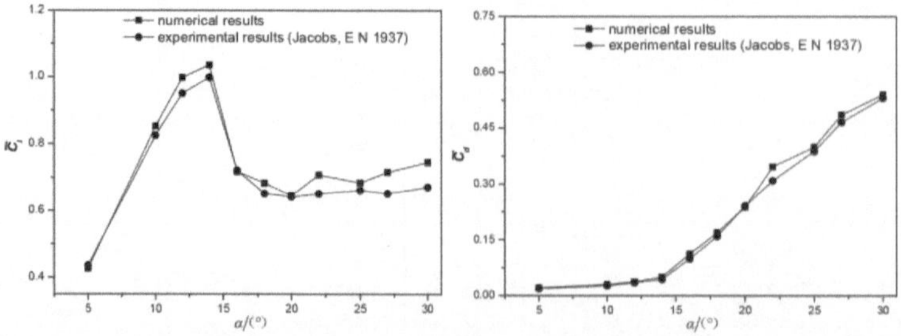

Figure 4. Comparisons of the time-average value of lift/drag coefficients with the experimental values when $Re_L = 1.63 \times 10^5$ and $t = 50$ s.

3.2 *Control of flow and drag/lift characteristic*

Firstly, use the DES model to simulate the flow characteristics in the absence of the Lorentz force about 20 *s*. Turn on the electromagnetic force at this instant and the flow becomes stable again at $t = 70\ s$. Fig. 5 illustrates the numerical results for the effect of the electromagnetic force on the vorticity distribution of the flow around the hydrofoil when $Re_L = 1.63 \times 10^5$, $t = 70\ s$ and $\alpha = 18°$. Obviously, the vorticity distribution characteristics in the near-wall region are changed and the vorticity fluctuations along the spanwise direction are weakened with the increasing electromagnetic force parameter. Once the interaction parameter reaches a certain critical value, the jet flow phenomenon behind the hydrofoil happens.

The effect of the electromagnetic force is not obvious and the vortex shedding becomes weaker when the interaction parameter value of $N = 0.25$. In the case of $0.5 \leq N < 1.5$, the flow separation point of the upper surface moves further to the trailing edge, and the flow separation is basically suppressed. There is no large-scale vortex pulsation in the upward direction and still a small amount of vortex in the wake flow field. The vorticity field of the wake is similar to that in the absence of electromagnetic force before the stall angle $\alpha = 12°$. There is no separation vortex around the flow field, and the electromagnetic force drives the fluid flow to produce a narrow laminar vortex structure when $N \geq 1.5$.

Fig. 6 shows the numerical results for the vorticity distribution of the middle spanwise section controlled by the electromagnetic force when $Re_L = 1.63 \times 10^5$, $t = 70\ s$ and $\alpha = 18°$. It can be seen that the effect of the electromagnetic force first changes the vorticity distribution at the leading edge, so that the position of flow separation moving to the trailing edge.

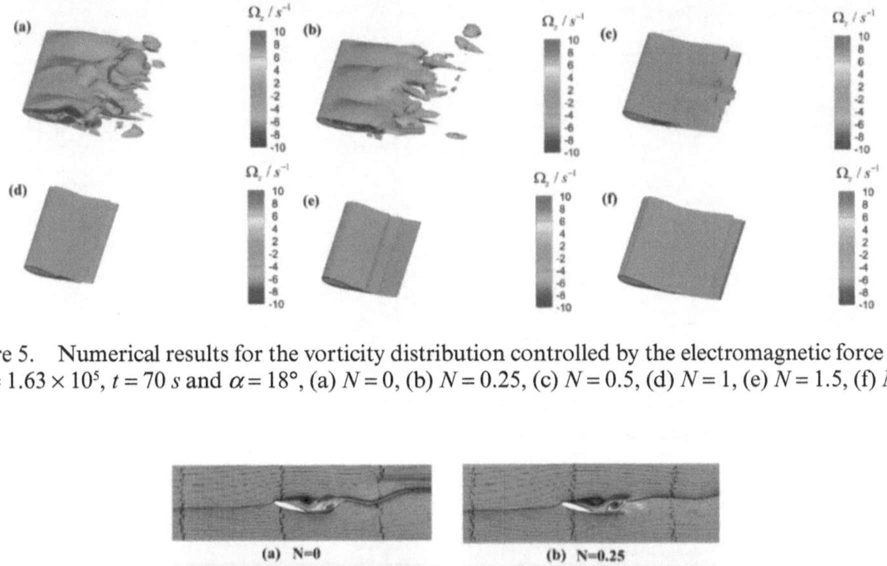

Figure 5. Numerical results for the vorticity distribution controlled by the electromagnetic force when $Re_L = 1.63 \times 10^5$, $t = 70\ s$ and $\alpha = 18°$, (a) $N = 0$, (b) $N = 0.25$, (c) $N = 0.5$, (d) $N = 1$, (e) $N = 1.5$, (f) $N = 2$.

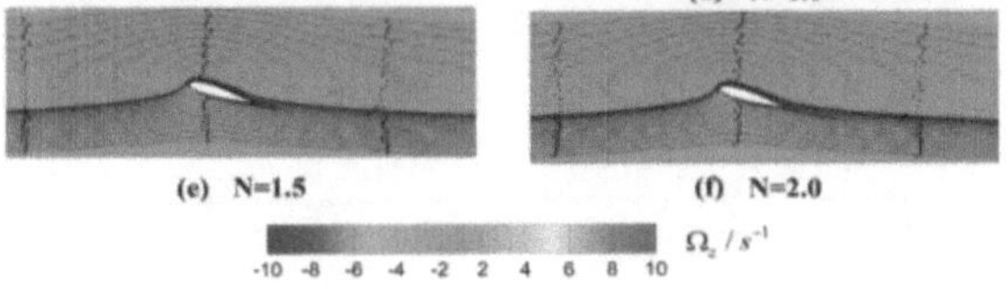

Figure 6. Numerical results for the vorticity distribution of the middle spanwise section controlled by the electromagnetic force when $Re_L = 1.63 \times 10^5$, $t = 70\ s$ and $\alpha = 18°$.

When the electromagnetic force increases, the separation position approaches to the tail edge. Then the separation and induced vortex are changed into laminar. The electromagnetic force accelerates the surface flow and the extended area of the narrow vortex structure increases with the increasing of the electromagnetic force.

When the hydrofoil is stalled, its lift coefficient is drastically reduced and the drag is significantly increased. Electromagnetic force changes the flow field and the drag/lift characteristics will inevitably change. The drag/lift characteristics are different due to the angle. Fig. 7 shows the numerical results for the time histories and the time-averaged value of the lift coefficients controlled by the electromagnetic force when $Re_L = 1.63 \times 10^5$, and $\alpha = 18°$. The results show that the mean value of the lift coefficient is only 0.68 and large scale pulsation in the wake field without electromagnetic force in the stall state at $\alpha = 18°$. The lift coefficient is also basically fluctuating up and down at the time-average value. The mean value of lift coefficient increase is not large when the electromagnetic force effect is small for $N = 0.25$. The mean value is significantly increased and no significant pulsation characteristics when $N = 0.5$. For the case of $N \geq 0.5$, the mean value of the lift coefficient shows a quadratic growth trend with the electromagnetic force.

Figure 8 shows the numerical results for the time histories and the time-averaged value of the drag coefficients controlled by the electromagnetic force when $Re_L = 1.63 \times 10^5$ and $\alpha = 18°$. The drag coefficients have a significant pulsating characteristic which are similar to the lift without electromagnetic force. The mean value of the drag coefficient increases significantly compared with the $\alpha = 12°$, which is about 0.168. The electromagnetic force decreases the drag, but the reduced amplitude is not large, and its temporal characteristics indicate that the drag coefficient fluctuates at the time-averaged value when $N = 0.25$. For the case of $N = 0.5$, the mean value of the drag coefficient is significantly reduced, and the value

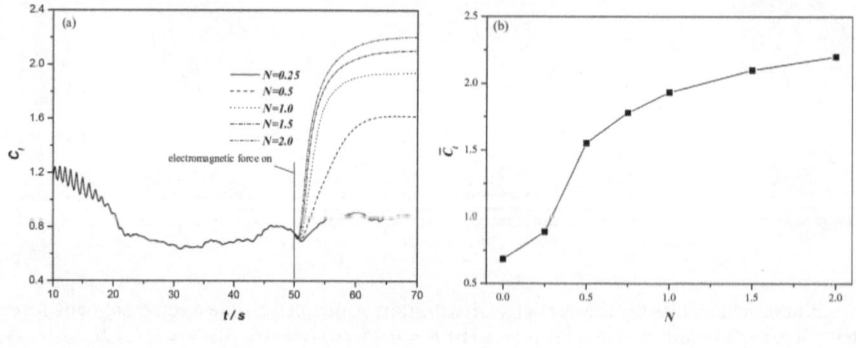

Figure 7. Numerical results for the time histories and the time-average value of the lift coefficients controlled by the electromagnetic force when $Re_L = 1.63 \times 10^5$ and $\alpha = 18°$.

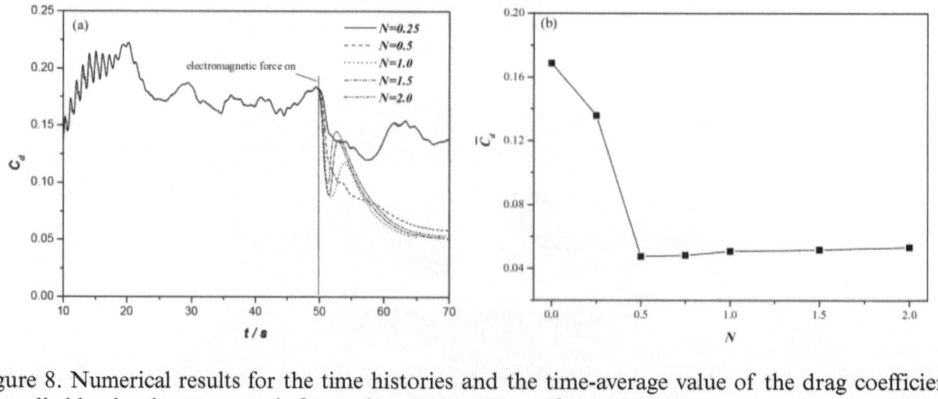

Figure 8. Numerical results for the time histories and the time-average value of the drag coefficients controlled by the electromagnetic force when $Re_L = 1.63 \times 10^5$ and $\alpha = 18°$.

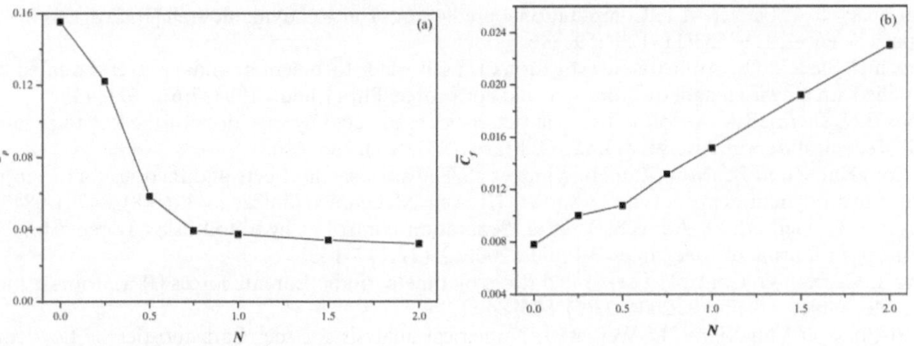

Figure 9. Numerical results for the time-average value of the pressure and friction coefficient controlled by the electromagnetic force when $Re_L = 1.63 \times 10^5$ and $\alpha = 18°$.

finally becomes stable with the increase of the numerical time. The mean value of the drag coefficient changes slowly and only slightly increases when $N \geq 0.5$.

The numerical results for the time-averaged value of the pressure and friction coefficient controlled by the electromagnetic force when $Re_L = 1.63 \times 10^5$ and $\alpha = 18°$ are illustrated in Fig. 9. The results imply that the pressure drag increases obviously and reaches more than 95% in the stall angle $\alpha = 18°$. The flow separation is the main reason for the increase in the total drag coefficient. The mean value of the pressure coefficient shows a quadratic attenuation and the friction coefficient increases linearly with the increasing of the electromagnetic force when $N \geq 0.25$. As the pressure and friction coefficient are in the opposite trending, the total drag is increasing.

4 CONCLUSIONS

The flow characteristics around a hydrofoil and the stall mechanism influenced by the electromagnetic force at a chord length Reynolds number $Re_L = 1.63 \times 10^5$ are simulated with the method of detached eddy simulation (DES). The regulation and mechanism of the three dimensional flow characteristics at the normal and stall attack angle were investigated.

The results show that the electromagnetic force can completely restrain the flow separation and the large scale eddies. The large scale vortex structures were changed to the laminar flow characteristics at the stall attack angle. Meanwhile, the electromagnetic force can increase the lift, reduce the drag, decrease the fluctuations of lift/drag, and raise the lift-drag ratio.

When N is less than the critical value N_c, the value of lift and drag coefficient change are not obvious. For the case of $N \geq N_c$, the lift coefficient shows a quadratic increase and the drag increases slowly and slightly. Moreover, the electromagnetic force can significantly delay the stall problem and increase the stall angle, improve the working performance.

REFERENCES

Albrecht T, Weier T, Gerbeth G, et al. Numerical and Experimental Investigation of Electromagnetic Separation Control Using Different Wave Forms [C]//5th AIAA Flow Control Conference, Chicago, IL, USA. 2010.

Chen YaoHui, Fan BaoChun, Chen ZhiHua, et al. Experimental and numerical investigations on the electro-magnetic control of hydrofoil wake [J]. Acta Mechanica Sinica, 2008, 57(2): 648–653.

Chen Yaohui, Fan Baochun, Chen Zhihua, et al. Influences of Lorentz force on the hydrofoil lift [J]. Acta Mechanica Sinica, 2009, 25(5): 589–595.

Cierpka C, Weier T, Gerbeth G. Synchronized force and particle image velocimetry measurements on a NACA 0015 in poststall under control of time periodic electromagnetic forcing [J]. Physics of Fluids, 2010, 22(7): 331–341.

Ekaterinaris J A, Platzer M F. Computational prediction of airfoil dynamic stall [J]. Progress in Aerospace Sciences, 1998, 33(11–12): 759–846.
Henoch C, Stace J. Experimental investigation of a salt water turbulent boundary layer modified by an applied streamwise magneto-hydrodynamic body force. Phys Fluids, 1995, 7(6): 1371–1383.
Jacobs E N, Sherman A. Airfoil section characteristics as affected by variations of the Reynolds number [J]. Technical Report Archive & Image Library, 1937, 8(2): 286–290.
Liu ZongKai, Zhou BenMou, Liu HuiXing, et al. The analysis of effects and theories for electromagnetic hydrodynamics propulsion by surface [J]. Acta Mechanica Sinica, 2011, 60(8): 84701-084701.
Mutschke G, Gerbeth G, Albrecht T, et al. Separation control at hydrofoil using Lorentz forces [J]. European Journal of Mechanics-B/Fluids, 2006, 25(2): 137–152.
Weier T, Gerbeth G. Control of separated flows by time periodic Lorentz forces [J]. European Journal of Mechanics—B/Fluids, 2004, 23(6): 835–849.
Yin Ji-Fu, You Yun-Xiang, Li Wei, et al. Numerical analysis for the characteristics of flow control around a circular cylinder with a turbulent boundary layer separation using the electromagnetic force [J]. Acta Phys. Sin. Vol. 63, No. 4 (2014) 044701.
Yin Ji-fu. Investigation on flow characteristics around blunt bodies controlled by electromagnetic forces [D]. Shanghai Jiaotong University, 2014.

Hydraulic Engineering V – Hong, Liu & Xie (Eds)
© 2018 Taylor & Francis Group, London, ISBN 978-0-8153-7471-8

ADAMS-MATLAB co-simulation of trailing suction hopper dredger under wave action

Jinwen Wang, Wei Shu, Guojun Hong, Jifu Yin, Zhongwei Zhou & Yiying Sun
National Engineering Research Center of Dredging Technology and Equipment, Shanghai, China

ABSTRACT: In view of the heave motion of Trailing Suction Hopper Dredger (TSHD) under wave action, Proportional-Integral-Derivative (PID) controller was designed and implemented to improve the heave compensator. It was co-simulated in both ADAMS and MATLAB software to study dynamic characters of TSHD under wave action. Simulation results show that the PID controller can effectively improve the stability of the drag head.

1 INTRODUCTION

The reliability of the TSHD during dredging is an important target to evaluate the performance of the TSHD. TSHD hull is connected with the seabed by lifting hanger, drag arm, drag head and other equipment. The TSHD will be subjected to the huge dynamic force by wind, wave and flow in the open ocean. The medium and long period wave seriously affects the construction of dredgers. Therefore, it is necessary to study the influence of wave and other external environment on the construction of TSHD.

As the most widely used mechanical system simulation software, ADAMS can be used to build and test virtual prototype on the computer. It can realize real-time online simulation and understand the movement performance of complex mechanical system. The ADAMS method was used to study the rule about the change of the mass center position of a TSHD's drag head with time when working in different soil hardness of seabed under the function of swell compensator. (Liu Zhi et al, 2006).

MATLAB has powerful control function, which can be used to establish a control scheme for complex mechanical system combined with ADAMS. (Zong Xiaoping et al, 2009, Liao C et al, 2015,Zhang Xiuqin et al, 2012, Çakan, Abdullah et al, 2016).

In view of the heave motion of the TSHD under wave action, co-simulation based on ADAMS and MATLAB software was introduced in this paper. It presented a detailed procedure for virtual testing of TSHD affected by heave compensator with PID controller.

2 MODELING

The research object is a TSHD owed by CCCC Shanghai Dredging Co., Ltd. The three-dimensional model of the TSHD's body and drag arms mechanism was established by CREO software, and saved as Parasolid format to ensure the integrity of the 3D solid Model. In order to facilitate the simulation, the hull and drag arm mechanism was simplified. Drag arm was divided into three parts: elbow, upper suction tube and lower suction tube. The drag head is included in lower suction tube. The TSHD model is showed as Picture 1.

The model was then imported into ADAMS. In ADAMS, a cuboid was built to simulate the seabed, which was located under the drag head. The angle between the TSHD hull and the drag arm was set to 30 degrees. The material and other basic information of the parts properties were redefined. The constraint relation of the parts was defined according to the physical prototype. In addition, the friction coefficient was set for each joint. The constraint type is listed in Table 1.

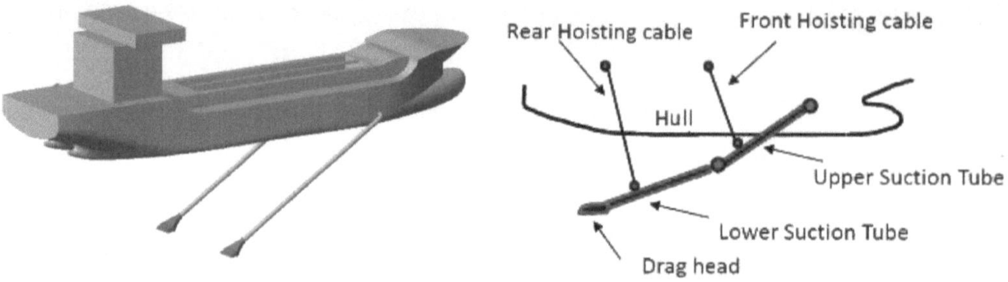

Figure 1. TSHD model.

Table 1. The constraint type of TSHD.

	Constraint type		
No.	Part 1	Part 2	Type
1	TSHD hull	elbow	revolute joint
2	elbow	upper suction tube	revolute joint
3	upper suction tube	lower suction tube	hooke joint
4	drag head	seabed	contact force

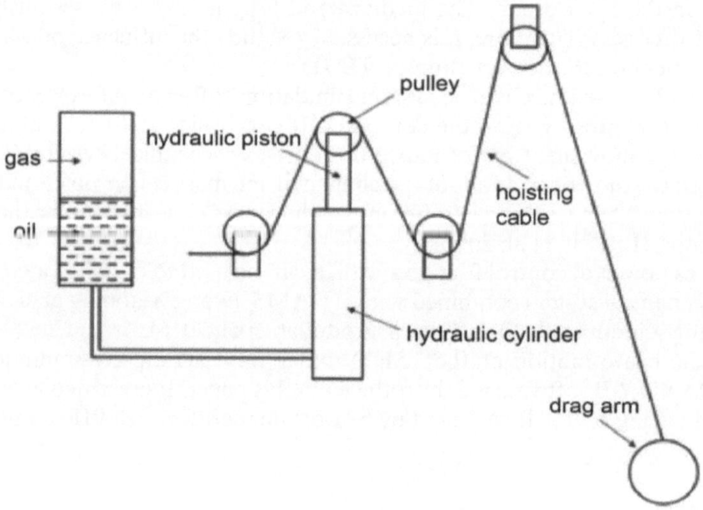

Figure 2. The schematic diagram of heave compensator.

A force F_1 was defined to simulate front hoisting cable between TSHD hull and upper suction tube, meanwhile, the corresponding stiffness and damping coefficient (k, c) were defined appropriately.

As the key research object, the heave compensator controls the lifting of drag head, which can make the drag head always cling to the seabed when the TSHD encountering to wave or uneven seabed. The schematic diagram is shown in Figure 2.

A spring mechanism was selected to simulate the heave compensator. (Lu Kuan, 2012) The stiffness coefficient, k, is made up by two parts. One is for the wave compensating system, k_1, which is defined as:

$$k_1 = \frac{A_P^2 \cdot n \cdot P_0}{V_0} \cdot \left(\frac{1}{1 - \frac{A_P \cdot x}{V_0}} \right)^{n+1} \tag{1}$$

where, A_P — piston surface area;

n — a constant;

P_0 — air pressure in equilibrium;

V_0 — the volume of air in equilibrium;

x — piston displacement.

The other one is for rear hoisting cable, k_2, which is defined as:

$$k_2 = ES_g / \left(L_g + x \right) \tag{2}$$

where, E — elastic modulus of hoisting cable;

S_g — cross sectional area of hoisting cable;

L_g — length of hoisting cable.

So, the stiffness coefficient can be expressed as:

$$k = \frac{k_1 \cdot k_2}{k_1 + k_2} \tag{3}$$

The damping coefficient is:

$$c = \frac{A_P^2 \cdot 128 \cdot \eta \cdot l_s}{\pi \cdot D_s^4} \tag{4}$$

where, D_s — catheter diameter;

η — dynamic viscosity;

l_s — catheter length.

The force of fluid at the speed of 1.5 m/s was loaded on the drag arm. We imported the spline function file which was calculated by Aqwa software under 12s periodic waves, and added spline function to the movement of TSHD hull in Z direction. The movement curve is shown in Figure 3.

The simulation model of TSHD under the environment of ADAMS is shown in Fig. 4.

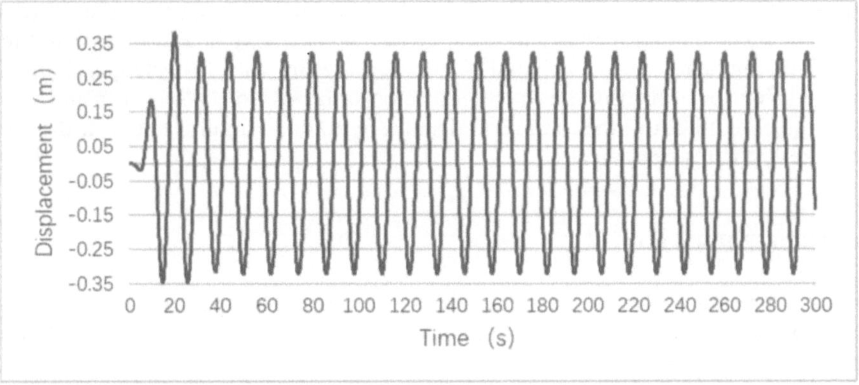

Figure 3. Movement displacement curve of TSHD.

Figure 4. Simulation model in ADAMS.

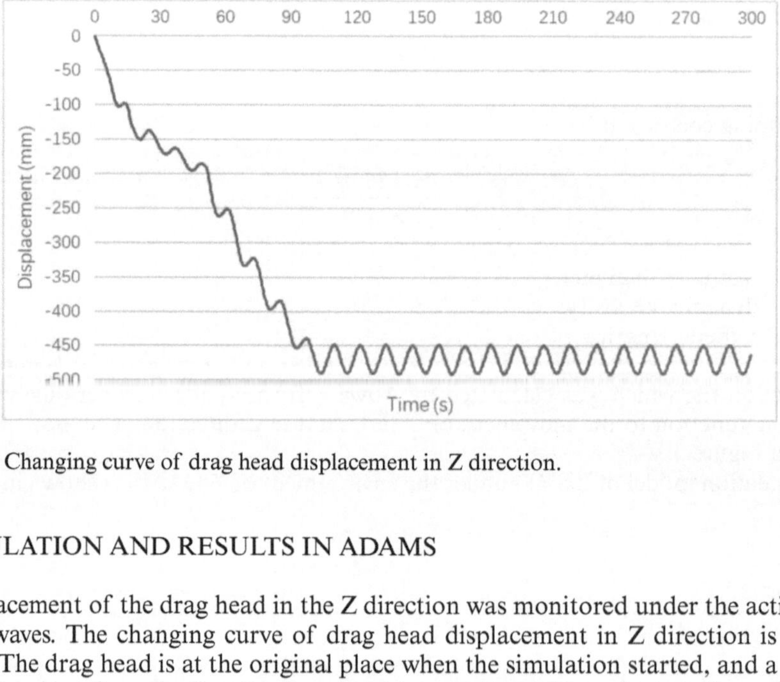

Figure 5. Changing curve of drag head displacement in Z direction.

3 SIMULATION AND RESULTS IN ADAMS

The displacement of the drag head in the Z direction was monitored under the action of 12s periodic waves. The changing curve of drag head displacement in Z direction is shown in Figure 5. The drag head is at the original place when the simulation started, and a period of time will be given for adjusting the drag head to certain position. As seen from the results, in the process of simulation, the drag head declines gradually, finally locates at the -469 mm position and fluctuates periodically.

4 CONTROL DESIGN IN MATLAB

In order to realize the control design of the heave compensator, the upper end of the spring mechanism was connected to a translational joint, and was set to drive along the Z direction, which was used to simulate the telescopic control by the winch on the rear hoisting cable.

In the co-simulation modeling, we assumed that the inputs of the system are displacement of the translational joint motion S_{i1} and displacement of TSHD hull S_{i2} which was provided

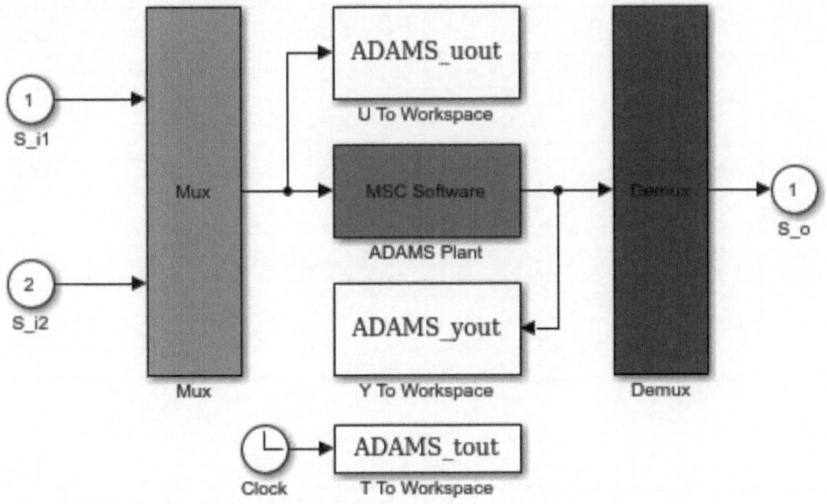

Figure 6. MATLAB model exported from ADAMS.

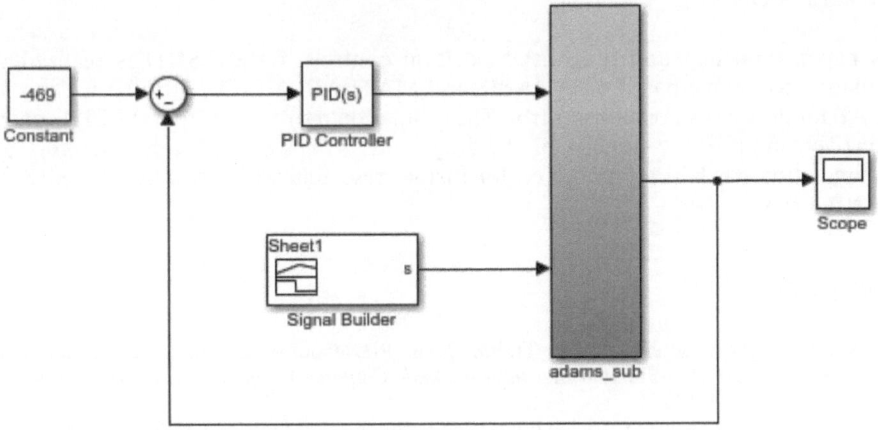

Figure 7. PID control block diagram of co-simulation model.

by the spline function file (seen in Figure 3), and the output is displacement of drag head in Z direction S_o. The co-simulation system was exported from ADAMS that is given in Figure 6. In this system, the displacement of the translational joint motion S_{i1} was used for controlling the springby manually tuned PID controller to stable the drag head position. Control block diagram is shown in Figure 7.

5 CO-SIMULATION RESULTS

The response of PID controlled drag head position in Z direction is shown in Figure 8. As seen from Figure 8, the drag head position in Z direction is decreased from 0 mm to –469 mm and stabilized in this desired position. From this result, the co-simulation system achieved successfully using manually tuned PID controller. Compared with the model without PID controller, the co-simulation system can stabilize the drag head better.

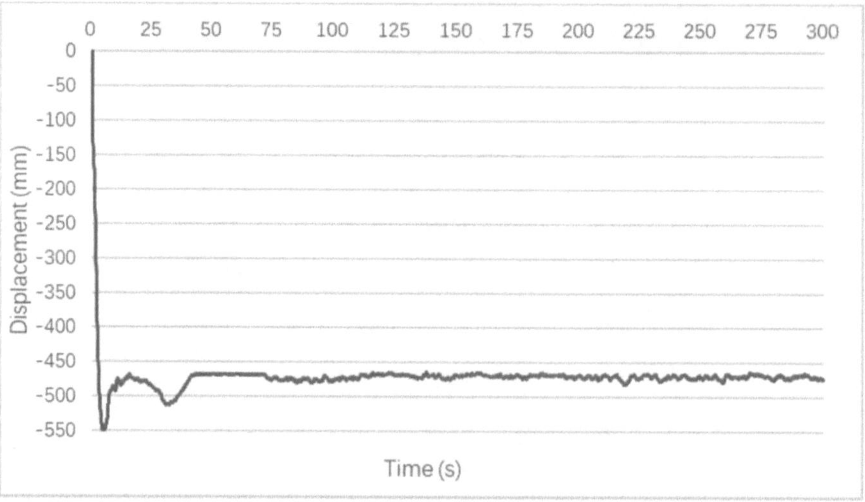

Figure 8. Drag head position of co-simulation model.

6 CONCLUSION

In this paper, modeling and drag head position control of the TSHD is realized using co-simulation technique based on ADAMS and MATLAB. A PID controller is designed in MATLAB for drag head position control. The simulation results show that the PID controller can effectively improve the stability of the drag head. What's more, the current work about the co-simulation has laid a foundation for further research, which can be applied to other control strategies.

REFERENCES

Çakan, Abdullah, F.M. Botsalı, & M. Tınkır. 2016. PID Control of Inverted Pendulum Using Adams and Matlab Co-Simulation. *International Conference on Control, Mechatronics and Automation ACM*:136–139.

Liao C, Zhang J, Chen W & Lv M. 2015. Union simulation on lower limbs rehabilitation robot based on MATLAB and ADAMS. *Industrial Electronics and Applications IEEE*:557–562.

Liu Zhi & Jiang Jibin. 2006. Virtual Simulation of Tshd Dredging Process Basing on Adams[J]. *NEW TECHNOLOGY & NEW PROCESS* 01:69–71.

Lu Kuan. 2012. Research on Numerical Simulation of Trailing Suction Hoper Dredger and its Dredging Accuracy Control[D]. *Tianjin University*.

Zhang Xiuqin, Yang Bo, Yang Chao & Xu Guanneng. 2012. Research on ABS of Multi-axle Truck Based on ADAMS/Car and Matlab/Simulink. *Procedia Engineering* 37(2):120–124.

Zong Xiaoping & Li Yueyue. 2009. The two degrees of freedom manipulator control simulation based on ADAMS and MATLAB. *MICROCOMPUTER INFORMATION* 35:29–30.

Hydraulic Engineering V – Hong, Liu & Xie (Eds)
© 2018 Taylor & Francis Group, London, ISBN 978-0-8153-7471-8

A method of measuring the critical jet pressure of clay cutting by a moving jet

L. Gu, F.S. Ni, Z.S. Li & L.Q. Xu
Engineering Research Center of Dredging Technology, Ministry of Education, Hohai University, Changzhou, Jiangsu, China
Faculty of Mechanical and Electrical Engineering, Hohai University, Changzhou, Jiangsu, China

ABSTRACT: To determine the appropriate jet parameters of a Trailing Suction Hopper Dredger (TSHD), it is necessary to measure the critical jet pressure of clay cutting by a moving water jet. In this study, a series of experiments were performed on clay samples prepared in a laboratory to investigate jet cutting using the window cutting method. First, the effects of the window width on the cutting depth were investigated. Then, the method for measuring the critical jet pressure of a moving jet was proposed. The results showed that the cutting depth increases linearly with the jet pressure and the fitting lines have an intercept on the jet pressure axis. For a certain strength of clay, the intercept is almost constant regardless of the change of the rest of the experimental conditions. This intercept indicates the critical jet pressure of clay cutting by a moving jet.

1 INTRODUCTION

The special drag with high-pressure water jet system (REACULA) developed by Belgium International Dredging Company for the "Jade River" dredger was the first application of the water jet in clay cutting (Lin 2002). It significantly improved the dredging yield and shortened the construction time (Vandycke 2002). Since then, the water jet system has gradually become the basic configuration of the modern trailing suction hopper dredger (TSHD). In particular, the system can improve the excavation efficiency and effectively reduce the adhesion in the clay dredging and therefore it is widely used. However, the jet parameters also depend on the experience in the dredging construction, which leads to the instability of construction effect. The fundamental reason is that it is impossible to accurately measure the critical jet pressure of the clay by the moving jet, making it difficult to reasonably set the jet system parameters.

At present, there have been many studies on the critical conditions of clay scouring by jet (Knapen et al. 2007). Mazurek et al. (2001) conducted a series experiments on clay scouring by jet with the scouring time of 44–188 h, in order to study the relationship between jet strength and scouring results. Li & Li (1997) analyzed the structure of a fixed jet and the causes of clay damage and derived an empirical formula of critical jet pressure through the experiment. In the experiments of Zong et al. (2014) & Ma Zhimin et al. (2015), the changes in the bed surface were considered as the standard critical condition. The perception methods include the visual inspection and B-ultrasound, and the critical states are divided into little movement and general movement. Many scholars have studied the method and apparatus of measuring the critical shear stress by a fixed vertical jet (Khanal et al. 2016, Wahl 2010, Hanson & Cook 2004).

It can be seen that most of the existing studies focus on the long-term scouring of fixed jets against clay. In contrast, the water jet rapidly moves with the vessel in the dredging with a velocity of 0.5–1.5 m/s. It is difficult to observe the changes in the bed surface using the above experimental method, and the experimental clay samples required for a moving jet also need a certain size. The measurement method of the critical jet pressure in view of clay cutting by a moving jet is explored in this study, and the feasibility of the developed method was tested with some experiments.

2 EXPERIMENTAL APPARATUS AND CLAY SAMPLES

The experiments were performed on the jet cutting platform (Figure 1) in the Engineering Research Center of Dredging Technology of Ministry of Education at Hohai University, China. The platform has a tank with length 9 m, width 1.8 m, and depth 1.5 m, together with a transparent glass and steel structure. A mobile carriage can move on the top track of the tank. The moving speed of the carriage can be controlled and maintained through a frequency conversion. The maximum speed can be up to 0.6 m/s. A nozzle is mounted vertically on the carriage and is supplied water by a multistage centrifugal pump with a maximum pressure of 300 bar. The flow rate is measured by an electromagnetic flowmeter with the accuracy grade of 0.5. All instruments are controlled by a computer, which is equipped with data-recording software for real-time operating parameters. The cutting depth is measured by a laser sensor with linear error within 0.26 mm.

In the tank, a soil pit (Figure 2) with length 2.2 m, width 0.4 m, and height 0.6 m is placed. The soil pit can prepare the clay sample through consolidation drainage. There are many holes on the steel plates around the soil pit for the drainage of soil matrix under load. The soil pit is equipped with some load chains to provide a total loading force of 18 t. Each chain is

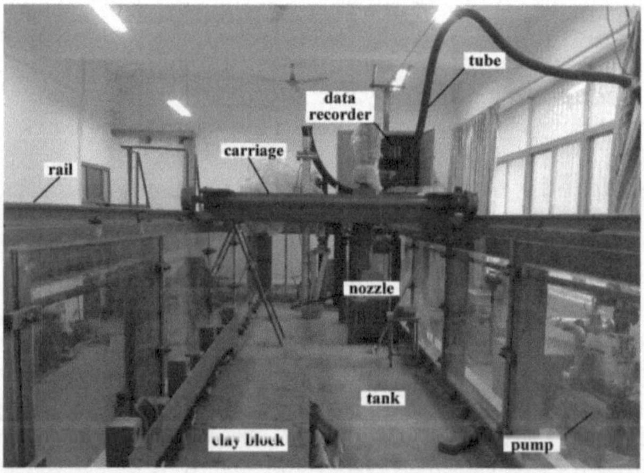

Figure 1. Jet cutting platform.

Figure 2. Preparation equipment of the clay sample.

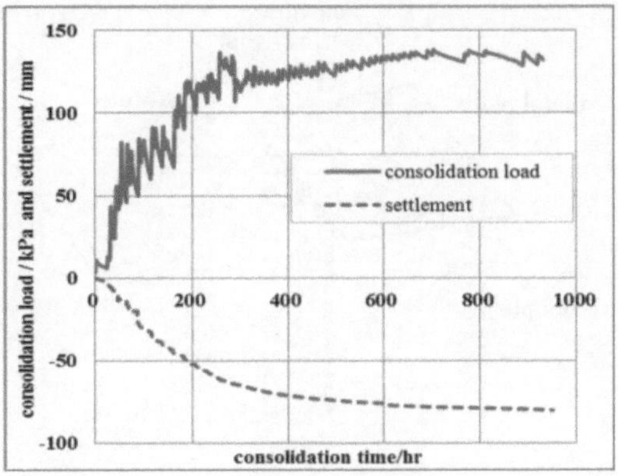

Figure 3. Curves of the consolidation load and the settlement.

Table 1. Characteristics of the experimental soil.

Liquid limit	Plastic limit	Dry density	Water content	Saturation	Medium diameter	Clay content	Permeability coefficient	Specific gravity
41.49%	20.56%	1.57 g/cm^3	26.86%	0.99	0.0128 mm	28.8%	2.45e-8 cm/s	2.72

connected to a tension sensor with the accuracy of 0.3%, which can gauge the load in real time. On the upper plate, some dial indicators are installed to measure the sedimentation value.

The loess is dried and the particles with size less than 2 mm are screened. After mixing with water, they are placed in the soil pit for consolidation drainage. Figure 3 shows the curves of the consolidation load and the settlement. According to the GB/T50123–1999, the basic physical properties of soil samples were measured, and the details of the parameters are presented in Table 1. It can be seen that the clay block is close to saturation and belongs to the low liquid-limit clay. The undrained shear strength of most soil in the middle area of the soil pit is about 16 kPa.

3 WINDOW CUTTING

The preparation process of the clay sample is complex and time-consuming. To ensure the repeatability of sample, the soil samples were prepared in one time. The water penetration rate of clay is very low, and the drainage path should not be too long in the consolidating drainage, so the soil sample size cannot be large. And the area required for the moving jet cutting is larger, so the homogeneous soil samples in the experiment are precious. From the perspective of saving soil samples, a method of window cutting was proposed in the experiment, as shown in Figure 4. A steel plate with window was installed above the clay sample, and the other areas of the soil sample were covered with metal plates. In this way, only the clay inside the window will be cutting in case of the jet moving above the soil sample.

However, such a cutting approach may bring a certain dimensional effect. For this consideration, some experiments were performed with the change of the window width (W). The cutting longitudinal cross-sectional shape along the moving direction of the jet at different window widths is shown in Figure 5. It can be seen that the slotted longitudinal section is in "V" shape in case of small window width, and a flattened bottom appears at the larger window width. During the jet cutting process, the jet affects the clay directly, destroys the soil, and consumes the energy. With the decline in fluid energy, the soil-breaking capacity of the jet decreases and the flow direction changes until it becomes smaller than the start-up condition of the clay particles (Zhang et al. 2016). Therefore, there is an effective range of the interaction between the jet

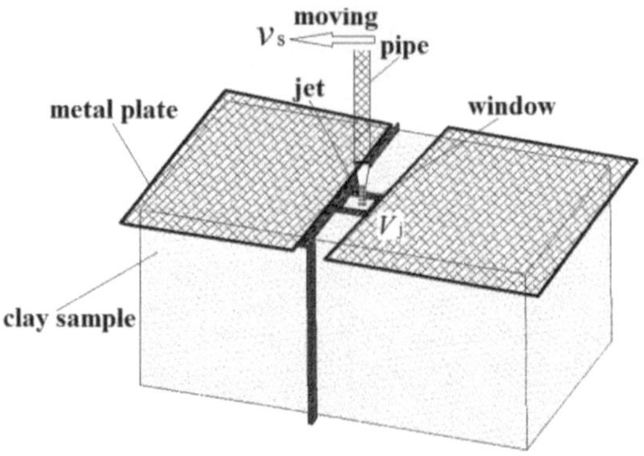

Figure 4. Sketch of the window cutting.

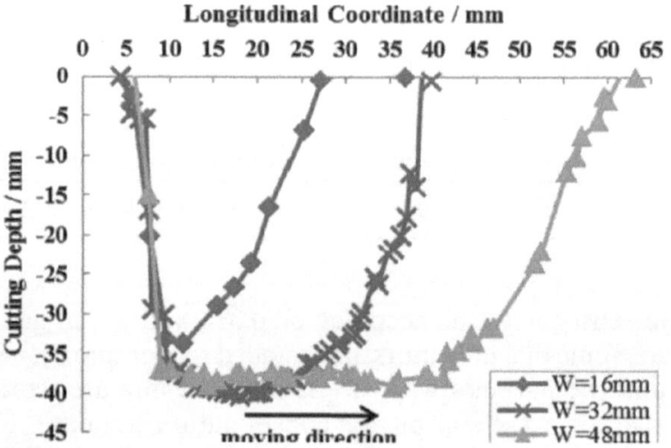

Figure 5. Cutting longitudinal section at different window widths.

and the clay. When the window width is less than this range, the interaction between the jet and the clay will be not sufficient ($W = 16$ mm in Figure 5), and the cutting depth will be shallow, in which case, the cutting results cannot reflect the jet cutting effect in the actual dredging. If the window width is greater than this range, then the jet can fully act with clay. The jet cannot produce cutting effect on the front side of the cutting bottom when moving to the back of the window and the flattened section will appear on the front side of the trench bottom. Then, the cutting depth will not change with further increase in the window width ($W = 32$ and 48 mm in Figure 5). The depth of the flattened section can represent the cutting depth of the dredging jet. In this study, W = 48 mm is used to avoid the influence of the window width.

4 CRITICAL CONDITIONS

In the research of rock breaking by jet, Matsumoto Koji (Japan) found that the breaking depth linearly increases with the jet pressure. The pressure at the zero breaking depth was considered as the critical pressure (Shen 1999). Whether the critical state of clay cutting can draw on this method needs study, but it must be clear (1) whether the cutting depth of clay by a moving jet will be affected by other factors and (2) the relationship between the cutting depth and the jet pressure.

4.1 *Nozzle diameter and moving velocity*

Nozzle diameter (d) and moving velocity (v_t) are the two important factors affecting the cutting results (Machin 2001). Under about the same jet pressure and soil strength, some tests were conducted with the nozzle diameters $d = 2$ and 4 mm and moving velocities $v_t = 0.1$, 0.2, and 0.4 m/s. In each test, the same area was cut several times and the cutting depth was recorded every time, which is shown in Figure 6. The abscissa is the accumulative cutting time $t = Nd/v_t$ (N is the number of repeated cutting trials), which indicates the characteristic time of the moving jet cutting at one point on the soil sample. It can be seen that, despite the change in the nozzle diameter and the moving velocity, the cutting depth only increases linearly with the accumulative cutting time t during a certain period of time ($t \leq 40$ ms). Therefore, under certain jet pressure, the cutting depth of the moving jet in the initial period can be regarded as the accumulation effect of the action time.

In the dredging project, the nozzle diameter is about 20 mm and the moving velocity during the construction is 0.5–1.5 m/s. It indicates that the accumulative cutting time in the project is less than 40 ms. Therefore, in this study, in order to obtain enough information about the cutting process, the experimental parameters used were as follows: nozzle diameter $d = 2$ mm, moving velocity $v_t = 0.4$ m/s, and the number of the repeated cutting trials $N = 8$. The cutting depth was recorded every time.

4.2 *Jet pressure*

Through a trial cutting test on clay, a value of jet pressure was obtained first at a visible cutting depth (3–5 mm). Then, on the basis of this value, the jet pressure progressively increased for the jet cutting experiments. The parameters of these experiments are shown in Table 2. The variation of cutting depth during each accumulative cutting time with jet pressure is shown in Figure 7. It can be seen that the changes are almost linear, which means that the cutting depth linearly increases with jet pressure. The fitting line of the data point has an intercept on the jet pressure axis. This intercept is called the critical jet pressure.

Interestingly, the intercepts of the fitting lines at different accumulative cutting times are almost the same. In other words, the critical jet pressure is almost constant and is not affected by the accumulative cutting time. And it can also be seen that as long as the accumulative time is equal, the cutting depth is same regardless of the change in the nozzle diameter and the moving velocity. Therefore, the critical jet pressure of clay does not have influence on the nozzle diameter and the moving velocity during a rapid moving jet cutting. As a result, the nozzle

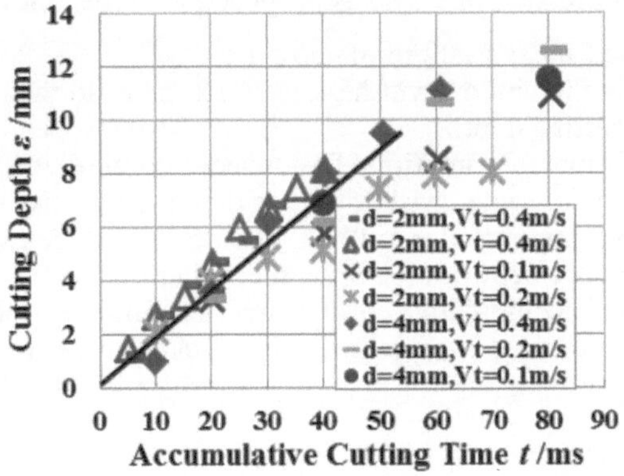

Figure 6. Variation of the cutting depth with the accumulative cutting time.

Table 2. Details of the test conditions.

Nozzle diameter d (mm)	Traversing velocity v_t (m/s)	Window width W (mm)	Undrained shear strength S_u (kPa)	Jet pressure P_j (kPa)
2	0.4	48	16	145, 200, 313, 365, 421

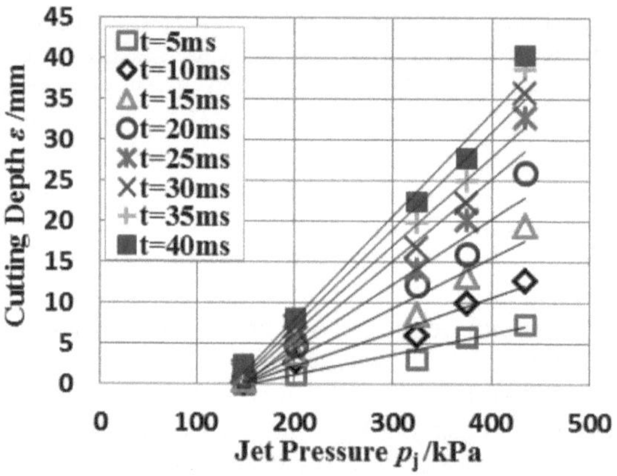

Figure 7. Variation of cutting depth with jet pressure.

diameter and the moving velocity can be selected according to the experimental conditions when measuring the critical jet pressure.

The repetitive experiments were performed using the prepared clay samples, and the experimental data were consistent with the above results. In addition, critical jet pressures of clay with different soil strengths were measured by the proposed method. The results show that the method can be used to measure the critical jet pressure for each strength of clay, indicating that this method can be used for the measurement of the critical jet pressure for clay cutting by a moving jet.

5 CONCLUSION

The effects of the cutting width, the nozzle diameter, and the moving velocity on the cutting depth were investigated by conducting experiments on clay cutting by a moving jet. The change in the cutting depth with jet pressure was analyzed. From the analysis, a method for measuring the critical jet pressure of clay was proposed with the following specific processes:

1. Measure the undrained shear strength of the clay;
2. Cut the clay into different window widths, and select the window width without considering the effect on the cutting depth;
3. According to the experimental conditions, first, select a nozzle diameter and a moving velocity and select the jet pressure with a visible cutting depth. Then, take three to five jet pressure values in steadily increasing order for multiple jet cutting, and measure each cutting depth;
4. According to the measurement results, draw the filling line of the cutting depth with the jet pressure in the same accumulative cutting time. The intercept of the fitting line on the jet pressure axis indicates the critical jet pressure of clay cutting by a moving jet.

ACKNOWLEDGMENTS

The authors thank Shuang Jiang and Hongbin Li for their help in the construction and maintenance of the experimental arrangement. This study was supported by the National Natural Science Foundation of China (NSFC-51709094).

REFERENCES

Hanson, G.J. & Cook K.R. 2004. Apparatus, test procedures, and analytical methods to measure soil erodibility in sity. *Applied Engineering in Agriculture*. 20(4) 455–462.

Khanal A., Fox G.A. & Al-Madhhachi A.T. et al. 2016.Variability of erodibility parameters from laboratory mini jet erosion tests. *Journal of Hydrologic Engineering*, 21(10): 04016030.

Knapen, A., Poesen, J. & Govers, G.G. et al. 2007. Resistance of soils to concentrated flow erosion: A review. *Earth-Science Reviews,* 80: 75–109.

Li, F.S., Li, J.H. 1997. Mechanism of soil broken by water jet and the engineer application. *Fluid Mechanics*, 25(2):26–29.

Lin, F. 2002. Using DRACULE system to dredging hard clay in trailing suction hopper dredger. *Shanghai Waterway Technology*, (4): 13 ~ 15. (in chinese).

Ma, Z., Zou, X & Zhao, X. et al. 2015. A method of measuring sediment incipient velocity based on B-scan ultrasound imaging technique. *Advances in Water Science*, 26(1): 59–65. (in chinese).

Machin, J.B., Messina, F.D. & Mangal J.K. et al. 2001. Recent research on stiff clay jetting. *Offshore Technology Conference*, OTC 13139. Houston, USA.

Mazurek, K.A., Rajaratnam, N. & Sego, D.C. 2001. Scour of cohesive soil by submerged circular turbulent impinging jets. *Journal of Hydraulic Engineering*, 127(7): 598–606.

Shen, Z.H. 1999. Theory and technology of water jet. *China Safety Science Journal*, 5(1): 63–65. (in chinese).

Vandycke, S. 2002. Dredging stiff to very stiff clay in the wielingen using the DRACULA system on a hopper dredger. *Terra et Aqua*: 3–8.

Wahl, T. 2010. A comparison of the hole erosion test and jet erosion test. *Joint Federal Interagency Conference on Sedimentation and Hydrologic Modeling.* Las Vegas, NV.

Zhang, S.S., Zhao, M. & TONG G. et al. 2016. Experimental research on trenching in stiff clay by submerged vertical traveling jets. *Journal of Coastal Research*, 32(2): 365–373.

Zong, Q.L., Xia, J.Q. & Zhang, Y. et al. 2014. Experimental study on scouring characteristics of cohesive bank soil in the Jingjiang reach. *Advances in Water Science*, 25(4): 567–57. (in chinese).

Hydraulic Engineering V – Hong, Liu & Xie (Eds)
© 2018 Taylor & Francis Group, London, ISBN 978-0-8153-7471-8

Experimental and numerical investigations on the ultra-high pressure jet of a TSHD

Shuai Jiang, Jinfu Yin, Guojun Hong & Liming Yin
Key Laboratory of Dredging Technology, Shanghai, China
National Engineering Research Center of Dredging Technology and Equipment, Shanghai, China

Hongjian Xu
Qingdao First Municipal Engineering Co. Ltd., Qingdao, China

ABSTRACT: This study investigates the submerged ultra-high pressure jet characteristics of a TSHD by experimental and numerical methods. The results indicate that the dynamic pressure undergoes exponential decay with the increasing jet distance. Moreover, it shows that the numerical results are in good agreement with the experimental ones. On the basis of the results, the study analyzes the jet flow characteristics of the ultra-high pressure jet. In the case of an ultra-high pressure jet, the moving speed does not have a significant change in the flow characteristics. The influence of the velocity field and the value of the velocity obviously increase with the increase in the jet pressure. Therefore, an increase in the jet pressure can improve the flow-field characteristics and the efficiency of the dredging process.

1 INTRODUCTION

There have been several publications on the capabilities of the jet technology in several industrial areas, such as cutting and machining (Hashish M, 1978; R. Kobayashi, 1998). In the dredging industry, the advanced trailing suction hopper dredgers are equipped with the high pressure jet system worldwide (Hong, 2008). High-pressure jet system is the necessary equipment for a TSHD, which plays an important role in the construction of cohesive soil and silt. At present, the pressure of the jet on the domestic dredgers is generally only 1–2 *MPa*, only for loose soil. When the soil with lower permeability coefficient is encountered, the effect of high pressure jet system is remarkably reduced due to the increase of soil shear resistance.

The high pressure jet can be applied to coal mining, oil drilling, metal cutting, and so on, when the pressure increases to above 10 and even 100 *MPa* (Rehbinder G, 1980; Agus M, 1993, J. Wang, 2003). The practical application of ultra-high pressure jet was realized more than 10 years ago in a large foreign dredging company. The dredging production increased by 20–30% using the ultra-high pressure jet. In the past 10 years, the research on ultra-high pressure jet system has produced huge economic benefits for the four major dredging groups (Warnier F, 2002; S Vandycke, 2005).

Because the construction technology and related theory are in development stages, the pressure on the jet system is relatively low for domestic dredgers. Particularly, in the construction process, the low-efficiency phenomenon has not been resolved due to the change in jet characteristics. Therefore, it is necessary to study the characteristics of submerged ultra-high pressure jet. On the basis of the experimental and numerical simulation methods, the effects of the submerged ultra-high pressure jet characteristic of the jet direction and the flow characteristics are analyzed.

2 EXPERIMENTAL AND NUMERICAL MODELS

2.1 *Experimental model*

Ultra-high pressure jet experiment and measurement devices are arranged on a multifunctional wave flume, which is 107 m long, 2 m wide, and 2.5 m high. Wave flume is mainly used to simulate the dredging equipment in working and navigation states under conditions of wind, wave, and flow. The flume has many experimental functions, including the effects of the waves and sediments on the various hydraulic structures under the action of the flow of water. A schematic diagram of the experimental model is shown in Figure 1.

The ultra-high pressure water jet equipment used for the test can provide a pressure <50 *MPa* and can be adjusted during experiment. The dynamic pressure characteristics of the water jet can be measured through the ultra-high pressure sensors. The experimental results will be transmitted to the data acquisition equipment for post-processing. Test sensors with no weld and impact resistance have a high-precision sapphire core. The cylindrical nozzle with ceramic core, which can prevent wear when the nozzle works for long time, is used for the test. The nozzle inlet diameter D_1 is about 0.008 m, and the outlet diameter D_2 is 0.0014 m. The outlet length L_1 is about 0.006 m, and the shrinkage length L_2 is 0.012 m.

2.2 *Measurement methods*

The ultra-high pressure jet system for test includes a jet control device, ultra-high pressure lines, and a jet nozzle. The jet pressure of the system can be controlled by a jet device, which can provide different pressures. The mobile bracket module can move in three dimensions along the length, width, and depth of the flume. The WIFI module is connected with the external control system to realize the three-dimensional movement of the moving bracket module. The relative position between the nozzle and the sensors is achieved by moving the bracket module. During the test, the position of the ultra-high nozzle is adjusted and the ultra-high pressure jet system is opened to measure the dynamic pressure under test conditions.

2.3 *Numerical model*

The numerical model is based on the real scale, which is the same as the nozzle used in the test. The numerical model adopts the three-dimensional numerical simulation method. The numerical grid is meshed with hexahedral, and the number of meshes is 2,300,000. The grid is shown in Figure 2. The numerical model boundary conditions mainly include the pressure inlet boundary of the nozzle inlet, the wall boundary of the nozzle wall, and the pressure outlet boundary in the jet direction. The numerical simulation time term adopts the second-order implicit sag, and the pressure term is separated by the second-order upwind scheme. The velocity and pressure are coupled by the SIMPLE method. The discrete equation of the momentum equation adopts the second-order upwind scheme on the boundary.

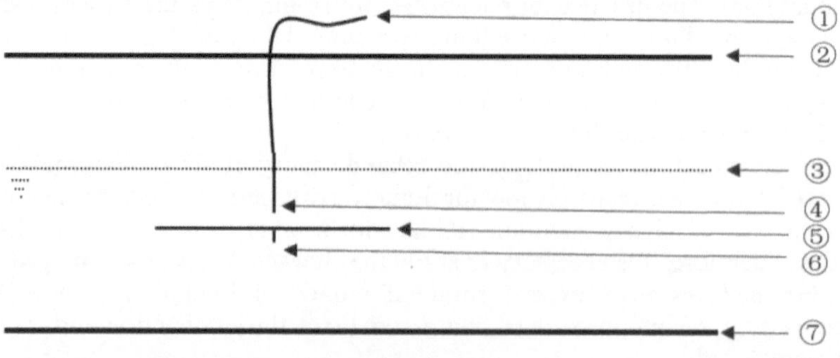

Figure 1. Schematic diagram of the experimental model.
(① ultra-high pressure pipeline, ② wave flume, ③ static water level, ④ ultra-high pressure nozzle, ⑤ test board, ⑥ ultra-high pressure sensor, ⑦ bottom of the flume).

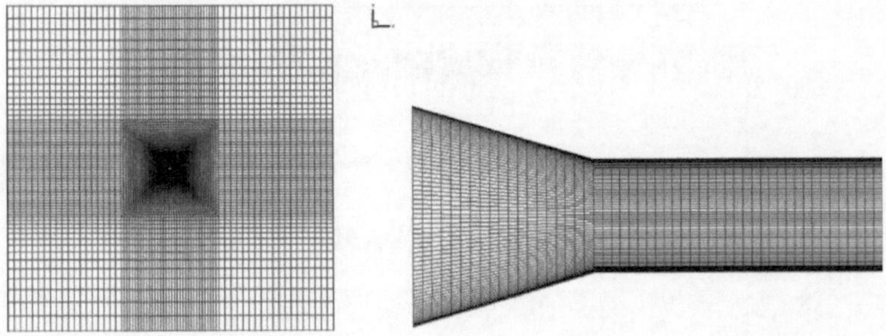

Figure 2. Grid diagram (xy plane and fractionated gain).

In this study, the two-equation turbulence model is used to simulate the characteristics of ultra-high pressure jet. Turbulence viscosity (μ_t), turbulent kinetic energy (k), and turbulence frequency (ω) are constrained in the turbulence model. The k–ω equations are shown in (1) and (2):

$$\frac{\partial \rho k}{\partial t}+\tilde{u}_i\frac{\partial \rho k}{\partial x_i}=P_k-\frac{\rho k^{3/2}}{l_{k-\omega}}+\frac{\partial}{\partial x_i}\left[(\rho v+\frac{\mu_t}{\sigma_k})\frac{\partial k}{\partial x_i}\right] \tag{1}$$

$$\frac{\partial(\rho\omega)}{\partial t}+\tilde{u}_i\frac{\partial(\rho\omega)}{\partial x_i}=C_\omega P_\omega-\beta_\omega\rho\omega^2+\frac{\partial}{\partial x_i}\left[(\rho v+\frac{\mu_t}{\sigma_k})\frac{\partial \omega}{\partial x_i}\right]+2\rho(1-F_1)\frac{1}{\sigma_{\omega 2}}\frac{1}{\omega}\frac{\partial k}{\partial x_i}\frac{\partial \omega}{\partial x_i} \tag{2}$$

where $\mu_t=\frac{\rho\alpha_1 k}{\max(\alpha_1\omega, SF_2)}$, P_k and P_ω are the turbulence generation factors, whose values can be found in Strelets M (2001), and the model constant value is 0.09.

3 RESULTS AND DISCUSSION

3.1 *Dynamic jet pressure test and numerical results*

Figure 3 shows the pressure signal on the jet centerline at different distances when the pressure is 10 *MPa*. It is evident from the figure that the pressure signal on the jet centerline fluctuates at a certain mean value. When the nozzle is very close to the sensor, the collected voltage signal increases obviously. As the jet distance increases, the voltage signal becomes weak. When the distance is 12 cm, the mean value of the voltage signal is about 128 mV.

The pressure signal is converted by the calibrated data. Figure 4 shows the dynamic pressure on the jet centerline at different jet distances. The results show that the experimental system can measure the dynamic pressure at the nozzle outlet. However, the dynamic pressure is too large and the sensor is easily damaged when it is close to the nozzle. Therefore, the experimental measurement distance is greater than $10D_2$.

The results show that when the jet distance is greater than $7D_2$, the jet dynamic pressure decreases sharply, and it reduces to 0.56 times of the inlet jet pressure when the jet distance is $10D_2$. The jet dynamic pressure exhibits exponential decay characteristics with the increase of the distance, especially the jet dynamic pressure is about 0.02 times of the initial pressure when the distance reaches $40D_2$. The experimental results are close to the numerical ones. When the jet distance is greater than $20D_2$, the values are slightly larger than the numerical results, but the trend is the same.

The closer the nozzle is to the mud surface, the greater is the dynamic pressure and the more favorable is the soil cutting, which is conducive to improving the dredging efficiency in the construction process. On the contrary, the farther away the mud surface from the nozzle is, the smaller the dynamic pressure and the lower the dredging efficiency will be.

Figure 5 shows the velocity on the jet centerline when the pressure is 10 *MPa*. V is the velocity of flow, V_{th} is the theoretical velocity, and L is the jet distance. The velocity on the jet

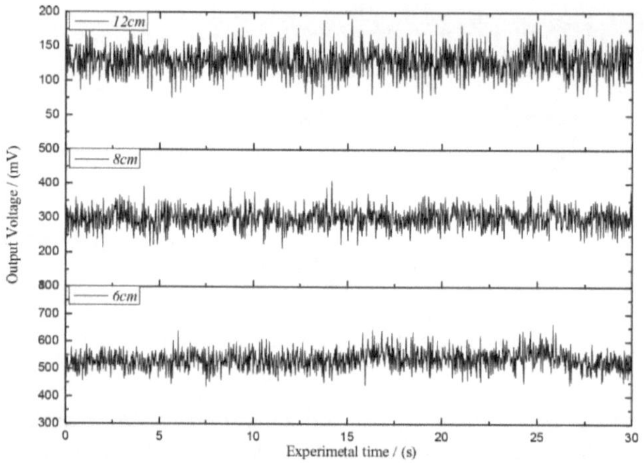

Figure 3. Pressure signal on the jet centerline at different jet distances when the jet pressure is 10 *MPa*.

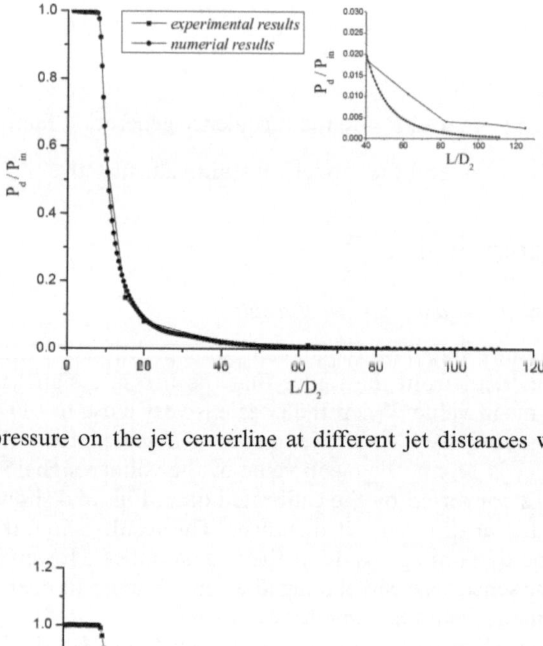

Figure 4. Dynamic pressure on the jet centerline at different jet distances when the jet pressure is 10 *MPa*.

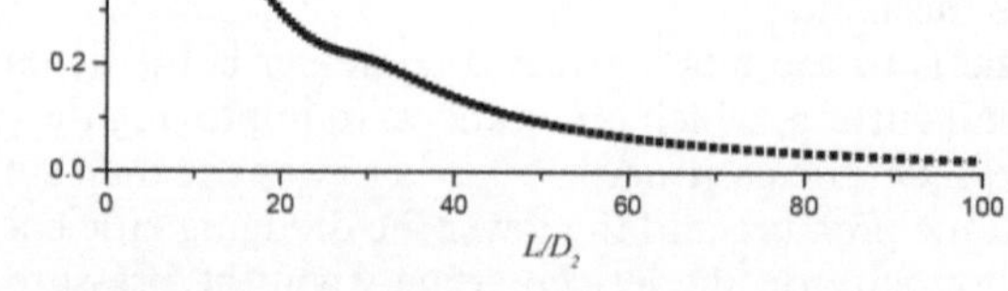

Figure 5. Velocity on the jet centerline when the jet pressure is 10 *MPa*.

centerline exhibits exponential decay distribution along the jet direction when the nozzle is stable. The velocity on the jet centerline decreases gradually with the increase of the distance. The velocity is almost zero when the distance from the nozzle outlet is approximately $100D_2$.

3.2 *Flow characteristics of the jet*

In order to observe the velocity characteristics at different cross sections of the ultra-high pressure nozzle, the velocity distribution at different distances at a pressure of 10 *MPa* is shown in Figure 6. As can be seen from the results, the maximum velocity on the jet centerline can reach the range of $7D_2$, which is consistent with the results of Figure 5. As the distance increases, the proportion of the maximum speed in the jet cross-sectional area becomes smaller. There is a significant velocity center on the jet cross section, but the value is significantly reduced when the jet distance is $10D_2$. When the distance is $20D_2$, there is no obvious jet center. The distribution of the velocity in the circumferential direction increases with the increase in the distance.

Figure 7 shows the velocity distribution at $7D_2$ section under different moving velocities when the pressure is 10 *MPa*. As is evident from the figure, the velocity characteristic does not change when the nozzle speed is 0.5 *m/s* in the $7D_2$ section. In the case of $u \geq 1.0$ *m/s*, the velocity field begins to shift on the opposite direction of the nozzle movement as a whole, but is not noticeable. Therefore, the effect of moving speeds on the speed field is not obvious at the jet distance of $7D_2$ under the 10 *MPa* jet.

In order to study the flow-field characteristics of different ultrahigh pressures, the velocity field characteristics at 10 and 30 *MPa* are shown in Figure 8. From the results, the influence of the velocity field and the value of the velocity obviously increase with the increase in the jet pressure. When the pressure exceeds 30 *MPa*, the influence of the velocity field is reduced

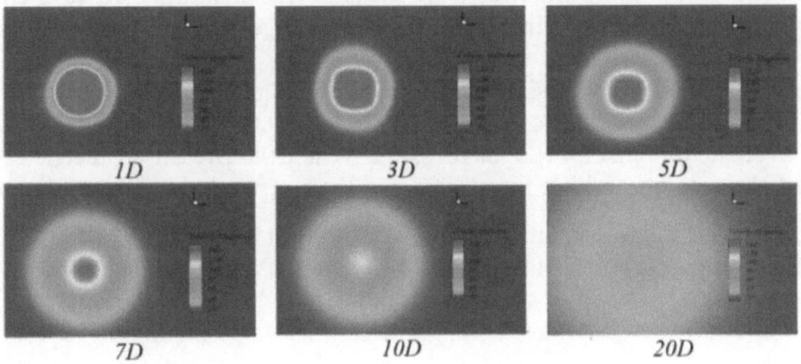

Figure 6. Velocity distribution at different distances when the jet pressure is 10 *MPa*.

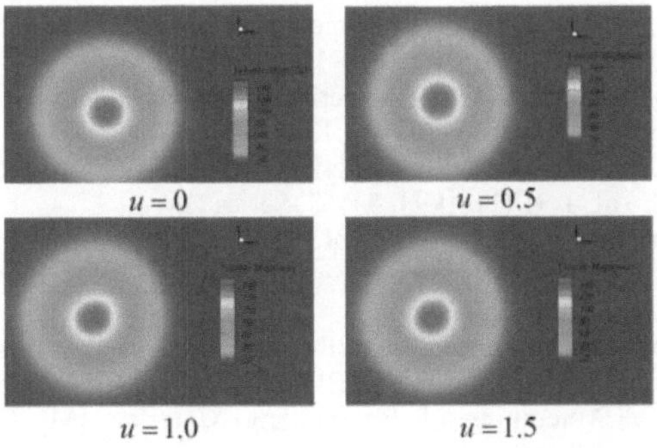

Figure 7. Velocity distribution at the $7D_2$ section under different moving speeds when the jet pressure is 10 *MPa*.

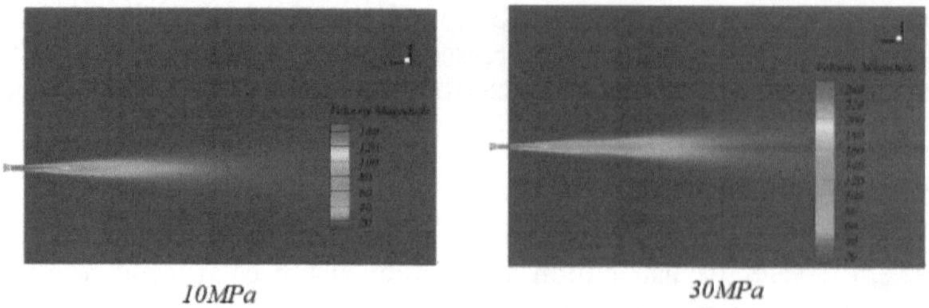

Figure 8. Velocity field characteristics at 10 and 30 *MPa*.

first and then expanded in the jet direction. As the pressure increases, the outwardly expanding area increases significantly. These show that at the initial stage, due to the larger spray resistance, spray area gradually reduces, but when the pressure exceeds a certain value, the flow field produces a secondary vortex phenomenon in the submerged jet. The influence of the velocity field will be further extended downstream of the jet.

4 CONCLUSION

This study investigated the characteristics of a submerged ultra-high pressure jet by experimental and numerical simulation methods. It analyzed the characteristics of the dynamic pressure and velocity distribution on the jet centerline. On the basis of this analysis, the study also analyzed the flow-field characteristics. From the analysis, the following conclusions can be drawn:

1. The experimental system could effectively measure the dynamic pressure characteristics of the submerged jet. The experimental results were in good agreement with the numerical ones.
2. The moving speed showed no significant change in the flow characteristic of the ultra-high pressure jet. The influence of the velocity field and the value of the velocity obviously increased with the increase in the jet pressure. When the pressure exceeded 30 *MPa*, the influence of the velocity field reduced first and then expanded in the jet direction.

Therefore, from the above conclusions, it is evident that an increase in the jet pressure can improve the flow-field characteristics and the efficiency of the dredging process.

REFERENCES

Agus M, Bortolussi A, Ciccu R, et al. The influence of rock properties on water jet performance [A]. In: The 7th American Water Jet Conference [C]. Washington: 1993, 427–442.

Hashish M, Duplessis M P. Theoretical and Experimental Investigation of Continuous Jet Penetration of Solids [J]. Journal of Engineering for Industry, 1978, 100(1):88.

Hong Guo-jun, Wang Jian, Lin Feng. Model experiments of draghead on training suction hopper dredger [J]. China Harbour Engineering, 2008(4):19–22.

Kobayashi, R., T. Arai, H. Yamada, Structure of high-speed water jet and the damage of metals in jet cutting technology, JSME Int. J. 31 (1), 1998, 53–57, Series II.

Rehbinder G. A theory about cutting rock with a water jet [J]. Rock Mechanics, 1980, 17(12):247–257.

Strelets M. Detached eddy simulation of massively separated flows [R]. Reno, Nevada: American Institute of Aeronautics and Astronautics, 2001.

Vandycke, S., M Van den Broeck, K Thomas. The DRACULA—system pushes the limits of hopper dredgers in cemented sand [C]. Proceedings of CEDA Dredging Days, 2005.

Wang, J. Abrasive Water jet Machining of Engineering Materials [M]. Trans Tech Publications, Switzerland, 2003.

Warnier F., Raes E, Vandycke S. Dredging stiff to very stiff clay in the wielingen using the DRACULA—system on a hopper dredge[C]//Institution of Engineers, 2002.

Hydraulic Engineering V – Hong, Liu & Xie (Eds)
 ISBN 978-0-8153-7471-8

Numerical simulation of the protection engineering effect in a new development district of Hangzhou Bay in Ningbo

W. Chen
School of Port and Transportation Engineering, Zhejiang Ocean University, Zhoushan, China

X.D. Mao
CCCC National Engineering Research Center of Dredging Technology and Equipment, Shanghai, China

L.Y. Chen
East Sea Information Center, SOA China, Shanghai, China

C.L. Hao
Second Institute of Oceanography, SOA China, Hangzhou, China

Y.L. Ni
School of Port and Transportation Engineering, Zhejiang Ocean University, Zhoushan, China

ABSTRACT: To protect the rich resources of tidal flats, a project involving two groins was set up to stabilize and promote tidal flat deposition in the southern coast of Hangzhou Bay. In this study, an unstructured mesh system was set up for numerical simulation to investigate the effect of the groins on hydrodynamics and morphology. A well-validated model was used to simulate the current. It was found that the groins were almost perpendicular to the current direction, directly affecting the movement of the current. Thus, the current significantly decreased between and outside the groins and increased at the heads of the groins. This characteristic finally caused a deposition of 0.25–0.60 m at both toes and between the groins and heavy erosion reaching 0.30–0.40 m at both heads of the groins. Finally, the simulation of the suspended sediment induced by riprap demonstrated that the suspended sediment was transported to an area larger than 29 km^2 with concentrations above 10 mg/L within and around the engineering region.

1 INTRODUCTION

The new development district of Hangzhou Bay is located in the northern region of Ningbo, Zhejiang Province, and the southern coast of Hangzhou Bay, as shown in Figure 1 (a). Hangzhou Bay has two international-standard harbors, namely Yangshan Harbor in Shanghai and Ningbo-Zhoushan Harbor in eastern Zhejiang Province.

The coast of this district is divided into a silting flat due to the morphology produced by Qiantang River and Hangzhou Bay (Feng & Bao 2006). According to the results of the survey conducted in the 1990s, tidal flat resources in this region had reached 416.73 million km^2, half of which is relatively stable and in good conditions for reclamation (IUCTFR 1988). Therefore, the government planned a protection program of groin construction to promote tidal flat deposition and stabilize the present tidal flats. Two groins with a total length of 6.606 km were designed to be constructed along the southern coast of Hangzhou Bay toward the right side of the Hangzhou Bay Cross-sea Bridge.

No. 1 groin with a length of 2.835 km is located 365 m east of the Hangzhou Bay Cross-sea Bridge, as shown in Figure 1 (b). The designed berm is 7.10 m with an upper width of 7.0 m, on which a 0.80 m high parapet wall is built. No. 2 groin in the east is about 3.0 km from

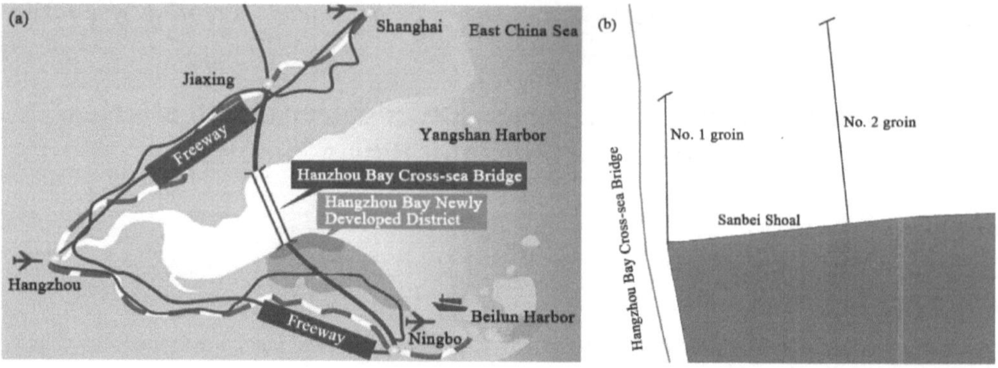

Figure 1. Geographical location of the study area.

No. 1 groin and has a length of 3.771 km from south to north. Its designed height is 6.40 m and the upper width is 8.8 m. The construction of the two long groins may affect the local hydrodynamics and sediment transport. In this study, we aim to investigate the hydrodynamic and deposition responses to the groins.

2 METHODOLOGY

Mike 21 Flow Model FM is based on a flexible mesh approach and the numerical solution of the 2D shallow water equations – depth-integrated incompressible Reynolds-averaged Navier–Stokes equations – invoking the assumptions of Boussinesq and of hydrostatic pressure (DHI 2011a).

2.1 *Hydrodynamic module and sediment transport module*

A brief introduction on the governing equations of hydrodynamic module and sediment transport module can be found elsewhere (DHI 2011a, DHI 2011b, Teisson 1991). The fine sediment transport module, Mike 21 Mud Transport Module (MT), is designed for simulating the transport of the cohesive sediment with grain size of less than 63 μm and fine sand with grain size of 63–125 μm (DHI 2011b). Because of the average D_{50} of about 14 μm, MT is used to simulate the suspended sediment transport.

2.2 *Bed morphology model*

According to the principle of sediment transport equilibrium, the sediment-carrying capacity of current, that is, S^* is defined as (Li et al. 2009):

$$S^* = k\frac{V^2}{gh} \tag{1}$$

where k is the coefficient of sediment carrying. When the actual sediment concentration is higher than S^*, the suspended sediment will be deposited. The decreased current will in turn decrease the sediment-carrying capacity. The bed morphology can be calculated based on this principle. Thus, the annual bed deposition/erosion is described by (Kong et al. 2014)

$$\Delta H = h_1 - h_2 = \frac{\alpha w_s S^* \Delta t}{\gamma_s'}\left(1 - \left(\frac{v_2}{v_1}\right)^2 \frac{h_1}{h_2}\right) \tag{2}$$

where v_1 and v_2 are the vertically averaged velocity under the pre-project and the post-project conditions, respectively; h_1 and h_2 are the water depth under the pre- and post-project conditions, respectively; α is the probability of settling, which is set at 0.5; and γ_s' is set at 801 kg/m^3.

When the deposition and erosion are in equilibrium, the water depth, current, and sediment are also under equilibrium, and therefore the final morphology can be calculated as

$$\Delta H_f = h_1\left[\left(\frac{v_2}{v_1}\right)^{0.667} - 1\right] \tag{3}$$

3 MODEL SETUP AND VERIFICATION

3.1 *Model setup*

A flexible triangular unstructured mesh system is established for hydrodynamic module and sediment transport module. The resolution of the mesh ranges from 5 to 8000 m, in which the groins are fined to reflect the accurate bathymetry and coastline, as shown in Figure 2. The bathymetry of large scale is the digitalized data from the chart published by Marine Navigation Department, and the bathymetry of the project region is interpolated by the measured data. In this study, the elevations are unified to the mean sea level (MSL).

The open boundary of FM module is forced with water level, which is predicted by harmonic constant obtained from the data set of OTIS in China Sea according to the forecast equation defined by Munk & Cartwright (1966):

$$\varsigma = A_0 + \sum_{i=1}^{11} H_i F_i \cos\left[\sigma_i t - (v_0 + u)_i + g_i\right] \tag{4}$$

where A_0 is the mean sea level; F_i and $(v_0 + u)_i$ are astronomical factors; σ_i is the angular frequency; H_i and g_i are the harmonic constants of a constituent, that is, amplitude and epoch, respectively; and 11 tidal constituents (M_2, S_2, N_2, K_2, K_1, O_1, P_1, Q_1, MF, M_4, and MS_4) are used to predict water level hourly.

According to the modification results, the Chezy coefficient is set at 0.015 m$^{1/2}$/s, and a Smagorinsky coefficient of 0.28 is used for the horizontal eddy viscosity. The Coriolis force varies in domain. The sediment settling velocity is 0.5 mm/s. According to the calibration

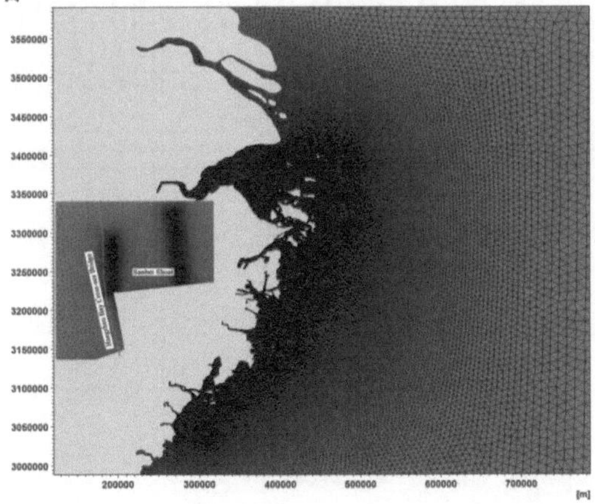

Figure 2. Triangular unstructured mesh system.

results, the critical shear stress for erosion applied in the model is set at 0.06 N/m^2, and the erosion coefficient (E) is calibrated to 1×10^{-5} kg/m^2/s.

3.2 *Model verification*

Field measurement is used for the validation of the model, including water levels of 2 gauge stations and tidal currents of 11 stations, as shown in Figure 3. The validation results of some selected stations are shown in Figures 4 and 5.

Simulated water levels are compared with measured data at LZW tide gauge station (Figure 4) for a typical period of 10 days, including spring and neap tides. The model tends to overestimate the water level at low spring tide level. In general, the simulated results have a high accuracy with the measured data for water level.

The hourly measured currents at L6 and L9 stations are utilized for the validation of currents during spring tide. The average error of current velocity is less than 0.09 m/s, and the simulated direction has a highly significant positive correlation with the measured data. That indicates that the model can satisfactorily reproduce the tidal current.

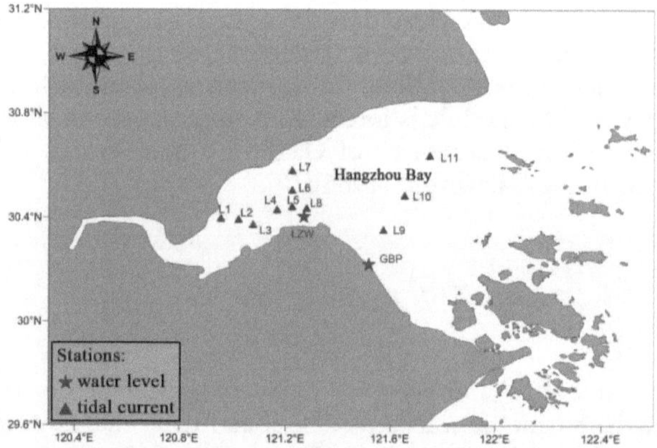

Figure 3. Location of gauge and tidal current stations for validation.

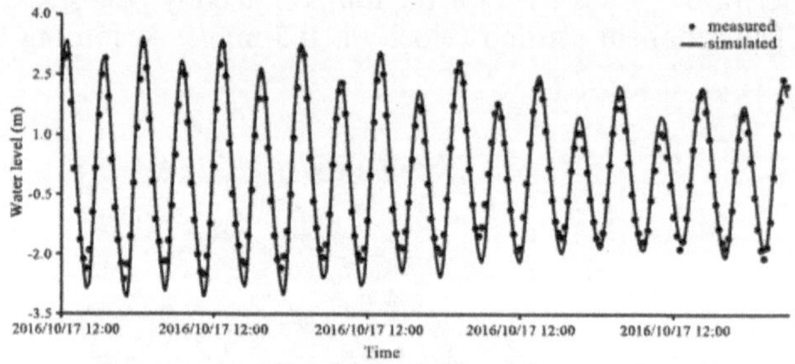

Figure 4. Comparison between the measured and simulated water levels at the LZW station.

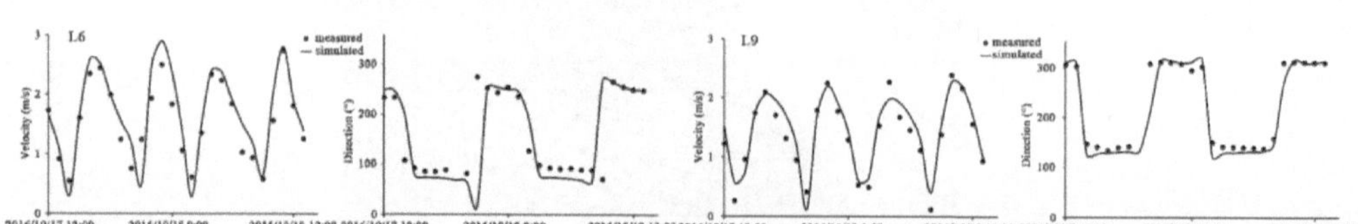

Figure 5. Comparison between the measured and simulated currents at the L6 and L9 stations.

4 RESULTS AND DISCUSSION

4.1 *Current response to the protection engineering*

The current vector response to the protection engineering is shown in Figure 6. As shown in Figure 6 (a), the groins disturb the flood movement of current, thus forming a counterclockwise current circulation in the western side of No. 2 groin, and the current flows along the groin. By contrast, there is a small change in current direction induced by No. 1 groin. This is because No. 2 groin is longer than No. 1 groin, which results in the latter being influenced by the former. The current vector in Figure 6 (b) indicates a larger-scale vector change at maximum ebb than at maximum flood, especially in the region between two groins. The shelter function of the two groins produces a significant change in the current, which results in the weak current in this region. It also causes a large current decrease toward the east of No. 2 groin. The protection engineering mainly changes the current in the region between the two groins and has a more significant effect during the ebb period.

The change of current velocity is also a key factor concerned, which is calculated for a spring tide in Figure 7. Its value ranges from 0.25 to 0.50 m/s at the head of No. 2 groin and from 0.15 to 0.30 m/s at the head of No. 1 groin during the flood period. Because of the shelter function of groins, the velocity reduces 0.15–0.30 m/s at the west of No. 1 groin and 0.25–0.50 m/s at the east of No. 2 groin in a large scale. A less than 0.10 m/s current velocity decrease occurs in the region between the two groins. During the ebb period, both value and scale of velocity change are larger than that during flood period. A velocity decrease of 0.60–1.00 m/s occurs in the region between the groins. The current velocity reduces 0.30–0.90 m/s outside the groins with a smaller decrease at the west side of No. 1 groin. However, there still is a large velocity increase of 0.30–0.50 m/s at both heads of the groins. It is concluded that the current velocity increases at the heads of the groins and decreases outside and inside the groins, generally showing a larger change and scale during the ebb period than during the flood period.

4.2 *Suspended sediment distribution due to groin construction*

The designed groins are constructed by riprap, which may cause bed suspended sediment. The suspended sediment is determined by the construction process and sediment source,

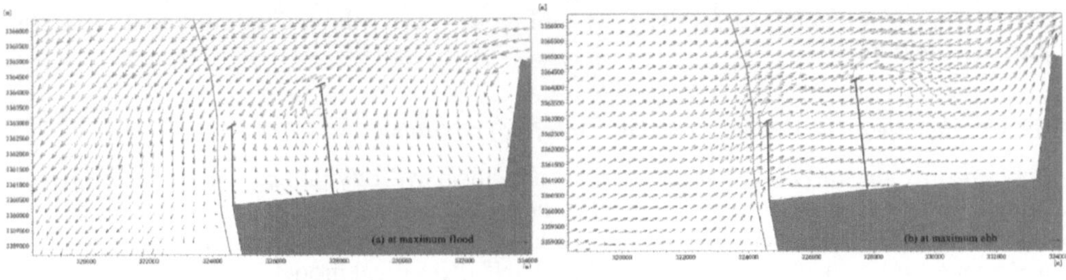

Figure 6. Tidal current fields under the pre- and post-project conditions.

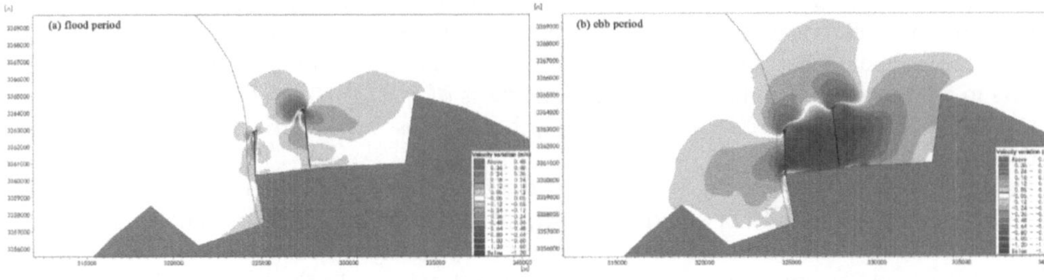

Figure 7. Average velocity change induced by protection engineering in a spring tide.

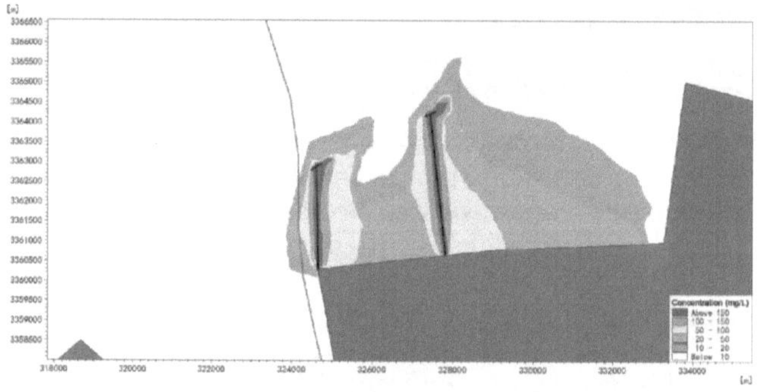

Figure 8. Maximum concentration distribution of the suspended sediment induced by riprap.

Table 1. Statistical data of areas with different isocline arrangements.

Concentration (mg/L)	10–20	20–50	50–100	100–150	>150
Area (km^2)	10.23	11.25	4.56	1.98	1.23

and it will affect the local environment. In this section, the most negative situation is taken into consideration, in which the source is set at a flow rate of 0.4 kg/s, and it is simplified as 12 point sources. The simulation time is 12 days, covering both spring tide and neap tide.

According to simulation results, the possible maximum concentration of the suspended sediment induced by riprap in an entire tidal period is shown in Figure 8, and the statistical data of areas with different isocline arrangements are provided in Table 1. The high sediment concentration is found along the groins, and the maximum value is larger than 150 mg/L. The suspended sediment is driven by tidal current; thus, the distribution scale is larger during the ebb period. The maximum envelope areas of 10–20, 20–50, and 50–100 mg/L are about 10.23, 11.25, and 4.56 km^2, respectively. The influence scale is limited within and around the engineering region.

4.3 *Morphology prediction under the post-project condition*

Sediment transport and morphology are driven by local hydrodynamics. The two long groins of the protection engineering inducing the regional current change will certainly cause morphology change. In this section, we aim to predict the morphology change under the post-project condition by using a half-empirical and half-theoretical model.

The predicted morphology 1 year after the establishment of protection engineering is shown in Figure 9 (a), where it is evident that the morphology trend approximately agrees with the current distribution. There is a significant deposition of 0.20–0.25 m at both toes of the groins, and the deposition gradually reduces to 0.10–0.15 m outside because of the current decrease. The deposition between the groins is about 0.15 m. The length of No. 2 groin results in a large scale of deposition with a thickness of 0.10–0.20 m. The increased current at the heads of the groins eroded 0.15–0.20 m, thereby possibly affecting the Hangzhou Bay Cross-sea Bridge.

As the protection engineering is located in the tidal flats with a relatively high elevation, the morphology will finally be in equilibrium, as shown in Figure 9 (b). The final deposition at both toes of the groins is about 0.50–0.60 m, and that between the groins is about 0.25–0.40 m. The erosion at both heads of the groins reaches 0.30–0.45 m, and the erosion of No. 1 groin (0.10–0.20 m) is close to the Cross-sea Bridge.

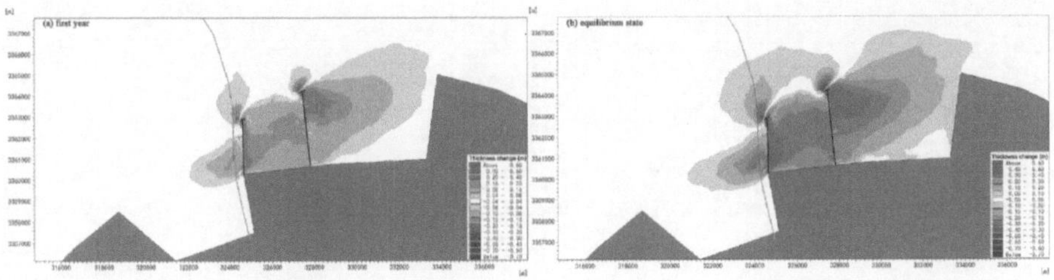

Figure 9. Bed morphology after the protection engineering.

5 CONCLUSIONS

This study used the numerical method to investigate the hydrodynamic response to the groins in protection engineering at the southern coast of Hangzhou Bay. The sediment distribution due to riprap in the construction process was analyzed and simplified by 12 point sources, and the morphology was predicted on the basis of a half-empirical and half-theoretical model.

The groins in the S–N orientation were almost perpendicular to the current direction, directly affecting the movement of the current. Therefore, the current between the groins and outside the groins had a significant decrease of 0.15–1.00 m/s, but a significant increase of 0.1–0.5 m/s at the heads of the groins. The larger change generally occurred around No. 2 groin. The response of the current determines the morphology after the protection engineering, resulting in a deposition thickness of 0.50–0.60 m at both toes of the groins, a deposition thickness of 0.25–0.40 m between the groins, and heavy erosion reaching 0.30–0.40 m at both heads of the groins. It was found that the 0.10–0.20 m erosion around the Cross-sea Bridge would not affect its service.

The suspended sediment induced by the groin construction was simplified as 12 point sources under the most negative condition with a flow rate of 0.4 kg/s. The simulation results demonstrated that the suspended sediment induced by riprap was driven by the local tidal current and was transported to a longer distance at the eastern side of No. 2 groin by a stronger ebb flow. The distribution area was larger than 29 km^2 with concentrations above 10 mg/L, in which about 7.77 km^2 area around the groins had concentrations above 50 mg/L.

ACKNOWLEDGMENTS

This study was financially supported by the Research Starting and Innovation Foundation of Zhejiang Ocean University and the Department of Education of Zhejiang Province.

REFERENCES

DHI 2011a. MIKE 21 & MIKE 3 FLOW MODEL FM: Hydrodynamic and Transport Module Scientific Documentation. Hørsholm, Denmark.

DHI 2011b. MIKE 21 & MIKE 3 FLOW MODEL FM: Mud Transport Module Scientific Documentation. Hørsholm, Denmark.

Feng, L.H. & Bao, Y.X. 2006. Coastal evolution process and tidal flat reclamation in Cixi City. Geography and Geo-Information Science 22(6): 75–78.

Investigations Unit of coastal and tidal flat resources of Zhejiang Province (IUCTFR) 1988. Comprehensive survey report on coastal and tidal flat resources of Zhejiang Province. Marine Press.

Kong, L.S. Cao, Z.D. Gu, F.F. 2014. Study on the erosion and deposition of the muddy sea bed. Zhejiang Hydrotechnics (195): 52–56.

Li, R.J. Luo, F. Zhou, H.M. 2009. Study on the flow induced sediment carrying capacity. Transactions of Oceanology and Limnology (1): 88–94.

Munk, W.H. & Cartwright, D.E. 1966. Tidal spectroscopy and prediction. Philosophical Transaction A259: 533–581.

Teisson, C. 1991. Cohesive suspended sediment transport: feasibility and limitations of numerical modeling. Journal of Hydraulic Research 29(6): 755–769.

Hydraulic Engineering V – Hong, Liu & Xie (Eds)

Adjustment of the flexible mattress with ecological cavities to near-bottom flow patterns

Ai-xing Ma
State Key Laboratory of Hydrology Water Resources and Hydraulic Engineering Science, Nanjing Hydraulic Research Institute, Nanjing, China

Min-xiong Cao
Nanjing Hydraulic Research Institute, Nanjing, China

Yuan Xu
Shanghai Waterway Engineering Design and Consulting Co. Ltd., Shanghai, China

Ya Deng
Hohai University, Nanjing, China

Dong-hai Huang
Shanghai Waterway Engineering Design and Consulting Co. Ltd., Shanghai, China

ABSTRACT: The classical flexible mattress has been widely used in waterway regulation engineering because of its advantage in protecting river beds. However, its disadvantage is the lack of ecological functions. Therefore, to optimize the classical flexible mattress, a new flexible mattress with ecological cavities was developed by waterway regulation designers. Flow patterns around the cavities were precisely measured by a 3D velocity meter in a flume. The results indicated that the velocity and turbulence intensity rapidly decreased when the fluid flowed into the cavities. Furthermore, the decay rate in the velocity-stable region reached 89% and the average vertical velocity in the cavity was only 10% of the bottom velocity in the upper surface. The profile of the vertical velocity was adjusted to the law of "large in the middle, small at the upper and bottom ends", and the peak value of the turbulence intensity was near the ceiling of the cavities. The bottom velocity of the cavity in the velocity-stable region was lower than the incipient velocity of the bed material load, thereby creating a favorable hydraulic condition and thus achieving the ecological benefits of little deposits. The findings of this study provide a reference for the design and application of the flexible mattress with ecological cavities.

1 INTRODUCTION

The flexible mattress of concrete interlocking block is widely used for bottom protection and beach protection in the middle and lower Yangtze waterway regulation (Hu 1998, Cao 2004, Zhang 2006, Chen 2016). This structure consists of geotextile and solid concrete blocks. The former is a cushion that can protect river beds, and the latter are ballasts that can maintain the structure stability. Because the geotextile can form isolating layers between the water and river beds, it produces an adverse effect on the habitat environment of the riverbed dweller. In order to reduce the negative effect on the ecological environment, the classical flexible mattress was improved by waterway designers (Xu 2016) on the basis of the idea of ecological restoration, and a new flexible mattress with ecological cavities was developed. In the new structure, solid concrete blocks were replaced by hollow ones, resulting in the decrease in the bottom velocity, forming a little environmental deposit and a microspace for the benthonic dweller.

When ecological flexible mattresses are laid on the bed, they caused the adjustment of near-bottom flow pattern and produced changes in flow velocity and turbulence features. In this study, we measure the flow vector field precisely near the new mattresses in a variable slope flume and analyze the adjustment to near-bottom flow. The findings have certain reference values on the design and application of new mattresses.

2 FLEXIBLE MATTRESS WITH ECOLOGICAL CAVITIES AND FLUME EXPERIMENT

2.1 *Structure of flexible mattress with ecological cavities*

The classical flexible mattress of concrete interlocking block consists of geotextile and solid concrete blocks. The size of the concrete block is 48 cm × 48 cm, with three different thicknesses, 12, 16, and 20 cm, and the net gap between the blocks is 2 cm.

Similar to the classical one, the flexible mattress with ecological cavities also concludes geotextile and ballast with concrete cavities. The geotextile is the same as the classical one; the size of concrete cavity is 60 cm × 60 cm × 40 cm (length, width, and height), which consists of an upper plate, a bottom plate, and four pillars. The thickness of the upper and the bottom plates are 8 and 10 cm, respectively. There are four holes in the bottom plate, each with a diameter of 12 cm. The cross profile of the pillars is square, with a uniform size of 8 cm × 8 cm. The layout of the cavities on the geotextile is symmetrical, with a net gap of 10 cm between the cavities and water resistance ratio of 51% in the near bottom. Figure 1 shows the structures of the classical and ecological flexible mattresses.

2.2 *Introduction of the flume experiment*

1. Experimental flume

Hydraulic experiment of the flexible mattress with ecological cavities was carried out in a variable slope flume in Nanjing Hydraulic Research Institute. The flume is 44 m long, 0.8 m wide, and 0.8 m deep, and the slope gradient of the experiment is 1%. The flow self-circular system is used in the design of the flume. The inlet flux is controlled by the electric valve and electromagnetic flowmeter and smoothed by a row of PVC piles. The outlet water level is altered through the grille tail-gate controlled by the stepper motor.

2. Model scale and model size of the cavity

In consideration of the cavity size and the dynamic conditions of the laboratory, the geometric scale of the model is normally set to 1:5.

The dimensions of the cavity model are 12 cm × 12 cm × 8 cm (length, width, and height); the thicknesses of the upper and bottom plates are 1.6 and 2 cm, respectively; the cross profile of the square pillar is 1.6 cm × 1.6 cm; the interior height of the cavity is 4.4 cm; and the net gap between cavities is 2 cm.

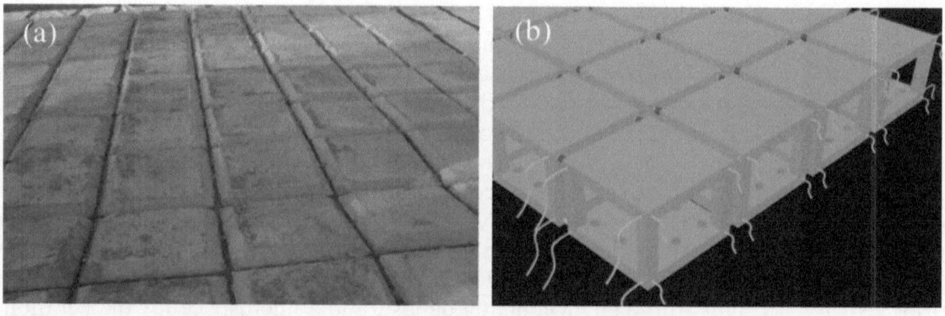

Figure 1. Flexible mattresses with (a) concrete interlocking block and (b) ecological cavity.

3. Flow conditions of the experiment
The flexible mattress with ecological cavities is used in the waterway regulation engineering at the left marginal beach of Fujiangsha in the lower Yangtze waterway regulation, where the median particle size of the bed material is 0.2 mm, the median diameter of the bed material load is 0.09 mm, and the incipient velocity (u_c) is 0.068 m/s.

The hydraulic conditions of the ebb current of the spring tide in the flood season were selected as the flow conditions in the laboratory experiment (Ma 2016): water depth, 8.21 m; average vertical velocity, 0.86 m/s. On the basis of the logarithmic distribution hypothesis of the vertical velocity in the near-bed region, the point velocity at 0.4 m above the riverbed, that is, at the top surface of the cavity, is 0.61 m/s.

Because of the large scale of the model, the condition of the prototype depth is hardly realized directly according to the scale. Therefore, the depth is reduced to 2 m; however, it is ensured that the point velocity at the upper surface of the cavity is same as the prototype. Finally, the experimental hydraulic conditions are a depth of 0.4 m and a point velocity at the cavity's upper surface of 0.27 m/s.

4. Layout of measurement instruments
There are 53 rows of model cavities along the flume with length of 7.4 m, and the corresponding prototype flexible mattress length is 37 m. A total of 15 vertical measured lines were arranged streamwise (Figure 2), of which 1 line (0#) was set at the distance of 10 m upstream from the front of the mattress and the remaining 14 lines (1#–14#) were arranged successively along the mattresses. A detailed measurement of the vertical velocity was conducted at the middle cavity of a row, with 9 measured points inside the cavity and 12 points from the upper surface of the cavity to the water surface.

Because of the narrow space of the interior cavity, a small hole on one side crosswise of the cavity upper surface, shown in Figure 3, was unfolded for the interior precise measurement after several repeated tests.

5. Velocity measurement instrument
The flow structure near the cavity is measured by the Acoustic Doppler Point Velocity Meter (Vectrino). The instrument can measure the spatial distribution of complex hydraulic elements, such as flow velocity and turbulence intensity near the structure, by measuring the instantaneous three-dimensional velocity. In the experiments, the velocity measuring frequency of the vertical point is 90 Hz, the measuring time is 40 s, and the velocity sample data of each vertical point is 3600.

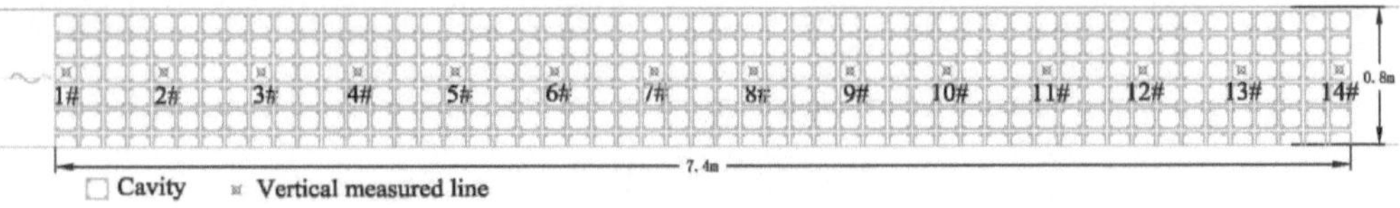

Figure 2. Layout of the vertical velocity measurement lines of the flexible mattress.

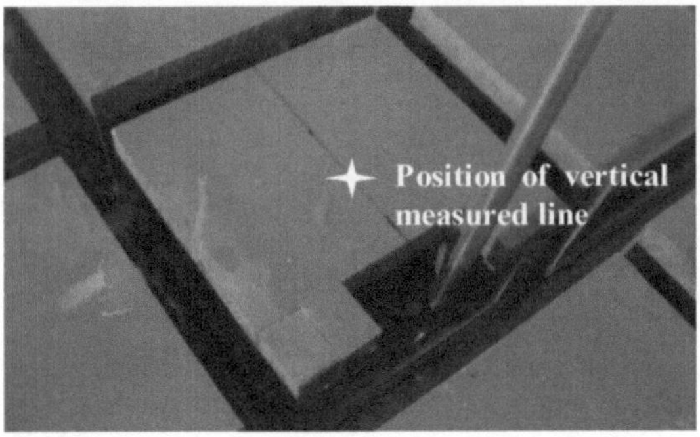

Figure 3. Velocity measurement of the interior cavity.

3 ANALYSIS OF EXPERIMENTAL RESULTS

The near-bottom flow pattern around the riverbed is altered by the ecological functions of the flexible mattress, such as the vertical velocity distribution and turbulence intensity. The geometric average of the instantaneous velocity along the flow direction (x-axis) is adopted as the time-averaged longitudinal velocity $(\bar{u})$, and the mean square root of the fluctuating velocity along the x-direction is adopted as the turbulence intensity (σ_x), that is

$$\bar{u} = \frac{\sum_{i=1}^{n} u_i}{n}, \quad \sigma_x = \sqrt{\frac{\sum_{i=1}^{n} \left(u_i - \bar{u}\right)^2}{n}} \tag{1}$$

where u_i is the *i-th* instantaneous velocity and n is the velocity sample data.

For convenience, the time-averaged longitudinal velocity (u) is used in the following. According to the model similarity law, for the undistorted model, the relationships among the measuring position (x_m), velocity (u_m), and turbulence intensity (σ_{xm}) of the model and the position (x_p), velocity (u_p), and turbulence intensity (σ_{xp}) of the prototype are given by:

$$x_p = \alpha_L x_m,\ u_p = (\alpha_L)^{0.5} u_m,\ \sigma_{xp} = (\alpha_L)^{0.5} \sigma_{xm} \tag{2}$$

where α_L is the geometric scale, that is, 5.

In order to facilitate comparative analysis, the velocity, the turbulence intensity, and the flow measuring vertical position are adopted as the parameters of the prototype.

3.1 *Vertical variation of the velocity and turbulence intensity*

The variation of the velocity and turbulence intensity along the relative depth, z/h (z is the height from the riverbed and h is the depth), of the vertical measured lines is shown in Figure 4. It shows that the distribution of the vertical velocity (measured line 0#) at the upstream of the mattress obeys the logarithmic law, with the small velocity at the bottom and the large one near the water surface. When the fluid flows into the flexible mattress, the vertical velocity distributions inside and outside the cavity are obviously different. The exterior distribution

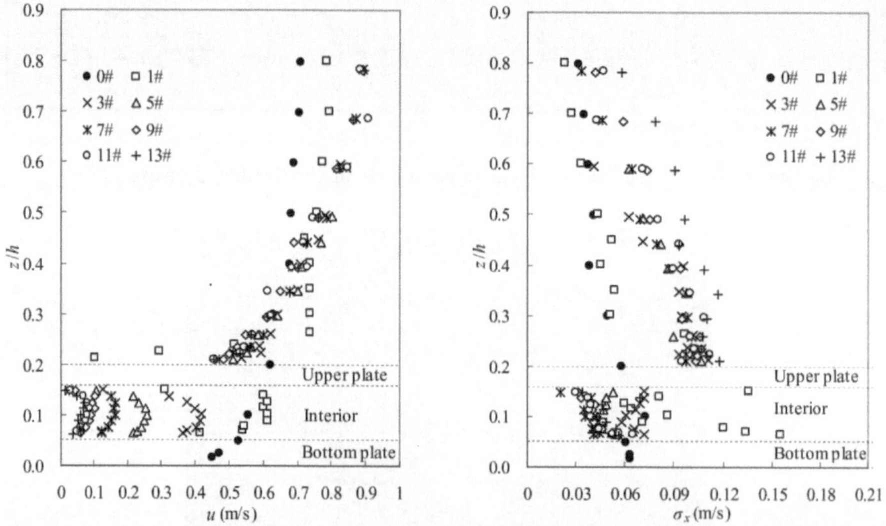

Figure 4. Variation of the vertical velocity (u) and turbulence intensity (σ_x) along the flexible mattress.

also obeyed the logarithmic rule, but the riverbed replaced by the upper surface of the cavity. The interior distribution is adjusted to the law of "large in the middle, small at the upper and bottom ends". The vertical velocity reduced rapidly and becomes smaller than that of the upstream flow because of the energy dissipation effect from the cavities.

The vertical distribution of turbulence intensity around the flexible mattress was also adjusted with the change of the vertical velocity, as shown in Figure 4. The turbulence intensity distribution of the cavity exterior is strong at the upper surface and weak on the water surface, whereas the maximum turbulence intensity appears near the top of the cavities. When the fluid flows into the first row of cavities, the interior turbulence peak increases obviously due to the sharp decrease in the cross section of the flow. The downstream turbulence intensity gradually becomes weaker, while the vertical distribution is weak around the upper surface and strong at the bottom, overall smaller than the upstream flow outside of the mattress.

3.2 *Variation of the characteristic velocity and turbulence intensity*

Velocity and turbulence intensity inside the cavities were adjusted when flexible mattress was paved on the riverbed. Figure 5 shows the variation of the characteristic vertical velocity and turbulence intensity (maximum, average, and near-bottom values) along the mattress inside the cavities. Velocity, turbulence, and amplitude of variation were high in the front of the mattress, whereas they decreased downstream the mattress and gradually became stable at the cavity of 22 m from the front, where both the vertical averaged velocity and the near-bottom velocity are lower than the incipient velocity of the bed material load (0.068 m/s) and slightly higher than the maximum velocity slightly (0.09 m/s).

3.3 *Variation of the relative velocity inside and outside the cavities*

For in-depth hydraulic analysis around the cavities, three indexes of the relative velocity were defined.

Velocity decay rate inside the cavity ($u_{in.}/u_{up.}$) is the ratio of the vertical averaged velocity ($u_{in.}$) inside the cavity to the upstream velocity ($u_{up.}$) outside mattress at the same position of the former, which represents the energy-dissipated effect on the inflow by the flexible mattress.

Vertical relative velocity inside and outside the cavity ($u_{in.}/u_{out.}$) is the ratio of the vertical averaged velocity ($u_{in.}$) inside the cavity to the bottom velocity ($u_{out.}$) around the upper surface of the cavity, which represents the hydraulic diversity around the mattress. In this study, $u_{out.}$ is the point velocity of 0.13 m from the cavity upper surface.

Relative incipient velocity of the near-bottom surface inside the cavity (u/u_c) is the ratio of the bottom velocity of 3 cm of the cavity to the incipient velocity (0.068 m/s) of the bed material load (d_{50} = 0.09 mm), which represents the depositing behavior of the mattress.

Figure 6 shows the variation of the velocity decay rate and the vertical relative velocity inside and outside the cavity along the flexible mattress. Both ratios decreased streamwise and gradually became stable at the distance of 22 m from the front of the mattress, behind

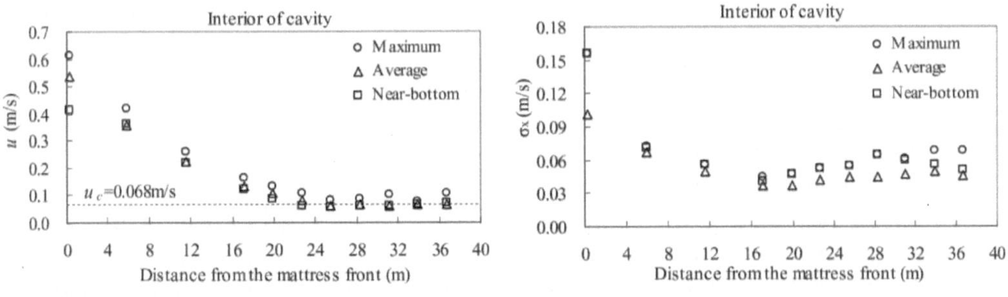

Figure 5. Variation of the interior characteristic velocity and turbulence intensity along the flexible mattress.

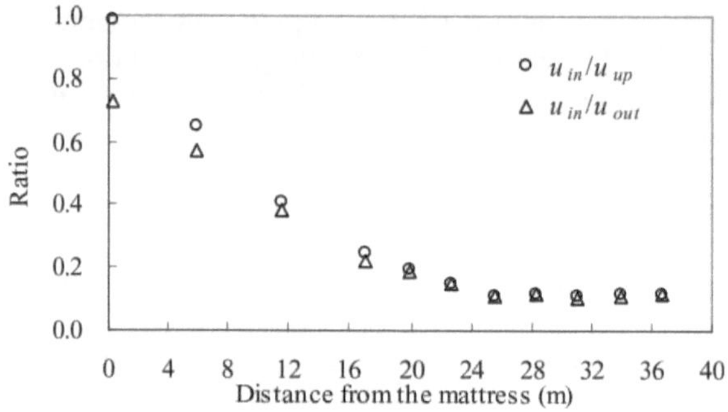

Figure 6. Variation of the velocity decay rate and the vertical relative velocity inside and outside the cavity along the flexible mattress.

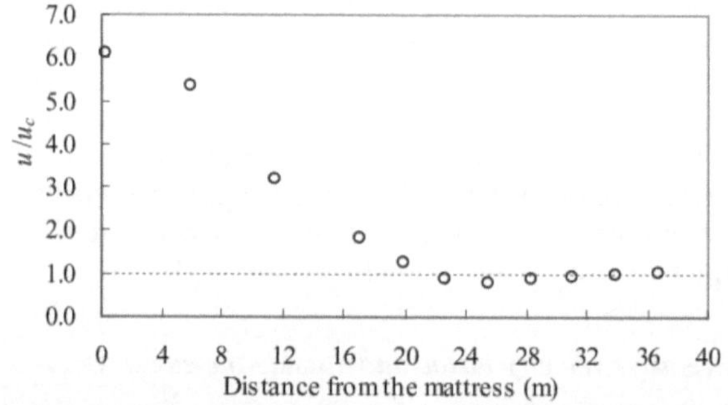

Figure 7. Variation of the relative incipient velocity of the near-bottom surface inside the cavity along the flexible mattress.

which the vertical relative velocity was maintained at around 0.1, with the corresponding velocity decay rate of 89%.

Figure 7 shows the variation of the relative incipient velocity of the near-bottom surface inside the cavity along the flexible mattress. The relative velocity reduced rapidly and gradually becomes stable at the distance of 22 m from the front of the mattress, behind which the ratio was maintained at 0.93.

Through the analysis of the hydraulic alteration around the cavities, the mechanism of producing ecological cavities along the flexible mattress could be summarized as follows: favorable hydraulic environment inside the cavities, where the dynamic conditions were weaker than the incipient ones of the sediment, was created by dissipating near-bottom flow energy by square pillars and separating the fluid between inside and outside of the cavity by the upper plate, thus achieving ecological benefits of little deposits.

4 CONCLUSION

This study carried out the hydrodynamic experiment on the flexible mattress with ecological cavities in a flume by setting the undistorted scale of the new mattress to 1:5. The main results can be summarized as follows:

1. The velocity and turbulence intensity were rapidly decreased when the fluid flowed into the cavities and then gradually became stable at the distance of about 20 m from the front of the mattress. The velocity decay rate in the stable flow region reached 89%, and the average vertical velocity in the cavity was only 10% of the bottom velocity in the upper surface.

2. The distribution of the vertical velocity inside the cavity no longer obeyed the logarithm law, but adjusted to the law of "large in the middle, small at the upper and bottom ends". The distribution of the turbulence intensity inside the cavity was weak at the upper surface and strong at the bottom, whereas the peak value of the turbulence intensity was near the ceiling of the cavities.
3. The bottom velocity of the cavity in the velocity-stable region was lower than the incipient velocity of the bed material load, thereby creating a favorable hydraulic environment and thus achieving the ecological benefits of little deposits.

ACKNOWLEDGMENTS

This study was financially supported by the National Key Research and Development Program of China (No. 2016YFC0402100) and the National Natural Science Foundation of China (Nos. 51309160 and 51579216).

REFERENCES

Cao, M. 2004. Application of flexible mattress in waterway regulation engineering of the Yangtze River. *Port & Waterway Engineering*, (9): 70–73.

Chen F. & Ma A.X. & Chen Y.H. 2016. Protection effect of soft mattress-permeable frame joint bottom protection downstream submerged breakwater. *Port & Waterway Engineering*, (4): 100–104.

Hu Y.L. 1998. Use of a soft drainage body (geotextile) in the bottom protection and beach stabilization. *Harbor Engineering*, (6): 33–36.

Ma A.X. 2016. Report of experiment study on the ecological bottom (beach) protection structure. Nanjing Hydraulic Research Institute.

Xu Y. & Huang D.H. & Ma A.X. 2016. Report of study on the ecological bottom (beach) protection structure. Shanghai Waterway Engineering Design and Consulting *Co. Ltd.*, Nanjing Hydraulic Research Institute.

Zhang J.M. 2006. Structural design of soft mattress in Yangtze estuary deepwater channel regulation project. *Port & Waterway Engineering*, (S2): 20–23.

Hydraulic Engineering V – Hong, Liu & Xie (Eds)
© 2018 Taylor & Francis Group, London, ISBN 978-0-8153-7471-8

Comparison and analysis of different critical velocity models about the pipeline transport for the medium and coarse sand

Runli Tao, Lu Zhang, Yuchi Hao, Yanchao Shen & Chaozhe Yuan
CCCC National Engineering Research Center of Dredging Techonology and Equipment Co. Ltd., China

ABSTRACT: One of the main concerns in slurry transport is the pipeline plugging. Plugging will not only cause engineering safety problems, but also cause economic losses. As a result, determination of critical velocity under certain working conditions is an important method to prevent blockage. Therefore, researching the critical velocity has great value on improving the efficiency of pipeline transport and optimizing the construction process. Currently, there are many definitions and formulas on the critical velocity, mostly empirical formulas, and it cannot be used in the dredging projects accurately. In order to obtain the role and value of these critical velocity models, this paper uses seven critical velocity models under the engineering condition of "Daxiaodeng fill and reclamation project" in Xiamen, China. In this paper, the calculated critical velocity is compared with the measured velocity. It is concluded that under the soil and working conditions in Xiamen, when the concentration is within 30%, the Durand model and IHC-MTI model are recommended to calculate the critical velocity. When the concentration is above 30%, the Durand model, IHC-MTI model and Wasp model are recommended to calculate the critical velocity.

1 INTRODUCTION

In dredging, the slurry is transported through the long pipeline from the dredged site to the storage site by the cutter suction dredger. Slurry is the mixture of sand and water with large discharge and high concentration in the pipeline, and the solids will gradually stratify and possibly deposit during the transportation. Meanwhile, due to the distance extension of the pipeline, the head loss increases and the slurry velocity in the pipeline decreases gradually. When the velocity is less than the critical velocity, pipeline plugging may happen. One of the main concerns in slurry transport is the pipeline plugging. Plugging will not only cause engineering safety problems, but also cause economic losses. Plugging can lead to frequent shutdowns, and a lot of time and manpower will be spent in solving this problem. As a result, determination of critical velocity under certain working conditions is an important method to prevent blockage. Therefore, researching the critical velocity has great value on improving the efficiency of pipeline transport and optimizing the construction process.

Currently, there are many definitions and formulas on the critical velocity, mostly empirical formulas, and it cannot be used in the dredging projects accurately. As a result, it needs to be selected carefully when using the critical velocity formula. The main Dutch and Belgium dredging companies use modified Durand & Condolios (1952) model, while the main companies in the USA use a modified Wilson et al. (1992) model (Miedema, 2014). Chinese dredging companies use "Dredging and reclamation engineering design specification Formula One" and "Dredging and reclamation engineering design specification Formula Two" to calculate the critical velocity. The IHC-MTI research institute uses Van den Berg (1998) equation. And professor Miedema of the dredging department of Delft University of Technology in the Netherlands recently proposed a new algorithm for the critical velocity, integrated in the DHLLDV model (Miedema, 2016). Since the above-mentioned critical velocity models are empirical formulas, in order to obtain the role and value of these critical velocity models, this paper uses seven critical velocity models under the engineering condition of "Daxiaodeng fill and reclamation project" in Xiamen, China. In this paper, the calculated critical velocity will be compared with the measured velocity, and the role and value of the above-mentioned critical velocities in engineering projects will be obtained.

2 CRITICAL VELOCITY MODELS

Most critical velocity models are one term one equation models, based on the idea that the critical velocity can be explained that way. So there are a lot of critical velocity calculation formulas. The following seven widely used critical velocity models will be introduced:

2.1 *Durand & Condolios (1952)*

Based on the experiments, Durand (1953) proposed the following equation for the critical velocity:

$$V_c = F_{DL}\sqrt{2gD(s-1)} \tag{1}$$

where V_c is the critical velocity, F_{DL} is the Durand velocity factor, which depends on the particle diameter and the concentration of solids. D is the diameter of pipe, S is the specific gravity of solids, and g is the gravity acceleration. The value of F_{DL} can be determined by Durand graph.

2.2 *Wilson (1992) model*

Wilson et al. (1992) published the sliding bed model as a nomograph, and the fit function of the critical velocity can be estimated in the following equation:

$$V_c = (2gD(S_s - S_f))^{1/2} \cdot \left(\frac{0.018}{f_f}\right)^{0.13} \tag{2}$$

where f_f is the friction factor. V_n is the critical velocity, F_{DL} is the Durand velocity factor, which depends on the particle diameter and the concentration of solids. D is the diameter of pipe, S_s is the relative density of solids, S_f is the relative density of slurry, and g is the gravity acceleration.

2.3 *Dredging and reclamation engineering design specification Formula One (dredging and reclamation engineering design specification, 2012)*

$$\mathrm{V_c} = 0.928C^{0.105}d_s^{\,0.056}\sqrt{2gD(\gamma_s - 1)} \tag{3}$$

where V_c is the critical velocity, C is the volumetric concentration, d_s is the average particle diameter, D is the diameter of pipe, γ_s is the specific gravity of solids, and g is the gravity acceleration.

2.4 *Dredging and reclamation engineering design specification Formula Two (design code for dredging and reclamation works, 2012)*

$$V_c = (90C)^{1/3} g^{1/4} D^{1/2} V_{ss}^{\,1/2} d_s^{\,-1/4} \tag{4}$$

where V_c is the critical velocity, C is the volumetric concentration, d_s is the average diameter, g is the gravity acceleration, D is the pipe diameter, V_{ss} is the terminal settling velocity.

2.5 *IHC-MTI model*

Van den Berg (1998) presented an equation for the critical velocity, which is used in IHC-MTI.

$$V_c = 0.298\left(5 - \frac{1}{\sqrt{1000 \cdot d}}\right) \cdot \left(\frac{C_v}{C_{v+0.1}}\right)^{1/6} \cdot \sqrt{2 \cdot g \cdot D_p \cdot R_{sd}} \tag{5}$$

where V_c is the critical velocity, C_v is the volumetric concentration, d is the average particle diameter, g is the gravity acceleration, D_p is the pipe diameter, R_{sd} is the relative submerged density.

2.6 *Wasp et al. (1977) equation*

Wasp et al. (1977) derived an equation similar to the Durand & Condolios (1952) equation. (Miedema, book)

$$V_c = 4 \cdot C^{1/5} \cdot \sqrt{2 \cdot g \cdot D_p \cdot R_{sg}} \cdot \left(\frac{d}{D_p}\right)^{1/6} \tag{6}$$

where V_c is the critical velocity, C is the volumetric concentration, d is the average particle diameter, g is the gravity acceleration, D_p is the pipe diameter, R_{sd} is the relative submerged density.

2.7 *Delft Head Loss & Limit Deposit Velocity model (DHLLDV)*

In Miedema (2016) paper, the definition of the critical velocity is the line speed below which there may be either a stationary bed or a sliding bed, depending on the particle diameter and the pipe diameter, but above which no bed (stationary or sliding) exists, the Limit Deposit Velocity. The DHLLDV framework starts with models for five fundamental flow regimes. The main problem of determining the critical velocity is to determine the Froude number F_L.

For very small particles, the Froude number F_L is:

$$F_{L,vs} = \frac{1.4 \cdot (\nu_l \cdot R_{sd} \cdot g)^{1/3} \cdot \sqrt{\frac{8}{\lambda_l}}}{(2 \cdot g \cdot R_{sd} \cdot D_p)^{1/2}} \tag{7}$$

For small particles, the Froude number F_L is:

$$F_{L,ss} = \alpha_p \cdot \left(\frac{\nu_l\left(1 - \frac{C_{vs}}{\kappa_c}\right)}{\lambda_l \cdot (2 \cdot g \cdot R_{sd} \cdot D_p)^{1/2}}\right)^{1/3} \tag{8}$$

with: $\alpha_p = 3.4 \cdot \left(\frac{1.65}{R_{sd}}\right)^{2/9}$, $\kappa_C = 0.175 \cdot (1 + \beta)$

If $F_{L,vs} > F_{L,ss} \Rightarrow F_{L,s} = F_{L,vs}$
If $F_{L,vs} \leq F_{L,ss} \Rightarrow F_{L,s} = F_{L,vs}$
For medium and large particles, the Froude number F_L is:

$$F_{L,r} = \alpha_p \cdot \left[\frac{\left(1 - \frac{C_{vs}}{\kappa_c}\right)^{\beta} \cdot C_{vs} \cdot \left(\mu_{sf} \cdot C_{vb} \cdot \frac{\pi}{8}\right)^{1/2} \cdot C_{vr,ldv}^{1/2}}{\lambda_l} \right] \tag{9}$$

where $C_{vr,ldv}$ is the bed fraction at the Limit Deposit Velocity.

The bed fraction at the Limit Deposit Velocity depends on the particle diameter to pipe diameter ratio:

$$d \le 0.015 \cdot D_p \qquad C_{vr,ldv} = \frac{0.0065}{2 \cdot g \cdot R_{sd} \cdot D_p}$$

$$d > 0.015 \cdot D_p \qquad C_{vr,ldv} = \frac{0.053}{2 \cdot g \cdot R_{sd} \cdot D_p} \cdot \left(\frac{d}{D_p}\right)^{1/2}$$

The resulting upper limit of the Froude number $F_{L,ul}$ value can be determined according to:
If $F_{L,vs} > F_{L,r} \Rightarrow F_{L,ul} = F_{L,s} \cdot e^{-d/d_0} + F_{L,r} \cdot \left(1 - e^{-d/d_0}\right)$
If $F_{L,vs} \le F_{L,ss} \Rightarrow F_{L,ul} = F_{L,s}$

$d > d_{rough} \Rightarrow F_{L,ul} = F_{L,r}$

With: $d_0 = 0.0005 \cdot \left(\frac{1.65}{R_{sd}}\right)^2$

The lower limit of the LDV is the transition velocity between the sliding bed regime and the heterogeneous regime, the Froude number $F_{L,ul}$ value can be determined according to:

$$F_{L,ll} = \frac{-B - \sqrt{B^2 - 4 \cdot A \cdot C}}{2 \cdot A \cdot \left(2 \cdot g \cdot R_{sd} \cdot D_p\right)^{1/2}} \tag{10}$$

The final resulting Froude F_L value can now be determined according to:

If $F_{L,ul} \ge F_{L,ll} \Rightarrow F_L = F_{L,ul}$

If $F_{L,ll} > F_{L,ul} \Rightarrow F_L = F_{L,ll}$

3 PROJECT INTRODUCTION

"The Daxiaodeng fill and land reclamation project" is located in the east of Xiamen Island, Xiang'an District, southeast direction. And the land reclamation area is about 7.58 km^2. The soil is mainly the medium and coarse sand. And the project related parameters are in Table 1.

Table 1. Xiamen land reclamation project related parameters.

Pipe Diameter	850 mm
Average Particle Diameter	0.72 mm
Specific gravity of Solids	2.65 t/m^3
Water density	1.025 t/m^3

4 COMPARISON AND ANALYSIS

The above-mentioned seven critical velocity models are applied under the condition of the land reclamation project in Xiamen with the medium and coarse sand, the respective calculated critical velocities with different concentrations are obtained in Figure 1. And they are compared with the measured velocity data. The measured concentration-velocity data is from the "Xin Haitun" Cutter suction dredger in August, 2016. And there are about 30 000 sets of data in total, the measured values in Figure 1 are obtained after averaging data.

4.1 *Comparison of different critical velocity models under low concentration condition*

Figure 1 shows the measured data and seven critical velocity models results under lower concentration which is 18% to 30%. In Figure 1, the black line is the measured velocity data, which is between 5 m/s and 6 m/s, and the measured velocity doesn't change a lot as the concentration changes. The red line is the Durand model result, it shows that the calculated critical velocity increases with the increase of the concentration, and the minimum velocity is 5.46 m/s at the concentration of 18%. The maximum velocity is 5.82 m/s at the concentration of 30%. The grey line is the Wilson model result. It shows that the calculated critical velocity remains 4.96 m/s with the change of the concentration under this engineering condition, because under this engineering condition, the friction factor is mainly affected by ε/D, and it has nothing to do with concentration. The purple line is the "Dredging and reclamation engineering design specification Formula One" result, the critical velocity increases with the increase of the concentration, and the minimum velocity is 4 m/s at the concentration of 18%. The maximum velocity is 4.2 m/s at the concentration of 30%. The blue line is the "Dredging and reclamation engineering design specification Formula One" result, the critical velocity increases with the increase of the concentration, and the minimum velocity is 7.76 m/s at the concentration of 18%. The maximum velocity is 9.2 m/s at the concentration of 30%. The green line is the IHC-MTI model result. It shows that the result of IHC-MTI model is very close to the Durand model. The dashed red line is Wasp model result, the critical velocity

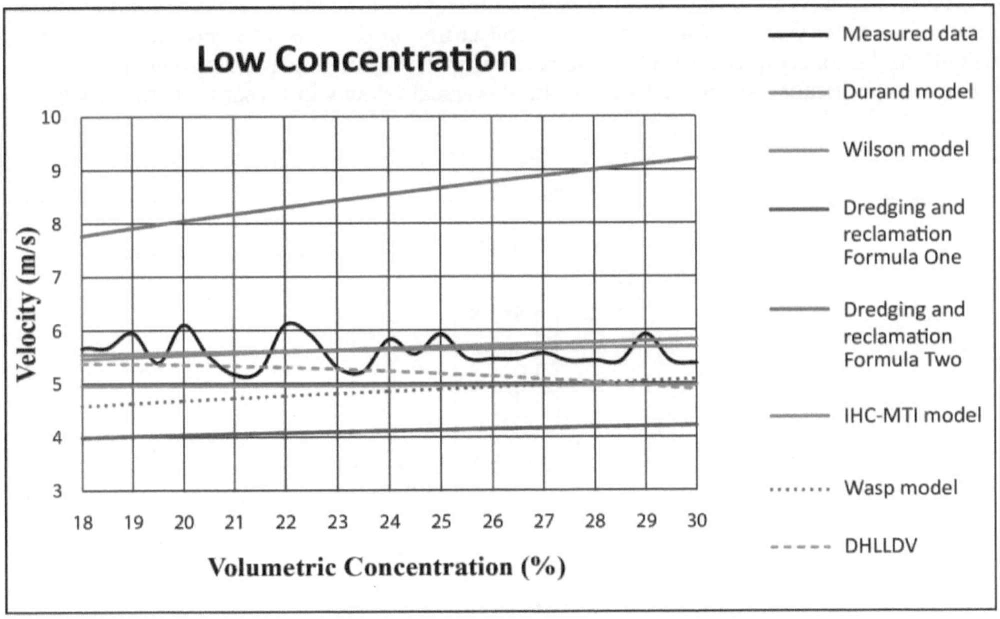

Figure 1. The calculated critical velocities of different models under different concentrations.

increases with the increase of the concentration, and the minimum velocity is 4.58 m/s at the concentration of 18%. The maximum velocity is 5.07 m/s at the concentration of 30%. The dashed green line is the DHLLDV model result, the critical velocity decreases with the increase of the concentration, and the minimum velocity is 5.015 m/s at the concentration of 30%. The maximum velocity is 5.36 m/s at the concentration of 18%.

For the seven models under low concentration, it shows that the calculated critical velocities of Durand model, Wilson model, IHC-MTI model, Wasp model and DHLLDV model are close. The Durand model result is similar to the IHC-MTI model, and these two model results are the closest to the measured velocity. The Dredging and reclamation engineering design specification Formula One and Formula Two deviate far from other models. The result of Dredging and reclamation engineering design specification Formula One is the smallest, and the result of Dredging and reclamation engineering design specification Formula Two is the largest, and the maximum velocity is 9.2 m/s. If Dredging and reclamation engineering design specification Formula Two is used to calculate the critical velocity, it may cause a waste of fuel consumption in dredging projects.

In general, under the soil and working conditions in Xiamen, when the concentration is within 30%, the Durand model and IHC-MTI model are recommended to calculate the critical velocity.

4.2 *Comparison of different critical velocity models under high concentration condition*

Figure 2 shows the measured data and seven critical velocity models results under high concentration which is 31% to 58%. In Figure 1, the black line is the measured velocity data, which is between 5 m/s and 6 m/s, and the measured velocity doesn't change a lot as the concentration changes. The red line is the Durand model result, it shows that the calculated critical velocity increases with the increase of the concentration, and the minimum velocity is 5.85 m/s at the concentration of 31%. The maximum velocity is 6.32 m/s at the concentration of 58%. The grey line is the Wilson model result. It shows that the calculated critical velocity remains 4.96 m/s with the change of the concentration under this engineering condition, because under this engineering condition, the friction factor is mainly affected by ε/D, and it has nothing to do with concentration. The purple line is the "Dredging and reclamation engineering design specification Formula One" result, the critical velocity increases with the increase of the concentration, and the minimum velocity is 4.22 m/s at the concentration of 31%. The maximum velocity is 4.52 m/s at the concentration of 58%. The blue line is the "Dredging and reclamation engineering design specification Formula Two" result, the critical velocity increases with the increase of

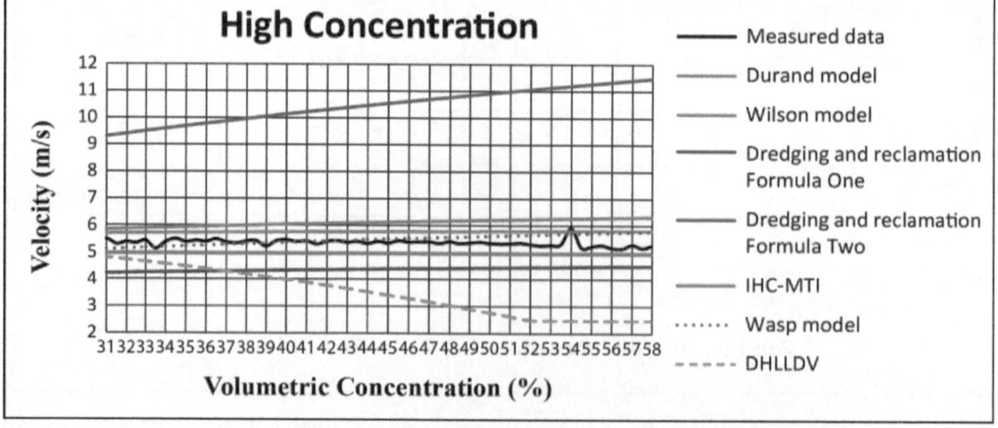

Figure 2. The calculated critical velocities of different models under different concentrations.

the concentration, and the minimum velocity is 9.31 m/s at the concentration of 31%. The maximum velocity is 11.54 m/s at the concentration of 58%. The green line is the IHC-MTI model result. It shows that the result of IHC-MTI model deviates from the Durand model further and further as the concentration increases. And the minimum velocity is 5.7 m/s at the concentration of 31%. The maximum velocity is 5.81 m/s at the concentration of 58%. The dashed red line is Wasp model result, the critical velocity increases with the increase of the concentration, and the minimum velocity is 5.1 m/s at the concentration of 31%. The maximum velocity is 5.8 m/s at the concentration of 58%. And it is closer to the IHC-MTI model result with the increase of concentration. The dashed green line is the DHLLDV model result, the critical velocity decreases with the increase of the concentration, and the minimum velocity is 2.5 m/s at the concentration of 58%. The maximum velocity is 4.6 m/s at the concentration of 31%.

For the seven models under high concentration, it shows that the calculated critical velocities of Durand model, Wilson model, IHC-MTI model and Wasp model are close in Figure 2. And these four model results are the closest to the measured velocity. Especially when the concentration increases, The Wasp model result is similar to the IHC-MTI model, The Dredging and reclamation engineering design specification Formula One and Formula Two deviate far from other models. The result of Dredging and reclamation engineering design specification Formula One is still the smallest compared with the low concentration, and the result of Dredging and reclamation engineering design specification Formula Two is still the largest. If Dredging and reclamation engineering design specification Formula Two is used to calculate the critical velocity, it may cause a waste of fuel consumption in dredging projects. Compared with Figure 1, it is worth noting that the changes of DHLLDV. In Figure 2, when the concentration is above 40%, the calculated DHLLDV model result deviates too far from the reality, which doesn't have reference significance. Actually in real dredging projects, when the concentration is too high, such as 50% or higher, the solids in the pipe start to deposite, there appears sliding bed. As a result, when the concentration comes to 50% or higher, the calculation of critical velocity often has no meaning.

In general, under the soil and working conditions in Xiamen, when the concentration is above 30%, the Durand model, IHC-MTI model and Wasp model are recommended to calculate the critical velocity.

5 CONCLUSIONS

Most of the existing models for the transport of settling solids in pipes are based on empirical experiments, and the models are generally optimized for one particular of the transport process. However, dredging companies work in a wide variety of situations, with different sizes dredgers, pipeline lengths ranging from tens of meters to tens of thousands of meters, and dredged materials ranging from silt and clay to cobbles and boulders. The existing models do not provide much guidance on how the slurry behavior changes as parameters such as the pipe diameter, available pumping power and particle size change. As a result, this paper compares seven widely used critical velocity models under the soil and working conditions of "Daxiaodeng fill and land reclamation project" in Xiamen, and the suitable critical velocity model for this soil and working condition can be obtained. And it can also provide guidance for other similar dredging projects.

For the seven models under low concentration, critical velocities of Durand model, Wilson model, IHC-MTI model, Wasp model and DHLLDV model are close. The Durand model result is similar to the IHC-MTI model. The Dredging and reclamation engineering design specification Formula One and Formula Two deviate far from other models. The result of Dredging and reclamation engineering design specification Formula One is the smallest, and the result of Dredging and reclamation engineering design specification Formula Two is the largest. If Dredging and reclamation engineering design specification Formula Two is used to calculate the critical velocity, it may cause a waste of fuel consumption in dredging projects. In general, under the soil and working conditions in Xiamen, when the concentration is within 30%, the Durand model and IHC-MTI model are recommended to calculate the critical velocity.

For the seven models under high concentration, critical velocities of Durand model, Wilson model, IHC-MTI model and Wasp model are close. Especially when the concentration increases, The Wasp model result is similar to the IHC-MTI model, The Dredging and reclamation engineering design specification Formula One and Formula Two deviate far from other models. The result of Dredging and reclamation engineering design specification Formula One is still the smallest compared with the low concentration, and the result of Dredging and reclamation engineering design specification Formula Two is still the largest. If Dredging and reclamation engineering design specification Formula Two is used to calculate the critical velocity, it may cause a waste of fuel consumption in dredging projects. Actually in real dredging projects, when the concentration is too high, such as 50% or higher, the solids in the pipe start to deposite, there appears sliding bed. As a result, when the concentration comes to 50% or higher, the calculation of critical velocity often has no meaning. In general, under the soil and working conditions in Xiamen, when the concentration is above 30%, the Durand model, IHC-MTI model and Wasp model are recommended to calculate the critical velocity.

REFERENCES

Berg, C.H. (1998). Pipelines as Transportation Systems. Kinderdijk, The Netherlands: European Mining Course Proceedings, IHC-MTI.

Durand, R., & Condolios, E. (1952). Etude experimentale du refoulement des materieaux en conduits en particulier des produits de dragage et des schlamms. *Deuxiemes Journees de l'Hydraulique*, 27–55.

Miedema, S.A., & Ramsdell, R.C. (2014). The Delft Head Loss & Limit Deposit Velocity Model. Hydrotransport (p. 15). Denver, USA: BHR Group.

Miedema, S.A., & Ramsdell, R.C. (2016). The Delft Head Loss & Limit Deposit Velocity Framework. Journal of Dredging Engineering, Vol. 15(2), 3–33.

Ministry of Transport of the People's Republic of China (2012). Design Code for Dredging and Reclamation Works. China Communications Press.

Wasp, E.J., Kenny, J.P., & Gandhi, R.L. (1977). Solid liquid flow slurry pipeline transportation. Transactions Technical Publications.

Wilson, K.C., Addie, G.R., & Clift, R. (1992). Slurry Transport using Centrifugal Pumps. New York: Elsevier Applied Sciences.

Hydraulic Engineering V – Hong, Liu & Xie (Eds)
 ISBN 978-0-8153-7471-8

Coastal responses and shoreline restoration after Holocene sea-level changes in Hong Kong, China

Y.Y. Sun & J. Xing
National Engineering Research Center of Dredging Technology and Equipment, Shanghai, China

Y.Q. Zong
The University of Hong Kong, Hong Kong

G.J. Hong
National Engineering Research Center of Dredging Technology and Equipment, Shanghai, China

ABSTRACT: Coastal changes and shoreline restoration after the mid-Holocene sea-level highstand are widely attributed to a combined effect of natural and anthropogenic factors. The mangrove forest wetland in the Mai Po Nature Reserve, Hong Kong had undergone several changes in the past, namely (1) coastal environmental changes influenced by Holocene transgression, (2) mangrove forest expansion caused by the generation of additional accommodation space in the past 100 years, (3) land use conversion driven by human activities and social development since the 1950s and wetland protection in recent decades, and (4) increased sedimentation rates induced by industrialization and urbanization in the Pearl River Delta.

1 INTRODUCTION

Coastal changes and shoreline restoration present a combined effect of sea-level changes, tectonics, sediment discharge balance, biological actions, and human activities. During the early Holocene, coastal change was predominantly due to the rapid rise of sea level, which changed the landscape of coastal regions. After the large-scale transgression, which resulted in mid-Holocene, coastal morphology underwent significant changes due to a number of controlling factors, including small-scale fluctuations in sea levels. In Hong Kong, which is a developed area, coastal and shoreline changes could determine the spatial patterns of urban development, human activities, and environmental conservation. In this study, we investigate the temporal and spatial changes of a wetland habitat to examine the response of the wetland to natural and anthropogenic factors.

2 SITE STUDY

The study site is located in the Mai Po Nature Reserve at the head of the Deep Bay, Northwest of Hong Kong (Figure 1), which is an integral part of the Deep Bay region. Mai Po covers approximately 380 hectares of tidal flats and mangrove forests. Over time, sea level has significantly changed the landscape and coastal environment of the Mai Po area. Since the 1970s, the inland part of the area has been converted into the so-called "Kei-Wais" system (trenches dug from the mangrove forest for shrimp farming), which is drained during low tides and receives seawater during high tides. The inflow and outflow are artificially controlled by means of sluices. The boundary of the area is covered by natural mangrove forests, consisting of *Kandelia candel* (Rhizophoraceae), *Aegiceras corniculatum* (Myrsinaceae), *Acanthus ilicifolius* (Acanthaceae), and occasional *Avicennia marina* (Avicenniaceae). In studies regarding

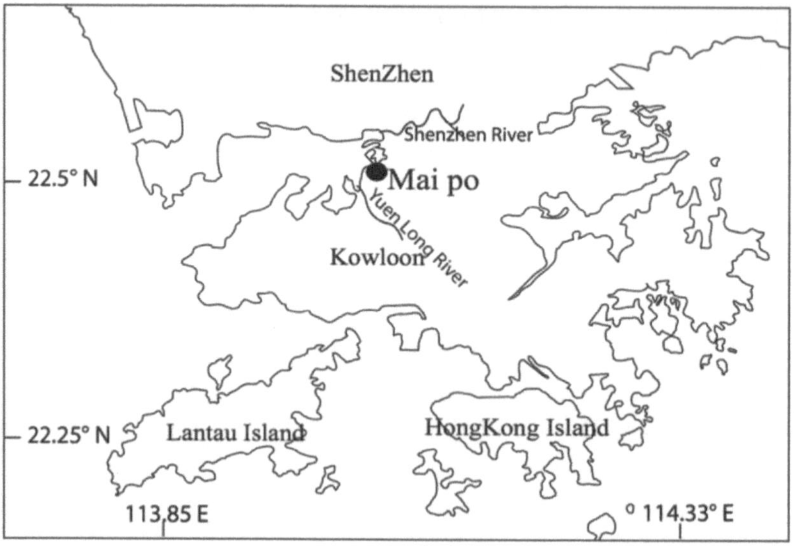

Figure 1. Mai Po nature reserve in Hong Kong.

coastal changes, mangrove forests are regarded as a biological factor that could protect coastal erosion and promote wetland progradation (Thampanya, Vermaat et al. 2006).

3 METHODOLOGY

3.1 *Laboratory methods for sediment analyses*

3.1.1 *Organic carbon analysis*

The stable organic carbon isotope ratio ($\delta^{13}C$) is a ratio of ^{13}C to ^{12}C isotopes measured from a sediment sample and calibrated to the established standard reference (VPDB). In aquatic environments, floating and benthic macrophytes and microalgae are important contributors of organic carbon. Freshwater macrophytes and algae tend to follow the Calvin Cycle pathway and have a $\delta^{13}C$ ratio similar to that of C3 plants. Marine macrophytes and algae, however, tend to follow the Hatch–Slack pathway and have a $\delta^{13}C$ ratio similar to that of C4 plants, but change only between −24‰ and −17‰ (Lamb, Wilson et al. 2006). Therefore, $\delta^{13}C$ ratios can help differentiate major sources of organic carbon and thus the strength of the marine influence (Zong, Lloyd et al. 2006). In addition, the ratio of total organic carbon to total nitrogen can help differentiate vascular plant organic matter from aquatic plant organic matter (Meyers 1994, Lamb, Wilson et al. 2006).

3.1.2 *Dating analysis*

According to radiocarbon dating (Arnold and Libby 1951), ^{14}C concentration in the atmosphere had been constant due to the constant passage of cosmic rays that affects the nucleus and protons of nitrogen to produce ^{14}C atoms in the atmosphere. When a plant or carbonate organism dies, ^{14}C within them will start to decay at a constant rate or half-life of a ^{14}C atom, that is, 5730 years. In this study, samples were sent to Beta Analytic for measurement. For marine shell samples, the calibration was based on CALIB611 with the Marine09 program (Reimer, Baillie et al. 2009) and a correction factor for marine reservoir effects (−128 ± 40, according to Southon et al. (2002)) for the samples collected from Mai Po wetlands. For organic matter, IntCal09 program (Reimer, Baillie et al. 2009) was used.

^{210}Pb and ^{137}Cs dating methods are based on short-life nuclides, because these two nuclides have facilitated the determination of ages of sediments in the past century. ^{210}Pb is continuously produced in the atmosphere (Krishnaswami and Lal 1978). Certain amount of ^{210}Pb

exists in the terrestrial sediment systems. By measuring the ratio of ^{210}Pb to ^{206}Pb in a column of sediments in relation to depth and assuming that the atmospheric flux of ^{210}Pb has remained constant, the time elapsed since the deposition of lead can be determined (Olsson 1986). In the natural systems, there is a certain but very low (close to non-detectable) level of concentration of ^{137}Cs. Two ^{137}Cs time markers from a down-core concentration curve can be used for age determination (Frignani and Langone 1991). First, the onset of increase in ^{137}Cs concentration can be dated back to the early 1950s, and the ^{137}Cs nuclides had started to accumulate in the atmosphere since 1954. The second time maker is 1964, because atmospheric testing of nuclear bombs had started to decline since the winter of 1963.

3.1.3 *Particle size analysis*

Particle size analysis is effective in determining the hydrodynamics of the sedimentary environment. In practice, a core sequence was subsampled for about 1 g each at certain intervals. Approximately 20 ml of 10% concentration hydroperoxide was added to each subsample to remove any organic matter. Then, approximately 20 ml of 1 mol/L hydrochloric acid was added to each subsample to remove any carbonate material. The reaction was accelerated by heating in both steps. After rinsing of the subsample with distilled water and neutralizing the subsample to pH 7.0, sodium metaphosphate solution was added to the subsample to disperse or separate sediment particles under ultrasonic vibration. Finally, each subsample was sent to a laser diffraction particle analyzer (Malvern Mastersizer 2000s) for measurement.

3.2 *GIA model*

Model simulations for glacial isostatic adjustment were performed in this study using a three-dimensional viscosity model (i.e., RF3 L20 ($\beta = 0.4$)) with radial and lateral variations. Two model grid resolutions, $2° \times 2°$ and $0.5° \times 0.5°$, were adapted. The land–ocean boundaries were verified and corrected in both the lower-resolution model and higher-resolution model, to ensure accuracy and precision because of spatial sensitivities in these models. During simulations, the coupled Laplace finite element method (Wang et al., 2006) was used to first tune the models to fit the paleo-sea-level observations from near-field sites, that is, former glaciated areas. Then, 3D viscosity models (RF3 L20 ($\beta = 0.4$)) were combined with different Ice models (ICE-4G, ICE-5G, and ICE-6G) to simulate the glacial isostatic adjustment worldwide. Outputs were obtained from a series of simulations, that is, RF3 L20($\beta = 0.4$)+ICE-4G, RF3 L20($\beta = 0.4$)+ICE-5G and RF3 L20($\beta = 0.4$)+ICE-6G with resolutions of $2° \times 2°$ and $0.5° \times 0.5°$, respectively. The computed results include the global distributed records in the past 18,000 years for RF3 L20($\beta = 0.4$)+ICE4G and RF3 L20($\beta = 0.4$)+ICE5G, and 19,000 years for RF3 L20($\beta = 0.4$)+ICE6G.

4 RESULTS AND ANALYSIS

4.1 *Early Holocene sea-level changes*

Before 18,000 years BP, *according* to the GIA models (Peltier 2004, Wang, Wu et al. 2009, Peltier, Argus et al. 2015), RSL in the deep bay area was nearly 120 m lower than the current value (Figure 2), and Mai Po was an inland upland area. Sea level had risen since the last glacial maximum and reached the level near present in 7000 years BP and submerged the Deep Bay area (Lambeck, Rouby et al. 2014, Sun, Wu et al. 2015, Zong, Huang et al. 2015).

4.2 *Results from cores in the mangrove forest*

After mid-Holocene, the sediments from Shenzhen River in north, Yuen Long River in south, in situ mangroves, and other sources increased deposition in the Deep Bay area, thereby leading to the wetland seaward expansion, which continues even today.

In core 14P4, three sedimentary units are recognizable (Figure 3). Unit 1 is sandy silt sediment with abundant marine shells. Unit 2 contains mainly silt without marine shells. Unit 3 is of clayey silt with an increased level of organic matter. Two time markers can be found in core

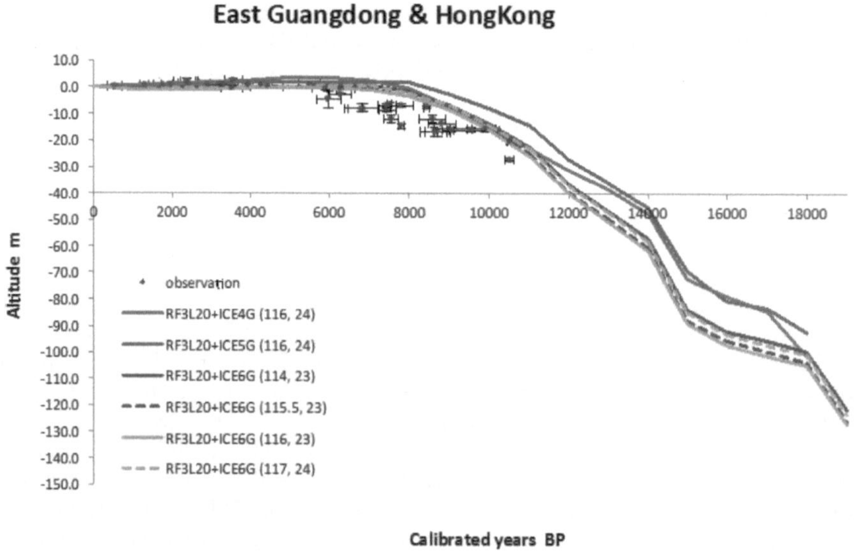

Figure 2. Sea-level changes predicted for East Guangdong and Hong Kong.

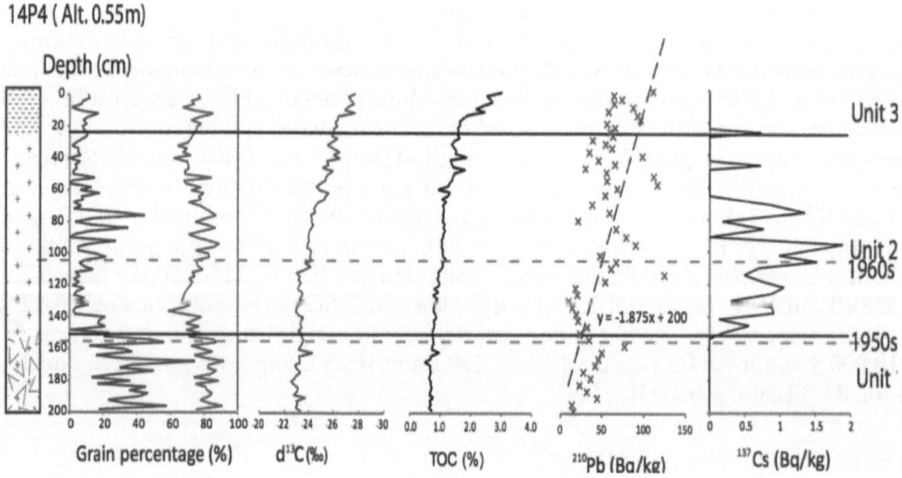

Figure 3. Results from core 14P1 (lithology, chronology, particle size, organic carbon isotopes, and total organic carbon).

14P4. First, the onset of increase of ^{137}Cs concentration was found at a depth of 150.5 cm, during 1950. Second, the ^{137}Cs maximum activity appeared at a depth of 106.5 cm in the core, during 1963. According to these two time markers, the depositional rates can be calculated as 33.8 mm/s between 1950 and 1963, and 20.9 mm/s between 1963 and 2014. These rates are higher than those in core 14P1 on the mudflat. The particle size results also show a sudden change at the depth of 150 m from more sandy sediment to mainly silt, implying a weakened hydrological condition starting from around 1950. Some fluctuations in particle size in the upper 150 cm are noted, which may reflect lateral migration of the tidal creek. In Unit 1, the δ^{13}C ratios vary slightly around −23‰, with organic content between 0.6% and 0.8%. In Unit 2, the δ^{13}C ratios change gradually from −23‰ to −26‰. This is associated with an increasing trend of organic content from 0.8% to 1.5%. These changes imply a change from the mudflat condition to the mangrove forest environment. In Unit 3, the δ^{13}C ratios change further to −28‰, and organic content over 3% suggests the establishment of the mangrove forest. Together with the chronology, the data from particle size, $\delta^{13}C$ ratios and organic content, suggest the following environmental changes at this site. First, the site was under oceanic conditions. By the 1950s, the sedimentation rate became very high. Subsequently, mangrove vegetation colonized the site by the early 1980s. Approximately 15–20 years ago, this site was covered by the mangrove forest.

4.3 *Coastal and land use changes from aerial images and remote sensing data*

From the limited aerial images, NDVI data, and historical records from WWF (Melville and Morton 1983), land use, vegetation, and shoreline changes of the Deep Bay area over time have been investigated in the recent decades (Figure 4). The major changes include coastal reclamation, development of the "Kei-Wais" system, paddy cultivation, human settlement, and spreading of the mangrove forest. In the 1920s, this area was covered by the mangrove forest and later replaced by a type of brackish water paddy. From the early 1940s to the 1970s until Mai Po was regarded as a Nature Conservation Area in 1975, large-scale reclamation of 1700 hectares in the Deep Bay area took place. First, the Kei-Wais system and fish ponds were widely constructed and replaced most parts of the mangroves. Second, cultivation was also carried out on the southern part of the wetland area. Third, from Tin Shui Wai District to southwest of Mai Po, after the long-time brackish paddy cultivation, soil had become sufficiently free of salt for freshwater paddy cultivation.

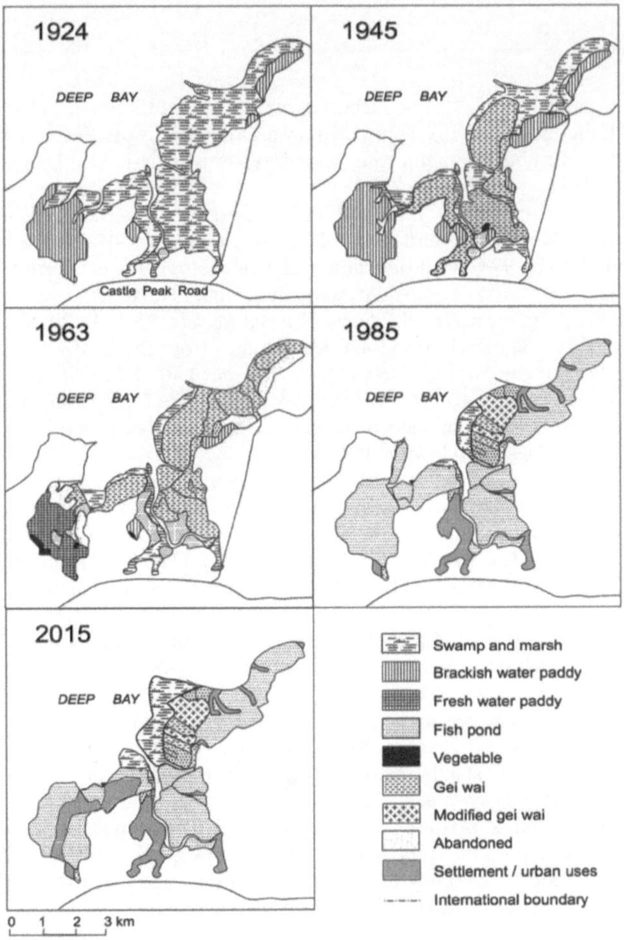

Figure 4. Land use changes in Mai Po (Irving and Leung 1987).

After it was regarded as a Nature Conservation Area, reclamation activities were stopped in Mai Po. At the same time, in other areas of the Deep Bay, such as Tin Shui Wai District, urbanization spread rapidly. The mangrove forest continued to expand toward mudflats in the past 20 years and formed two wide bands near Shenzhen River and Yuen Long River estuaries.

5 CONCLUSION

1. From the early to mid-Holocene, Mai Po had been regarded as a subtidal area under deep water and having significant tidal water exchange with the ocean. This environmental condition facilitated the growth of marine shells on the mudflats. This implies that in the past, Mai Po was possibly connected to the ocean not only through the Deep Bay, but also through the low-lying channel of Teun Mun.
2. Over the past 100 years, the most significant coastal change in Mai Po has been the expansion of the mangrove forest. The mangrove front was located at the current border fence in the 1960s. The front might have expanded seaward and reached the lower part of the mangrove forest by the 1980s.
3. High sedimentation rate has been recorded for the past 50 years, which is possibly due to human activities such as industrialization and urbanization in Hong Kong and Shenzhen, which may have increased the sediment yield. Over the past decades, sea-level rise has generated additional accommodation space, which has also led to rapid sedimentation.

REFERENCES

Arnold, J.R. and W.F. Libby (1951). "Radiocarbon dates." *Science* **113**(2927): 111–120.

Frignani, M. and L. Langone (1991). "Accumulation rates and 137 Cs distribution in sediments off the Po River delta and the Emilia-Romagna coast (northwestern Adriatic Sea, Italy)." *Continental Shelf Research* **11**(6): 525–542.

Irving, R.T.A. and K.-w. Leung (1987). *Land-use and land-use change in the reclaimed coastal areas of Deep Bay*, Centre of Urban Studies and Urban Planning, University of Hong Kong.

Krishnaswami, S. and D. Lal (1978). Radionuclide limnochronology. *Lakes*, Springer: 153–177.

Lamb, A.L., et al. (2006). "A review of coastal palaeoclimate and relative sea-level reconstructions using δ13C and C/N ratios in organic material." *Earth-Science Reviews* **75**(1–4): 29–57.

Lambeck, K., et al. (2014). "Sea level and global ice volumes from the Last Glacial Maximum to the Holocene." *Proceedings of the National Academy of Sciences* **111**(43): 15296–15303.

Melville, D. and B. Morton (1983). *Mai Po Marshes*, World Wildlife Fund Hong Kong.

Meyers, P.A. (1994). "Preservation of elemental and isotopic source identification of sedimentary organic matter." *Chemical Geology* **114**(3): 289–302.

Olsson, I.U. (1986). Radiometric dating. Handbook of Holocene palaeoecology and palaeohydrology.

Peltier, W., et al. (2015). "Space geodesy constrains ice age terminal deglaciation: The global ICE_6G_C (VM5a) model." *Journal of Geophysical Research: Solid Earth* **120**(1): 450–487.

Peltier, W.R. (2004). "GLOBAL GLACIAL ISOSTASY AND THE SURFACE OF THE ICE-AGE EARTH: The ICE-5G (VM2) Model and GRACE." *Annual Review of Earth and Planetary Sciences* **32**(1): 111–149.

Reimer, P.J., et al. (2009). "IntCal09 and Marine09 radiocarbon age calibration curves, 0–50,000 yeats cal BP."

Southon, J., et al. (2002). "Marine reservoir corrections for the Indian Ocean and Southeast Asia." *Radiocarbon* **44**(1): 167–180.

Sun,Y.Y., et al. (2015). "Holocene sea-level reconstution for GuangDong Coast and a Comparison with Model Outputs." *Quaternary Sciences* **35**(2): 281–290.

Thampanya, U., et al. (2006). "Coastal erosion and mangrove progradation of Southern Thailand." *Estuarine, Coastal and Shelf Science* **68**(1–2): 75–85.

Wang.H.S., et al. (2009). "Glacial isostatic adjustment model constrained by geodetic measurements and relative sea level." *Chinese Journal of Geophysics (Acta Geophysica Sinica)* **52**(10): 2450–2460.

Zong, Y., et al. (2015). "Late Quaternary tectonics, sea-level change and lithostratigraphy along the northern coast of the South China Sea." *Geological Society, London, Special Publications* 429.

Zong, Y., et al. (2006). "Reconstruction of Holocene monsoon history from the Pearl River Estuary, southern China, using diatoms and carbon isotope ratios." *Holocene* **16**(2): 251–263.

Hydraulic Engineering V – Hong, Liu & Xie (Eds)
© 2018 Taylor & Francis Group, London, ISBN 978-0-8153-7471-8

Application of the Coupled Eulerian–Lagrangian (CEL) method to the modeling of rock-breaking by a high-pressure waterjet

Zhongwei Zhou, Wei Shu, Jifu Yin & Guojun Hong
National Engineering Center of Dredging Technology and Equipment, Shanghai, China

Fangmei Su
Shanghai Electric Power Engineering Company, Shanghai, China

ABSTRACT: Numerical modeling of rock-breaking by a waterjet is a complex problem involving various nonlinear and coupled phenomena such as high velocity, large strains and fluid–structure interaction. This paper introduces the coupled Eulerian–Lagrangian (CEL) method, in which the water is described by the Eulerian formulation and the rock by the Lagrangian formulation. Furthermore, it adopts the modified Drucker–Prager (D-P) constitutive model to consider the effects of strain hardening, damage evolvement of the rock material, and rational failure criterion. This paper investigates three jet velocities to analyze the formation of the breaking hole, the distribution of stress, the depth of the broken hole, and the shape of the water flow, which are generally consistent with the experimental and numerical data in the literature. This result indicates that the application of the CEL method provides the possibility of modeling geotechnical problems of rock-breaking by a high-pressure waterjet.

1 INTRODUCTION

High-pressure waterjet technology has been widely used in precision cutting, crushing, and processing of metal, rock, and composite materials. So far, the mechanism of rock-breaking using a high-pressure waterjet has not been fully understood, increasing the difficulty of theoretical studies. In addition, transient features of rock-breaking and low pellucidity have complicated experimental studies. Rock-breaking with jet also involves large deformation and fluid–solid interaction, and the conventional finite element method could not solve the problem. In this situation, it is of great importance to propose a reliable and effective analytical model to study the behavior of breaking rock with a waterjet. At present, the widely used numerical methods are the arbitrary Lagrangian–Eulerian (ALE) method and the smooth-particle hydrodynamic (SPH) method.

For the ALE method, large deformation analysis is realized by remeshing grids and interpolation technique. Remeshed grids have the same topology and connectivity as the original grids. That is, the new grid nodes are obtained by moving the original grid nodes without changing the element connectivity. Chen et al. (2009) used the ALE method to simulate the coal damage caused by high-pressure jet and investigated the effect of failure criterion. Liu et al. (2012) simulated breaking submerged rock with a waterjet using the ALE method and analyzed the parameter sensitivity by the range method. The drawback of the ALE method is its limited ability to overcome mesh distortion. For large displacement of the structure, for instance, very deep broken hole, mesh distortion would produce and affect the computational accuracy. Besides, grids are often remeshed in every few steps, so that the capture of interface is not sufficiently accurate.

The SPH method is a meshless method developed over the past 20 years. The continuous fluid (or solid) is described by interactional particle group. The mechanical behavior of the whole system is obtained by solving the dynamic equation of particle group and tracking trajectories of every particle. Zhao et al. (2017) described water and abrasive material by the SPH method, described rock by the FEM method, and compared the numerical erosion results with their experimental ones. Wang et al. (2017) studied the rock-breaking efficiency influenced by the jet velocity and the particle volume fraction using the SPH–FEM method and showed the shapes of the broken hole and the damage distribution.

The CEL method is different from the ALE and SPH methods in that it does not need to remesh grids so that no grid distortion exists. Even though the depth of the broken hole is very large, the fluid–structure interface can still be obtained accurately and timely and the numerical computation is precise. In the CEL method, water, being described by the Eulerian formulation, can avoid certain problems caused by water granulation compared with the SPH method. The CEL method can also analyze various Eulerian media effectively, such as breaking rock with a waterjet under submerged conditions. Therefore, it is obvious that the CEL method is appropriate to simulate breaking rock with a waterjet. Under the impact of the high-pressure waterjet, the effects of strain hardening and damage evolution in rock become significant. In this study, the modified Drucker–Prager (D-P) constitutive model is adopted, and the shear and tensile failure criterion is introduced to simulate rock failure.

2 THE PRINCIPLE AND RESEARCH STATUS OF THE CEL METHOD

In the contact problems of fluid–structure coupling, fluid materials are described by the Eulerian formulation and structures are described by the Lagrangian formulation. The interface can be defined by the boundary face of the Lagrangian part, which will change with the structure deformation and destruction. The contact constraints between the Eulerian part and the Lagrangian part can be described by a general contact based on the penalty function method. The general contact algorithm calculates automatically and tracks the interface between the Lagrangian and Eulerian parts. Thus, although the two types of grids overlap, the two types of material will never overlap in the physical space. The coupled method combines the advantages of the two types of grids, namely the grids are not distorted and they can calculate the material interface accurately. In the CEL method, the flow state of material can be obtained by calculating the Eulerian volume fraction (EVF) in Eulerian grids, which is a measure of how an element is filled of material.

In recent years, the method has been increasingly applied in machinery, rock–soil, and civil engineering fields for studying large deformation and fluid–structure coupling problems. Ducobu et al. (2016) described metal using Euler formulation for investigating the cutting process by cutter and compared with the ALE method. Su et al. (2016) described the soil by the Eulerian formulation for investigating the ultimate bearing capacity of anchor plates in the clay. However, there have been only a few studies on the rock-breaking process using the CEL method.

In this study, the erosion process of breaking rock with a waterjet is simulated. First, the modeling process using the CEL method is introduced. Then, three types of waterjet velocities are simulated. The impact results of rock and flow shapes of water are analyzed. The results show that the CEL method has an excellent ability and opens new possibilities of modeling rock-breaking by a waterjet.

3 NUMERICAL MODEL

A new model of breaking rock with a waterjet is shown in Figure 1. In order to illustrate the CEL method clearly, the selected cases are all representative and common in the literatures for high-pressure waterjet problems. For saving computing resources, the quarter symmetry 1/4 model was established. The jet flow had a diameter (D) of 2 mm and initial velocities (V) of 150, 200, and 250 m/s, and was fixed to be at a distance of 5 mm from the rock target. The shape of the 1/4 geometric model for rock surrounded by air was a cuboid of size $20 \times 20 \times 30$ mm. In order to eliminate the marginal effect, the target width was set at $10D$. The modeling of the jet/target interaction was carried out by the CEL method, which is only implemented in three dimensions.

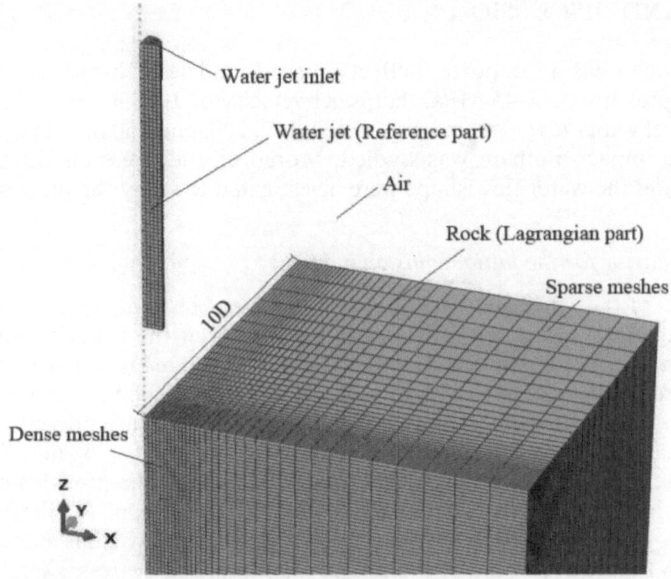

Figure 1. 3D model of rock-breaking with a waterjet.

Table 1. Main parameters of the rock.

Density (kg/m³)	Young's modulus (GPa)	Poisson's ratio	Angle of friction (°)	Dilation angle (°)	Compression strength (MPa)	Tensile strength (MPa)
2370	57.62	0.23	30	10	73	9.3

Here, the Mie-Grüneisen equation of state was adopted as the constitutive equation of water material, which is suitable to model materials at high pressure. Us-Up type was selected in the equation of state (EOS) in our model.

Under the impact of the high-pressure waterjet, the rock bears characteristics such as instant strong dynamic load, flexible impact, large deformation, and strain hardening and softening. In order to describe these characteristics reasonably, the extended Drucker-Prager model was adopted. It is an elastic–plastic model, and the hardening curve can be set. The damage is decided by the plastic strain, which is related to the softening curve. When the value becomes 1, the material will fail and the grids should be removed. Because the main damage styles are shear damage and ductile damage for rock, the shear and ductile damage criteria were introduced into our model. The main parameters of the rock are provided in Table 1.

For the CEL method, zone of Eulerian meshes should be large enough to cover the whole zone that the fluid material could flow through. If water goes out of meshes, the material will be ineffective. Rock is described by the Lagrangian meshes, which coincide with the Eulerian meshes. It is noted that the Eulerian meshes are not shown in Figure 1 and the meshed instance of the waterjet is just a reference part for calculating the EVF of water in the initial state. The volume fraction tool was used to assign the initial waterjet distribution in the Eulerian grids, and the waterjet inlet was set. Because the main damage zone concentrates on the center of rock, the meshes were distributed uniformly for reducing mesh quantity. The rock was fixed at a spatial location far away from the initial waterjet outlet for 5 mm. The interaction between the water and the rock was a general contact, and a contact pair must be set up because the Lagrangian meshes would fail and be removed.

4 RESULTS AND DISCUSSION

The waterjet velocity has an important effect on erosion. The high-pressure waterjet refers to the pumping pressure of 5–45 MPa, that is, jet velocity of 100–300 m/s. Therefore, in this study, three typical velocities (V) were investigated: 150, 200, and 250 m/s. The damage behavior for the initial impact moment was studied. Moreover, the stress distribution, the rock-breaking depth and the water flow shape were investigated to study the proposed model.

4.1 *Damage behavior for the initial collision moment*

The initial impact of the waterjet on the target is a complex behavior, which generates a shock wave that propagates forward in the target and rearward in the waterjet (Lu et al. 2015). In this study, the numerical results show that the impact center is not the first area where rock materials are damaged and that a ring-shaped crater is formed around the hump at the impact center on the target surface, as shown in Figure 2. The figure presents the rock damage field on the target when the waterjet impinges on the target ($V = 250$ m/s, $T = 20.5$ μs). In order to illustrate the relationship between damage and stress, the previous stress distributions are present in Figure 3 ($V = 250$ m/s, $T = 20$ μs). Thus, we calculated the axial and radial stresses, (S_{xx}), (S_{yy}), and (S_{zz}), along the red line at target surface, which is marked by the start and end points in Figure 2. It could be found that the compressive stress is small at the impact center and that the tensile stress appears at the distance of 1.3 mm from the original point. Thus, the superior shear stress could be produced, and the zone could be destroyed primarily. Figures 4 and 5 show the results of the experimental and numerical tests of the broken shape on the target surface at the initial impinging moment conducted by Mabrouki et al. (2000). By comparing Figure 2 with Figures 4 and 5, we can consider that our results are consistent with those of Mabrouki et al. (2000). They also studied the stress distribution on the target surface at the initial impinging moment. Four points on target surface were selected to investigate the variety of stresses and revealed the same principle with Figure 3.

4.2 *Hole shapes and stress distributions*

Figure 6 shows that the material is peeled off layer by layer under the impact of the waterjet, leading to the formation of a hole similar to a horn mouth. The destruction of the rock and the formation of the hole occur in the timescale of milliseconds. The hole wall is relatively smooth and regular. Lan (2007) conducted experiments to study the shape of the breaking hole, and in order to show it more clearly, he drew sketch maps of elevation view and side view, as shown in Figure 7. Figure 8 shows the breaking hole produced in the experiment conducted by Wang et al. (2017). Comparing the experimental results with our numerical results,

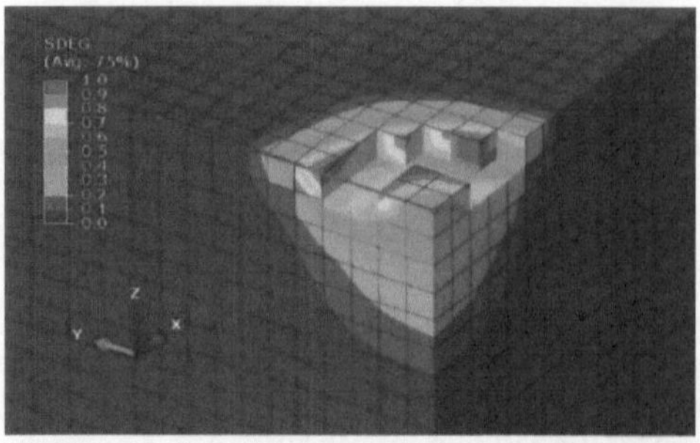

Figure 2. Damage degree distribution (V = 250 m/s, $T = 20.5$ μs) on the target for the initial impact moment and a hump.

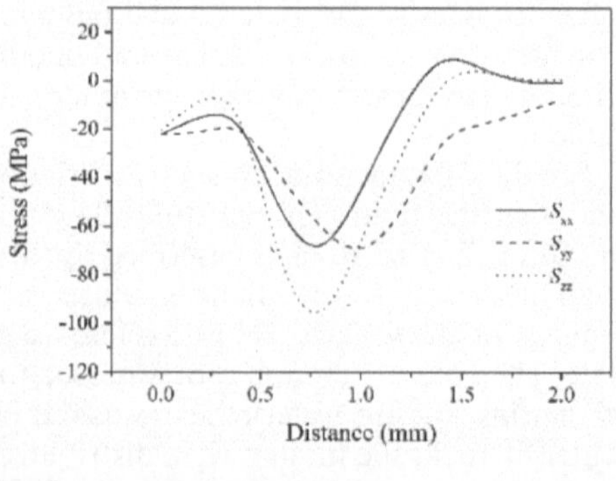

Figure 3. Stress–distance curve along the y-axis on the upper surface of the target (V = 250 m/s, $T = 20$ μs).

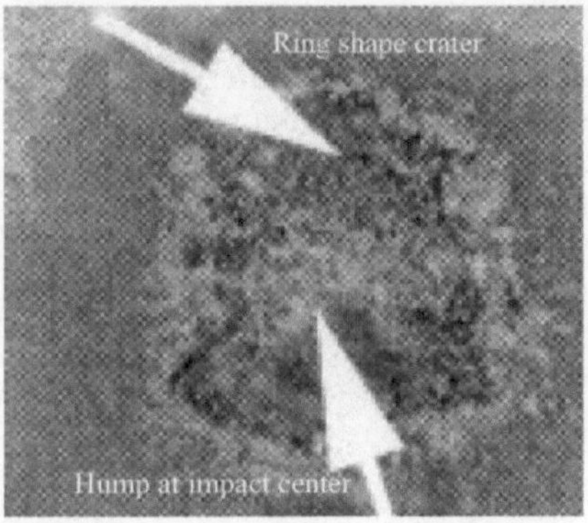

Figure 4. Experimental result of breaking the shape on the target surface conducted by Mabrouki et al. (2000).

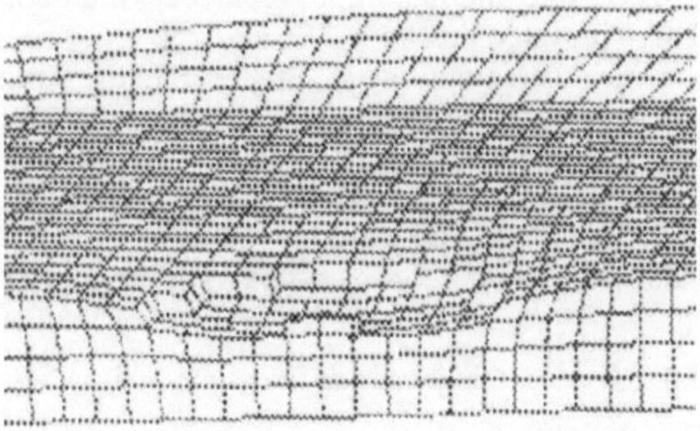

Figure 5. Numerical result of breaking the shape on the target surface conducted by Mabrouki et al. (2000).

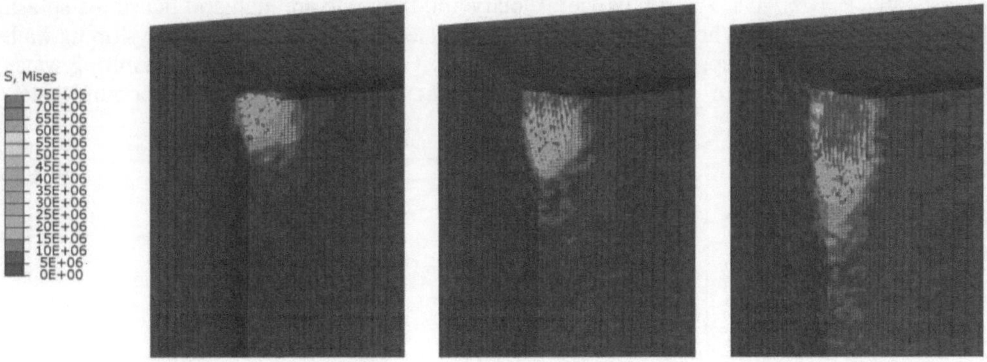

Figure 6. Hole shapes and stress distributions at different times ($V = 250$ m/s).

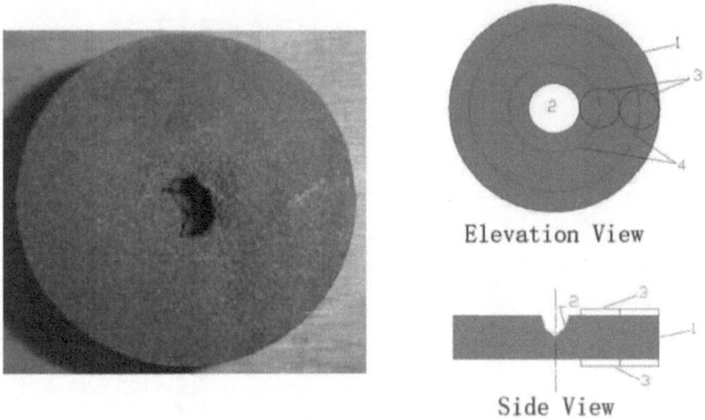

Figure 7. Image of the breaking hole and its sketch maps (Lan, 2007).

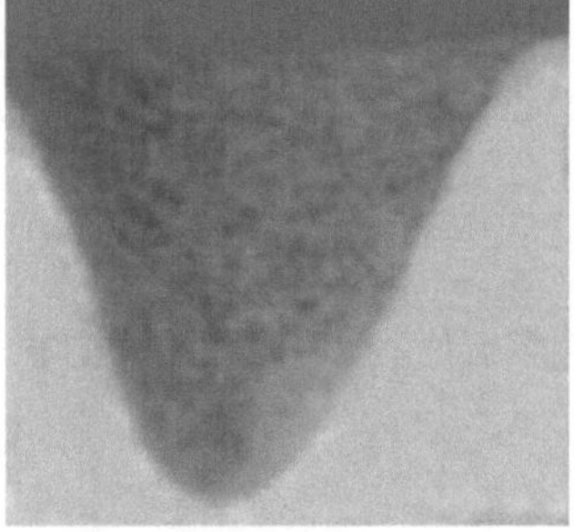

Figure 8. Breaking hole (Wang et al., 2017).

it is obvious that the shapes of the breaking holes are very similar. Our model accuracy can be verified to a certain extent. Three typical moments of case $V = 250$ m/s are selected to present Mises stress distributions in Figure 6, which illustrates that the stress is high near the contact interface and decays quickly from the interface.

4.3 *Variation of rock-breaking depth at different waterjet velocities*

Erosion rate and maximum rock-breaking depth are different at different waterjet velocities. Three jet velocities, 150, 200, and 250 m/s, are investigated in this study. Figure 5 shows that the various regulations of erosion rate are similar for different jet velocities, that is, the erosion rate is high initially and then becomes low, and finally, the rock-breaking depth does not increase anymore. The maximum rock-breaking depth increased with the increase in the waterjet velocity, and its corresponding dwell time decreases with jet velocity. It is noted that the curves in Figure 9 are not smooth enough because the waterjet is influenced by rebound water and could not affect the target for certain moments.

4.4 *Waterjet shapes at different moments*

For the CEL method, the water volume is decided by the Eulerian volume fraction (EVF). Figure 10 shows the evolution of the waterjet impact, rebound, and splash, which are more realistic than those of the SPH method. The legend of the figure presents the values of EVF. EVF = 1 means that the Eulerian grid is full of water. In the case of $V = 150$ m/s, $T = 100$ μs, the rock-breaking depth is shallow, so the water nearly spreads on the upper surface of the rock. For the last case ($V = 250$ m/s, $T = 200$ μs), jet velocity and flow rate are high and hence the splash effect is very significant. When the formed hole wall is steep, water will slide off the sloping wall. When the hole is very deep, water will rebound almost linearly and affect the resulting water flow. Therefore, the shape of the hole plays an important role in the movement of rebound water.

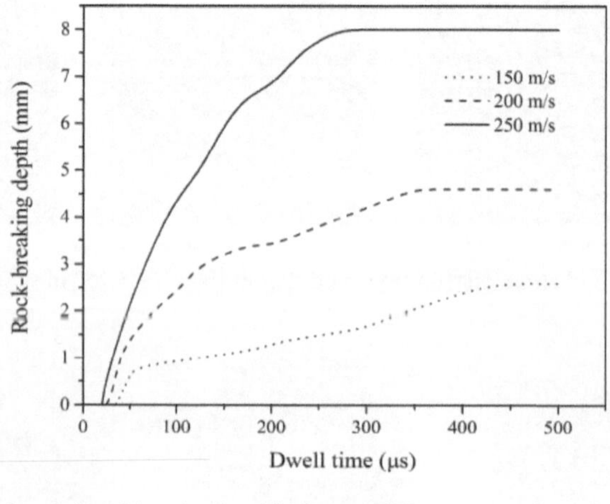

Figure 9. Variation of rock-breaking depth with time at different waterjet velocities.

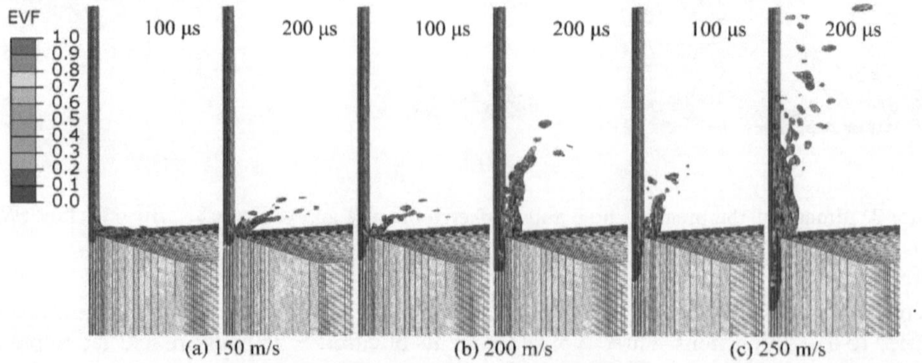

Figure 10. Water shapes for different jet velocities at 100 and 200 μs.

5 CONCLUSION

The aim of this study was to present a new numerical model to investigate the interaction between a high-velocity waterjet and the rock target. The numerical computation was carried out using the CEL method, which is advantageous over the widely used ALE and SPH methods in analyzing large deformation and fluid–structure coupling. The following conclusions can be drawn from this study:

1. The calculation results were verified by relevant experiments, which indicate that the coupled Eulerian–Lagrangian method can reflect the actual physical process of rock-breaking by the high-pressure waterjet. Therefore, the proposed numerical model can be used to analyze the damage evolution and failure mechanism of rocks.
2. The numerical results indicate the jet stagnation zone present at the impact center, and consequently the formation of a hump-shaped damage around the impact center. The formation of a breaking hole, similar to a horn mouth, was also observed under the impact of the high-pressure waterjet. The stress was found to be high near the contact–impact interface, which decayed quickly from the interface.
3. Furthermore, the rock-breaking depth increased nonlinearly with dwell time. The impact efficiency and the maximum depth were significantly influenced by the jet velocity. The shapes of the water flow were simulated more realistically by the CEL method than by the SPH method. The movements of rebound and splash water were closely related to the jet velocity and the depth and shape of the hole.

REFERENCES

Chen J.H., Liang Y.P., Cheng G.Q. 2009. Numerical analysis on coal-breaking process under high pressure waterjet. *Journal of Coal Science & Engineering*, 15(3): 289–294.

Ducobu F., Riviere E., Filippi E. 2016. Application of the Coupled Eulerian-Lagrangian (CEL) method to the modeling of orthogonal cutting. *European Journal of Mechanics*, 59: 58–66.

Lan H.J. 2007. Experimental and theoretical investigation into rock erosion of cavitating waterjets. *PhD thesis*, Chongqing University, Chongqing, China.

Liu J.L., Si H., Zhang H. 2012. Study on breaking rock efficiency of submerged waterjet. *China Safety Science Journal*, 22(11): 23–29.

Lu Y.Y., Huang F., Liu X.C., Ao X. 2015. On the failure pattern of sandstone impacted by high-velocity waterjet. *International Journal of Impact Engineering*, 76: 67–74.

Mabrouki T., Raissi K., Cornier A. 2000. Numerical simulation and experimental study of the interaction between a pure high-velocity waterjet and targets: contribution to investigate the decoating process. *Wear*, 239: 260–273.

Su F.M., Liu H.X., Li Z. 2016. Analysis of ultimate bearing capacity of plate anchors in clay using a coupled Eulerian-Lagrangian method. *Rock and Soil Mechanics*, 37(9): 2728–2736.

Wang F.X., Wang R.H., Zhou W.D., Chen G.C. 2017. Numerical simulation and experimental verification of the rock damage field under particle waterjet impacting. *International Journal of Impact Engineering*, 102: 169–179.

Zhao J., Zhang G.C., Xu Y.J., Wang R.H., Zhou W.D., Han L.X., Zhou Y. 2017. Mechanism and effect of jet parameters on particle waterjet. *Powder Technology*, 313: 231–244.

Hydraulic Engineering V – Hong, Liu & Xie (Eds)

Evaluating the effect of sulfate, alkalinity, and chloramine on iron release from iron pipes and the characteristics of iron corrosion using the response surface methodology

Fan Yang, Shaofeng Zhang & Weiyu Zhang
College of Engineering and Technology, Tianjin Agricultural University, Tianjin, China

Jianbo Guo
Tianjin Key Laboratory of Aquatic Science and Technology, Tianjin Chengjian University, Tianjin, China
School of Environmental and Municipal Engineering, Tianjin Chengjian University, Tianjin, China

Xinyuan Liu & Nan Wu
College of Engineering and Technology, Tianjin Agricultural University, Tianjin, China

ABSTRACT: This study used the Response Surface Methodology (RSM), particularly the Box–Behnken design model, to evaluate the effect of sulfate, alkalinity, and chloramine on iron release from iron pipes historically transporting groundwater, as well as the characteristics of iron corrosion during a changing water quality experiment. The RSM experiment showed that the amounts of Fe_3O_4, amorphous iron oxide, and intermediate iron products ($FeCO_3$ and Green Rust (GR)) increased in pipe section reactors carrying chloramine-treated water (NH_2ClRs), but their α-FeOOH contents decreased and β-FeOOH contents disappeared. In water with high LR (>1.4), the thin and non-protective scale layers containing lower amounts of amorphous iron oxide, Fe(II), derived from newly produced Fe_3O_4, or $FeCO_3$ or GR could be easily penetrated by oxidants and sulfate ions, resulting in higher iron release. However, the same unstable scale layers did not affect the rates of iron release in water with low LR (≤1.4). The RSM results indicated that there was a close relationship between iron release from these unstable scale layers and water quality (Larson ratio).

1 INTRODUCTION

Because of groundwater depletion and regulatory constraints, many countries have decreased groundwater utilization and increased the utilization of surface water from other sources for drinking purposes. The difference between the water qualities of these alternate supplies and old groundwater source might disrupt the existing distribution scale on pipes, which had adverse effects on water quality distribution.

Red water issue occurred in some areas that had been supplying local groundwater earlier in a northern city of China soon after the local groundwater was replaced by water from long-distance introduced source from a neighboring province in October 2008 (Yang et al., 2012; Li et al., 2010). Larson ratio (LR) has been used to evaluate the corrosiveness of water against iron-based pipes, which was developed from the relative corrosive behaviors of chlorides and sulfates to the protective properties of bicarbonate for the corrosion of iron pipes, and an LR value of above 0.5 indicates corrosive water (Schock et al., 1999). Compared with groundwater (LR 0.22–0.4), water from the new source had a higher LR (1.7), which lies in the corrosive range. According to previous studies, the thin corrosion mostly discovered in pipes historically transporting groundwater did not contain magnetite, exhibited an M/G ratio of <1.0, and contained higher contents of unstable ferric components such as γ-FeOOH, β-FeOOH, and amorphous iron oxide. Thus, it is concluded that the pipes historically transporting groundwater were less corroded (Yang et al., 2012).

Alkalinity, sulfate, and disinfector are primary water quality parameters influencing iron corrosion. To prevent the red water cases resulting from a source water switch, pipe section reactors harvested in groundwater treatment plant service areas were used for the RSM experiment to determine acceptable water quality for controlling iron release. New experimental waters with adjusted sulfate, alkalinity, and chloramine concentration designed by the RSM experiment were supplied to pipe section reactors for effluent total iron release. After 11d-RSM experiment, scale samples of pipe section reactors with different iron release rates were collected for XRD and amorphous iron analysis to investigate the physiochemical changes of the scale. The conclusions drawn from this study could help develop the water quality regulation method for less corrosive pipes historically transporting groundwater.

2 MATERIALS AND METHODS

2.1 *Preparation of iron pipe section reactors*

The cast-iron pipe was approximately 20 years old and harvested from the Beijing drinking water distribution system (DWDS) area, denoted as PipeB-GW3. In the past, PipeB-GW3 was transporting local groundwater (denoted as GW3) treated by groundwater treatment plant 3 (GWTP3). It was cut into 10 cm short sections with relatively little damage to the pipe scale layers, and the exposed metal surfaces at both ends of each section were coated with epoxy to prevent water contact. Short pipe sections were installed with polyvinyl chloride (PVC) plate covers and magnetic stirrers into pipe section reactors (Figure 1). The corrosion of the pipe section reactor returned to the original condition by receiving the old water source of GW3 for 2 months in continuous flow-through mode. Then, the pipe section reactors filled with GW3 water were carefully transported to the laboratory avoiding any scale damage by collision.

2.2 *Preparation of experimental solution*

The characteristics of the old source water of PipeB-GW3, the base solution used, and the Box–Behnken Design (BBD) experimental water are listed in Table 1. As shown in the table, the old source water had low sulfate and DO and high alkalinity, whereas the base solution and experimental water had higher sulfate and DO.

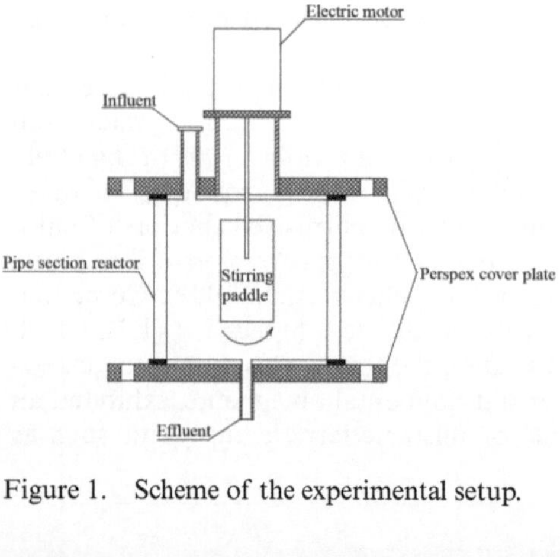

Figure 1. Scheme of the experimental setup.

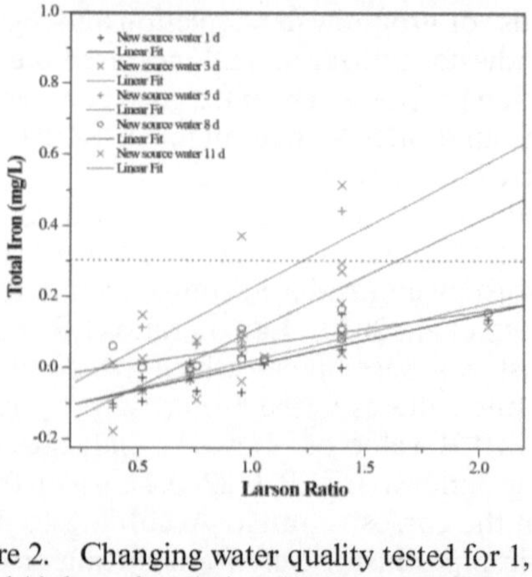

Figure 2. Changing water quality tested for 1, 3, 5, 8, and 11 days: the relationship between effluent total iron concentrations in pipe section reactors carrying chloramine-treated water and the Larson ratio.

Table 1. Water quality characteristics of the original source water of PipeB-GW3 and BBD experimental water (units: alkalinity in $mg \cdot L^{-1}$ as $CaCO_3$; conductivity in $\mu s \cdot cm^{-1}$; others in $mg \cdot L^{-1}$).

Water quality	Original source water	BBD experimental water	Analysis equipment
Alkalinity	194.5	Variables (130, 190, 250)	Acid base titration
Conductivity	408	Variables (540–780)	Multiparameter meter sension156
DO	6.5	8.1	HACH Co., USA
pH	7.3	Variables (7.7–8.0)	
Cl^-	23.1	40.2	Ion Chromatography System-1000
SO_4^{2-}	24.7	Variables (40, 120, 200)	Dionex Co., USA
LR[a]	0.26	Variables (0.39–2.0)	Calculated by equation
Chloramine	1.0 $LR = \frac{[Cl^-]+2[SO_4^{2-}]}{HCO_3^-}$	Variables (0–2.0)	DPD titration by HACH DR2800

[a]LR, Larson ratio, where $[Cl^-]$, $[SO_4^{2-}]$, $[HCO_3^-]$ are expressed in moles per liter.

2.3 *Batch mode study and water quality measurements*

Before the BBD experiments, pipe section reactors were first filled with GW3 water and run for 3 days to acquire blank values of iron release in each pipe section. According to the BBD experimental design, each pipe section reactor (marked by En) was then filled with new experimental water for 11 days. The new experimental water was renewed every day after unloading of the 1-day-run solution. During the period of changing water quality, the effluent total iron concentration tested at the end of 1, 3, 8, and 11 days for a given pipe section reactor was the actual measured data reduced by its corresponding blank value. Water samples were collected from the reactor outlet at the end of 1-day run and measured according to Standard Methods (Clesceri et al., 1998). The monitored parameters of experimental water included pH, dissolved oxygen (DO), conductivity, alkalinity, sulfate, and chloride (Table 1).

2.4 *Response surface methodology and statistical analysis*

A BBD model for the above three variables, each at two levels (i.e., low level and high level), was adopted as the experimental design model (Table 2). A total of 15 experiments were conducted to calculate the coefficients of the polynomial equation that was fitted on the experimental data (Box et al., 1987). The experimental design was generated using Design-Expert 7.0, and graphics using MATLAB software.

Table 3 provides the details of the independent variables (sulfate, alkalinity, and chloramine concentration), coded value of the variables, and the observed output response (total iron release). Each experimental run was carried out with the maintained process variable conditions as given in the design matrix. The effluent iron release presented in Table 3 was the average value of the tested values of 1, 3, 8, and 11 days. A first-order polynomial model fitted to the experimental data obtained from the experimental runs conducted using the BBD model is explained by the following equation:

$$Y = \beta_0 + \sum \beta_i x_i \qquad (1)$$

where Y is the total iron release, β_0 is the offset term, β_i is the first-order main effect, and x_i are the experimental variables studied, including sulfate, alkalinity, and chloramine concentration. The fit quality of the model was determined by the coefficient of determination (R^2) and the analysis of variance (ANOVA) (Table 4).

2.5 *Characteristics of iron corrosion*

At the end of 11 day batch mode study, the corrosion samples were removed from pipe section reactors and milled immediately in an anaerobic glove box to avoid spontaneous oxidation by oxygen. Then, the wet powder samples and small scale pieces were freeze-dried

Table 2. Experimental factors and the range and level of variables ($mg \cdot L^{-1}$).

	Low level	Medium level	High level
Variables	(–1)	(0)	(–1)
Alkalinity	130	190	250
Sulfate	40	120	200
Chloramine	0	1	2

Table 3. Box–Behnken design matrix for three variables together with the observed response (units: alkalinity in $mg \cdot L^{-1}$ as $CaCO_3$; others in $mg \cdot L^{-1}$).

	Variables						Variables				
No.	Alkalinity	SO_4^{2-}	Chloramine	LR	Total iron	No.	Alkalinity	SO_4^{2-}	Chloramine	LR	Total iron
E1	190	200	2.0	1.40	0.078	E9	130	120	2.0	1.40	0.118
E2	190	40	2.0	0.52	–0.024[a]	E10	190	120	1.0	0.96	0.079
E3	190	120	1.0	0.96	0.032	E11	130	120	0	1.40	0.098
E4	250	120	0	0.73	–0.026	E12	190	200	0	1.40	0.292
E5	130	200	1.0	2.04	0.129	E13	250	40	1.0	0.39	–0.063
E6	250	200	1.0	1.06	0.024	E14	190	40	0	0.52	0.067
E7	190	120	1.0	0.96	0.039	E15	250	120	2.0	0.73	–0.005
E8	130	40	1.0	0.76	–0.001						

[a]The total iron and turbidity values were the average values tested during the 11-day experiment minus their corresponding blank values obtained from the 3-day baseline period. Thus, there may be negative values.

and analyzed by X-ray powder diffraction (XRD) and free and amorphous iron extraction, respectively. A Rigaku (D/max-rA) X-ray powder diffractometer (Rigaku-USA Inc., Danvers, MA) with Cu Kα radiation at 40 kV and 100 mA was used for XRD studies. Crystalline phase identification and composition were determined by Jade XRD software and contrasted parameters of the intensity method (Chung, 1975). The dithionite-citrate-bicarbonate (DCB) technique described by Tan (Tan, 1996) was used to determine free and amorphous iron concentration, and the total iron was pretreated by microwave digestion and determined by an inductively coupled plasma optical emission spectrometer (ICP-OES, PerkinElmer Optima 2000, USA).

3 RESULTS AND DISCUSSION

3.1 *Investigation of iron release using the response surface methodology*

3.1.1 *Iron release in the RSM experiment*

Table 3 presents the coded value of the variables together with the average iron release and turbidity. NH_2ClR of E12 had the highest iron release, whereas NH_2ClR of E13 had the lowest iron release. The relationship between the effluent total iron concentrations during the tested days (1, 3, 5, 8 and 11 days) and the LR for NH_2ClR is shown in Figure 2. After switching to experimental water, the iron release in each pipe section reactor had a positive relationship with their appointed LR. For NH_2ClR, the effluent total iron concentrations of 1 and 3 days showed an increasing trend, but those of 5, 8, and 11 days showed a decreasing trend, and only a few detected data exceeded the drinking water standard.

3.1.2 *Fitting of the first-order polynomial equation and statistical analysis*

The application of the RSM based on the estimates of the parameters and the experimental results indicated an empirical relationship between the response and the input variables expressed by the following fitted regression model, which is a first-order polynomial equation (2):

$$C_{totalIron} = 0.15 - 8.64e^{-4}C_{alkalinity} + 8.52e^{-4}C_{sulfate} + 0.03C_{chloramine} \tag{2}$$

Both the statistical significance of the ratio of mean square due to regression model and the ratio of mean square due to residual were tested using ANOVA, which revealed that equation (2) suitably represents the actual relationship between the response and the variables. A very low probability value (p-value) of 0.003 (NH_2ClR) obtained from the Fisher *F*-test demonstrated a very high significance for the first-order equations in this study. The p-values of sulfate and alkalinity were lower than 0.1, whereas the p-value of chloramine was greater than 0.1. Therefore, the effect of sulfate and alkalinity on iron release was significant. The equation and ANOVA results were consistent with the results shown in Figure 2 that the LR (i.e., sulfate and alkalinity) had a positive relationship with iron release.

3.2 *X-Ray diffraction (XRD) analysis of corrosion*

According to effluent total iron concentration, the tested pipe section reactors can be classified into three different levels: E5, E12 (high-iron-release reactors); E1 and E14 (medium–iron-release reactors); and E2, E3, and E8 (low-iron-release reactors). At the end of the RSM experiment, scale samples from the Cl_2Rs were collected for composition and structure analysis, respectively.

The box figures illustrated in Figure 3 show the chemical and structural differences for scale of site collection samples and experimented pipe reactor samples. The aforementioned pipe section reactors were grouped into three categories on the basis of their iron release and iron oxide percentage used below was average values of multisamples in box figures.

Freshly obtained scale layers of PipeB-GW3 in the collection site contained Fe_3O_4 (18%), α-FeOOH (64%), $FeCO_3$ (9%), β-FeOOH (8%), and γ-FeOOH (5%); in addition, their

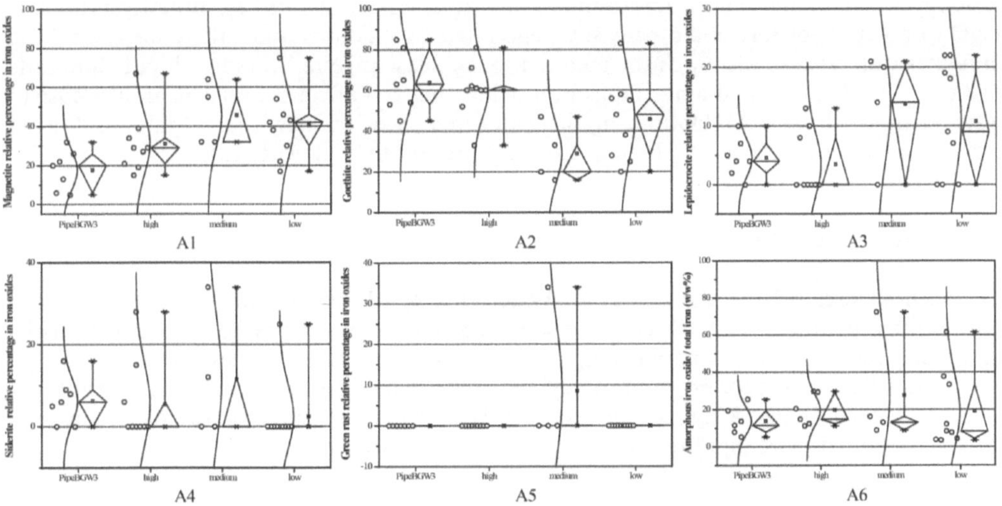

Figure 3. Variation of scale composition in pipe section reactors carrying chloramine-treated water (A1–A6) during the changing water quality experiment. PipeB-GW3 represents the scale samples in the collection site, "high" represents the scale samples in high-iron-release reactors, "medium" represents the scale samples in medium-iron-release reactors, and "low" represents the scale samples in low-iron-release reactors.

averaged amorphous iron oxide content was 13.6%. After changing the water quality, the percentage of Fe_3O_4 scale samples in NH_2ClRs increased, whereas that of α-FeOOH scale samples decreased and β-FeOOH samples disappeared. Many studies found that β-FeOOH might be bacterial reduced to transform into GR (Erbs, 2004), Fe_3O_4, and $FeCO_3$ (Kukkadapu et al., 2005). The medium iron release NH_2ClRs had the highest Fe_3O_4 (45%), γ-FeOOH (14%), $FeCO_3$ (12%), GR (8.5%), and amorphous iron oxide (27.6%) contents, but the lowest α-FeOOH content (29%). However, the high-iron-release and low-iron-release NH_2ClRs had lower Fe_3O_4, γ-FeOOH, $FeCO_3$, and amorphous iron oxide amounts. Previous studies have shown that Fe_3O_4 were the thermodynamically stable Fe oxides and that the dense magnetite tightly fixed on the metal surface possesses protective and anticorrosive properties (Sarin et al., 2004). The Fe_3O_4 content in NH_2ClRs increased during the 11-day experiment, but the scale layer was very thin and Fe(II) of newly formed Fe_3O_4 was still unstable and could be transformed into Fe(III), which resulted in the red water case.

In water with high LR (>1.4), the thin and non-protective scale layers containing lower amorphous iron oxide, Fe(II), derived from newly produced Fe_3O_4, or $FeCO_3$ or GR could be easily penetrated by oxidants and sulfate ions and had higher iron release, such as E5 and E12. However, the same unstable scale layers in E2, E3, and E8 did not affect the rates of iron release in water with low LR (≤1.4).

4 CONCLUSIONS

The influence of several key water quality parameters on the iron release from pipes historically transporting treated groundwater and their thin scale stability was investigated using the RSM in a changing water quality experiment. The results of the experiment indicated that the iron release from pipe section reactors was positively related to the LR; for example, sulfate concentration was positively related to iron release and alkalinity was negatively related to iron release. Furthermore, there was no significant relationship between iron release and chloramine. The RSM experiment showed that the high-iron-release and low-iron-release NH_2ClRs had lower amounts of Fe_3O_4, γ-FeOOH, $FeCO_3$, and amorphous iron oxide, whereas the medium-iron-release NH_2ClRs had higher amounts of Fe_3O_4, γ-FeOOH, $FeCO_3$, GR, and amorphous iron oxide. In water with high LR (>1.4), the thin and non-protective scale layers containing lower amorphous iron oxide, Fe(II), derived from newly produced Fe_3O_4, or $FeCO_3$ or GR could be easily penetrated by oxidants and sulfate ions, resulting in higher iron release. Although the medium-iron-release pipe reactors had higher amounts of Fe_3O_4, $FeCO_3$, GR, and amorphous iron oxide, they did affect the rates of iron release in water with low LR (≤1.4). Optimizing the water quality of the new source water and using reasonable water purification measures can help prevent the red water case.

ACKNOWLEDGMENTS

The authors acknowledge the Center for Water Quality Monitoring of Beijing Waterworks Group for providing access to sampling sites, sharing the water quality monitoring results of water treatment plants, and allowing us to publish those results. They also thank Weiyu Zhang and Jianbo Guo for providing analytical support and helpful instructions. The authors thank Shaofeng Zhang, Xinyuan Liu, and Nan Wu for their assistance in sample processing. All the authors declare that they have no conflict of interest.

FUNDING INFORMATION

This study was financially supported by the National Natural Science Foundation of China (No. 51308392), the Research Fund of Tianjin Key Laboratory of Aquatic Science and

Technology, and the Natural Science Foundation of Tianjin (China) (Nos. 14 JCYBJC43700 and 16 JCQNJC08200).

REFERENCES

Box, G.E.P., Draper, N.R., 1987, Empirical Model Building and Response Surface, John Wiley and Sons, New York.

Chung, F.H., 1975, Quantitative interpretation of X-ray diffraction patterns of mixtures. III. Simultaneous determination of a set of reference intensities, J. Appl. Crystallogr., 8(1): 17–19.

Clesceri, L.S., Greenberg, A.E., Eaton, A.D., et al, 1998, Standard method for the examination of water and wastewater (20th edition), American Public Health Association, Washington.

Erbs, M., 2004, Formation and redox reactions of green rusts under geochemical conditions found in natural soils and sediments. Doctor Dissertation. Swiss Federal Institute of Technology.

Kukkadapu, R.K., Zachara, J.M., Fredrickson, J.K., Kennedy, D.W., Dohnalkova, A.C., Mccready, D.E., 2005, Ferrous hydroxyl carbonate is a stable transformation product of biogenic magnetite, Am. Mineral., 90: 510–515.

Li, D., Li, Z., Yu, J.W., Cao, N., Liu, R.Y., Yang, M., 2010, Characterization of bacterial community structure in a DWDS during an occurrence of red water, Appl. Environ. Microbiol. 76: 7171–7180.

Sarin, P., Snoeyink, V.L., Lytle, D.A., 2004, Iron corrosion scales: model for scale growth, iron release and colored water formation, J. Environ. Eng., 130(4): 364–373.

Schock, M.R., Lytle, D.A., 1999, Internal corrosion and deposition control. Chapter 17 in Water quality and treatment: a handbook of community water supplies, 5th ed, AWWA and Edzwald J.K. McGraw-Hill Companies, New York, NY.

Tan, K H., 1996, Soil Sampling, Preparation, and Analysis, CRC Press, Marcel Dekker, New York.

Yang, F., Shi, B.Y., Gu, J.N., Wang, D.S., Yang, M., 2012, Morphology and composition characteristics of iron corrosion scales formed under different water source histories in a drinking water distribution system, Water Res., 46: 5423–5433.

Hydraulic Engineering V – Hong, Liu & Xie (Eds)
© 2018 Taylor & Francis Group, London, ISBN 978-0-8153-7471-8

Application of the DNDC model in simulating nitrate leaching due to irrigation and precipitation

Y.Y. Sun, J.W. Wang & G.J. Hong
National Engineering Research Center of Dredging Technology and Equipment, Shanghai, China

ABSTRACT: Nitrate leaching can lead to numerous environmental hazards; therefore, research on balancing water input, fertilization, N leaching and crop yield is necessary. Previous research has shown the ability of the DNDC model to accurately predict N leaching, which was a result of embedding the details of N transformation processes in the model. In this study, the N-leaching data obtained from previous experiments conducted in No. 3 field (Bush Estate) were used as validation data. This field consisted of 12 plots treated with different management scenarios from 1989 to 1992. After the calibration of crop growth parameters according to field measurements, the DNDC model was applied in this field for simulating nitrate leaching. Both the predicted and measured results indicate the value of 90–150 kg N ha^{-1} as the appropriate amount for fertilization to avoid N leaching. The comparison of the simulation and validation data shows that the simulated results were totally dependent. The simulated and observed RMSEs of the accumulated N-leaching amount in different plots for two sampling dates were in the range of 1.14–1.97. However, there were some problems that affected the accuracy of the DNDC model, including the underestimation of the N fertilizer effect and the overestimation of the water input effect.

1 INTRODUCTION

As an integral part of the agricultural system, nitrogen cycle includes import, removal, and transport of nitrogen within system, as well as export to external environment. It widely interacts with atmosphere and water environment by comprehensive exchanges (Li et al., 2002). The nitrogen in soil moves down with water after it reaches the depth under the root activity layer without being absorbed by crops. This process is called nitrate leaching, through which nitrate could enter groundwater and diffuse to the whole water environment. It results in water pollution, fauna illness, algal blooms in coasts, and eutrophication in catchments (Addiscott, 1996). Among all the factors contributing to N leaching, water input (precipitation and irrigation) and N fertilizers play significant roles (Addiscott, 1996; Bergström and Brink, 1986; Singh and Sekhon, 1978; Dise and Wright, 1995). Therefore, it is necessary to carry out studies on balancing the relationship between N fertilizer, water input, N leaching, and crop yield in the agricultural system.

The study site of this project is No. 3 field, Scottish Centre of Agricultural Engineering, Bush Estate, Midlothian. It has sandy loam soil developed from partially sorted glacial till. Before 1989, arable cultivation had been in practice in the site for many years. This site was divided into 12 plots and underwent different treatments between 1989 and 1992 (Vinten et al., 1992). Drained plots of 12,250 m^2 were isolated up and down the slope with trenches backfilled with gravel and lined with polythene on the downslope edge (Vinten et al., 1994).

2 DATA SOURCES AND METHODOLOGY

2.1 *Data sources*

The climate data used in this study are provided by the meteorological station near No. 3 field, including the daily maximum temperature (C°), minimum temperature (C°), and precipitation (cm) from 1989 to 1992. The atmospheric background concentrations of carbon dioxide and ammonia keep default values in the DNDC model (350 ppm and 0.06 μg N/m^3, respectively). The soil and field management data are obtained from previous studies (Vinten et al.1992; Vinten et al., 1994; Vinten, 1999) conducted in No. 3 field. The crop yield refers to spring barley records in Scotland (O'Sullivan et al., 1982) or the calculation from the component analysis.

The validation data are obtained from the previous experiments conducted in No. 3 fields. The drainage fluxes of each plot were measured by using tipping bucket flowmeters (Vinten et al., 1992). A simple flow-weighted sampling device was used to measure the concentrations of nitrate and ammonium in drainage water (Vinten et al., 1994). Then, the N-leaching data could be calculated. However, at the same time, the time intervals of two sampling dates are varied, and the N-leaching data are the accumulated results between two sampling dates. The detailed observed N-leaching data of the sampling dates between 7 November 1989 and 24 December 1992 are provided by SAC.

2.2 *DNDC model*

DeNitrification-DeComposition (DNDC) is a biogeochemistry model used for simulating carbon and nitrogen cycles in agro-ecosystems (Li et al., 2003, 2006, 2007), which has advantages in predicting N leaching in both site and region scales. There are two components in the DNDC model, one composed of climate, soil, crop growth, and decomposition submodels, and the other composed of nitrification, denitrification, and fermentation submodels. This model serves as a bridge between C and N cycles and the basic ecological drivers (Li, 2007). The model operation includes the setup of crop growth parameters and the input of climate, soil, and management parameters. Published yield data (Vinten et al., 1992) are used here for calibrating the crop growth parameters and testing the modified model. After the calibration, the model is used for the prediction of nitrate leaching.

3 RESULTS AND DISCUSSION

3.1 *Daily N leaching*

3.1.1 *Result of daily N leaching*

The observation data include the accumulated amount during two sampling dates. In order to perform more comparison and analysis with the observation data, the simulation data during two sampling dates are also accumulated in this study. The prediction curves and validation curves during the period of 7 November 1989 to 24 December 1992 are plotted in Figure 1 (a–h). Plots with similar management schemes are represented in the same chart and share the same prediction curve (a, b, d, and e in Figure 1).

3.1.2 *Statistics of daily N leaching*

On the one hand, it could be summarized from the curves in Figure 1 that the prediction curves fit the validation ones fairly and represent the total trend very well. The accumulated daily amounts of N leaching (the accumulated amount during two sampling dates) based on RMSE between prediction and validation (Figure 2.) are between 1.14 and 1.97, and no large RMSE appears in the statistic result.

On the other hand, water input (precipitation and irrigation) is a main factor that could affect the movement of soil water and have a significant influence on N leaching in No.3 field. The relationship between N leaching and water input is analyzed in Figure 3 using the method of correlation analysis. The result of sensitivity analysis (Table 1) show water input as the most sensitive factor, with a sensitivity of 2.429. The DNDC model supposes that the sensitivity of water input to N leaching is quite high; however, it can be stated that the model slightly overestimated the influences of water input.

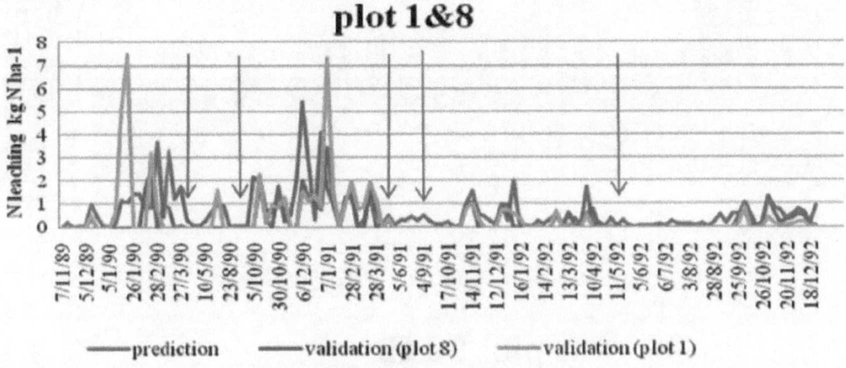

(a) Crop: SB/SB/G/G; Fertilizer: 128/150/0/0 kg N ha^{-1}
*SB=spring barley G=grassland
*arrowsindicate fertilization

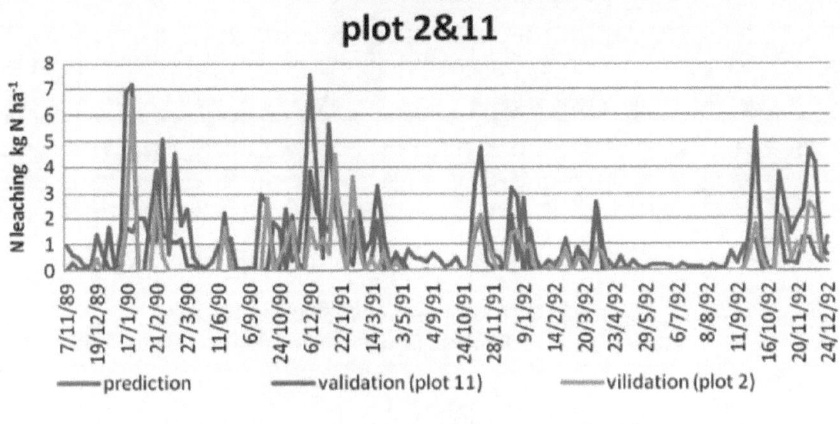

(b) Crop: SB/SB/SB/SB; Fertilizer: 128/150/150/150 kg N ha^{-1}

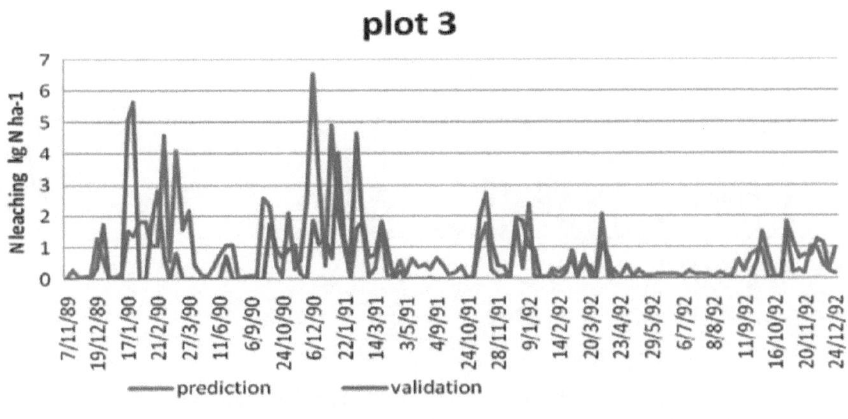

(c) Crop: SB/SB/SB/SB; Fertilizer: 128/0/0/0 kg N ha^{-1}

Figure 1. (*Continued*).

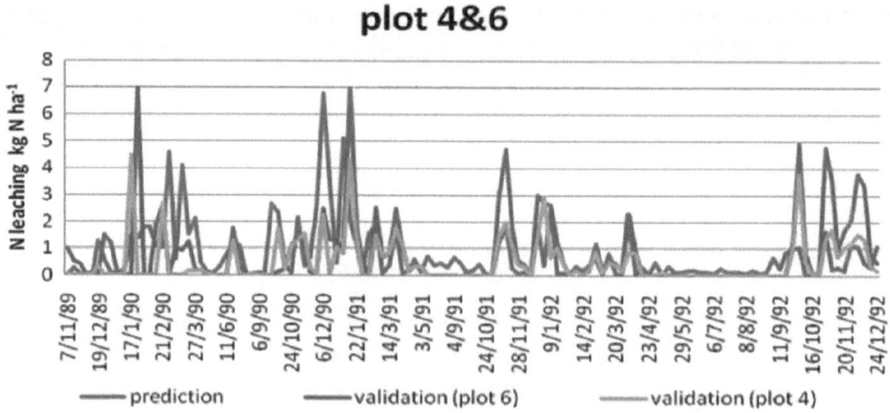

(d) Crop: SB/SB/SB/SB; Fertilizer: 128/120/120/120 kg N ha^{-1}

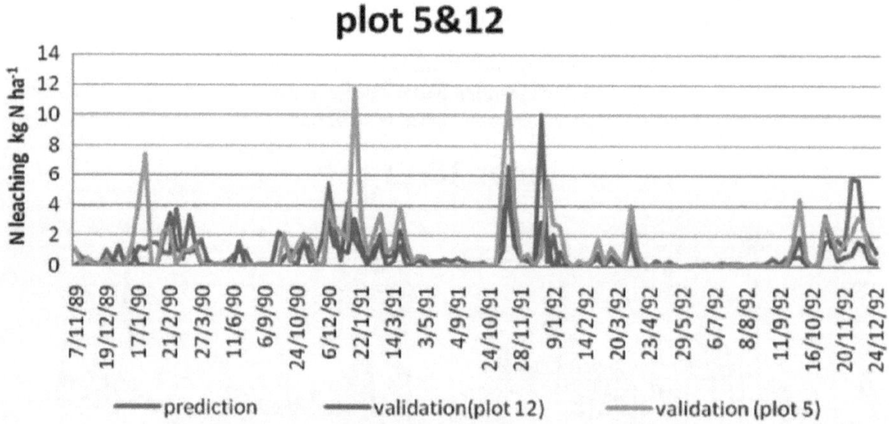

(e) Crop: SB/SB/BF/BF; Fertilizer: 128/150/0/0 kg N ha^{-1}
*BF=Bare fallow

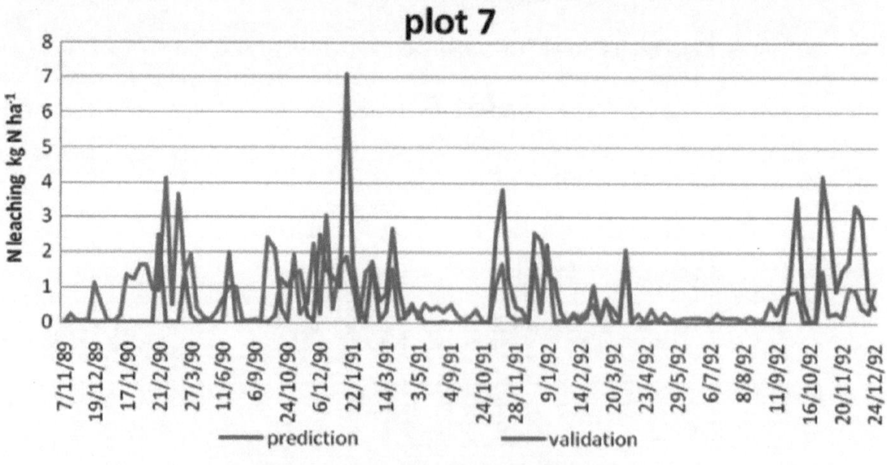

(f) Crop: SB/SB/SB/SB; Fertilizer: 128/90/90/90 kg N ha^{-1}

Figure 1. (*Continued*).

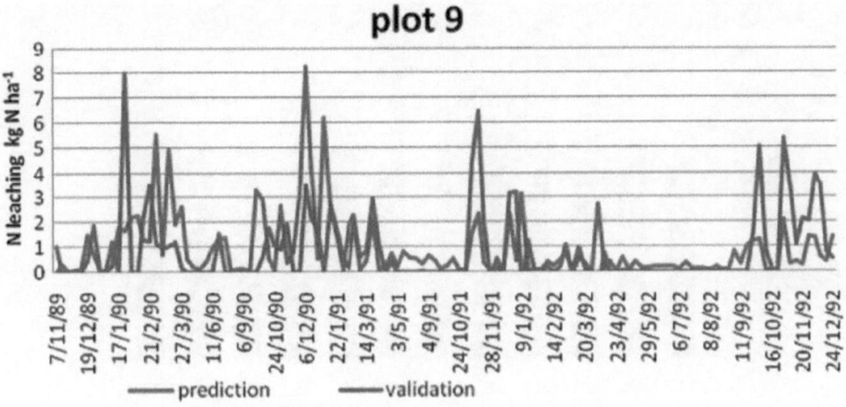

(g) Crop: SB/SB/SB/SB; Fertilizer: 128/180/180/180 kg N ha^{-1}

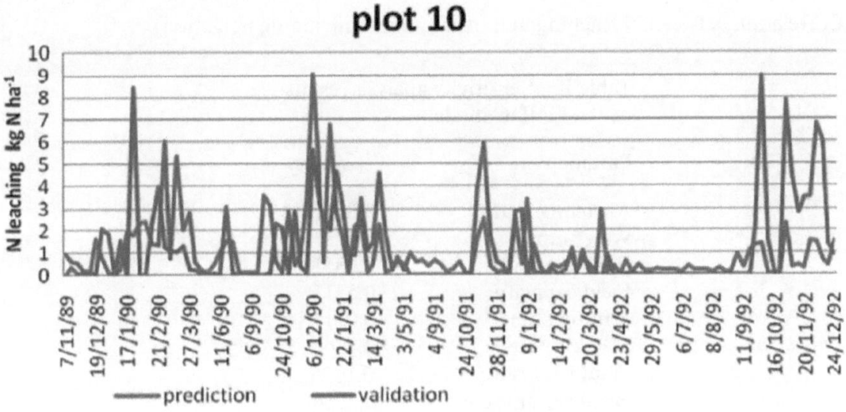

h) Crop: SB/SB/SB/SB; Fertilizer: 128/210/210/210 kg N ha^{-1}

Figure 1. Accumulated daily summary of N leaching in different plots.

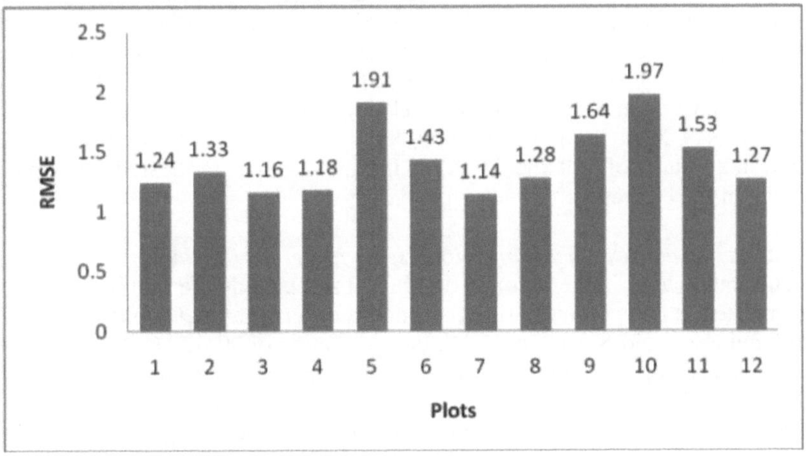

Figure 2. RMSE between prediction and validation in different plots.

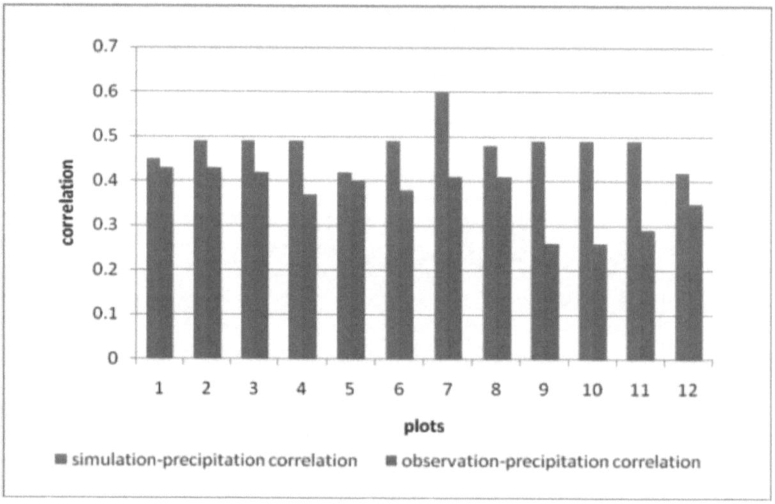

Figure 3. Correlation between N leaching and precipitation (including irrigation).

Table 1. Sensitivity analysis results of the DNDC model.

Parameters	Sensitivity
Air temperature	0.0122
Water input	2.429
Bulk density	0.6351
Field capacity	0.5511
Wilting point	0.0111
SOC content	0.7346
Plant C/N ratio	−0.1796
Fertilizer amount	0.5254

3.2 *Annual N leaching with respect to fertilizer amount*

3.2.1 *Result of annual N leaching with respect to fertilizer amount*

In 1989, all the 12 plots received the same treatments (spring barley, 128 kg N ha^{-1}). The observed amount of average annual N leaching is 16.17 kg N ha^{-1}, whereas the simulated amount is 13.4 kg N ha^{-1}. The RMSE between the observed and simulated amounts is 7.52. In the period from 1990 to 1992, different treatments (crops and fertilization) were applied in those plots. With particular emphasis on the fields with spring barley, both the simulated and observed amounts of N leaching for different N fertilizer application are shown in Figure 4. For fertilization amounts of 120 and 150 kg N ha^{-1}, two sets of replicated experiment results are also shown in the figures.

3.2.2 *Discussion on annual N leaching with respect to fertilizer amount*

In the agricultural system, if the amount of N fertilizer exceeds the requirement, N leaching could be aggravated. It is necessary to find the amount of fertilizer that could not only satisfy the need of crops but also avoid N leaching. With the application of 90–150 kg N ha^{-1} of N fertilizer, the amount of N leaching is found to be low in both observation and simulation data. Thus, this range could be regarded as the appropriate amount for fertilization to avoid N leaching.

From 1990 to 1992, the lowest simulated amount of N leaching occurred on the plots with application of 90 kg N ha^{-1} of fertilizer N, which then increased slowly and steadily. The observed N leaching amount usually undergoes inconspicuous changes with the applica-

(a)

(b)

(c)

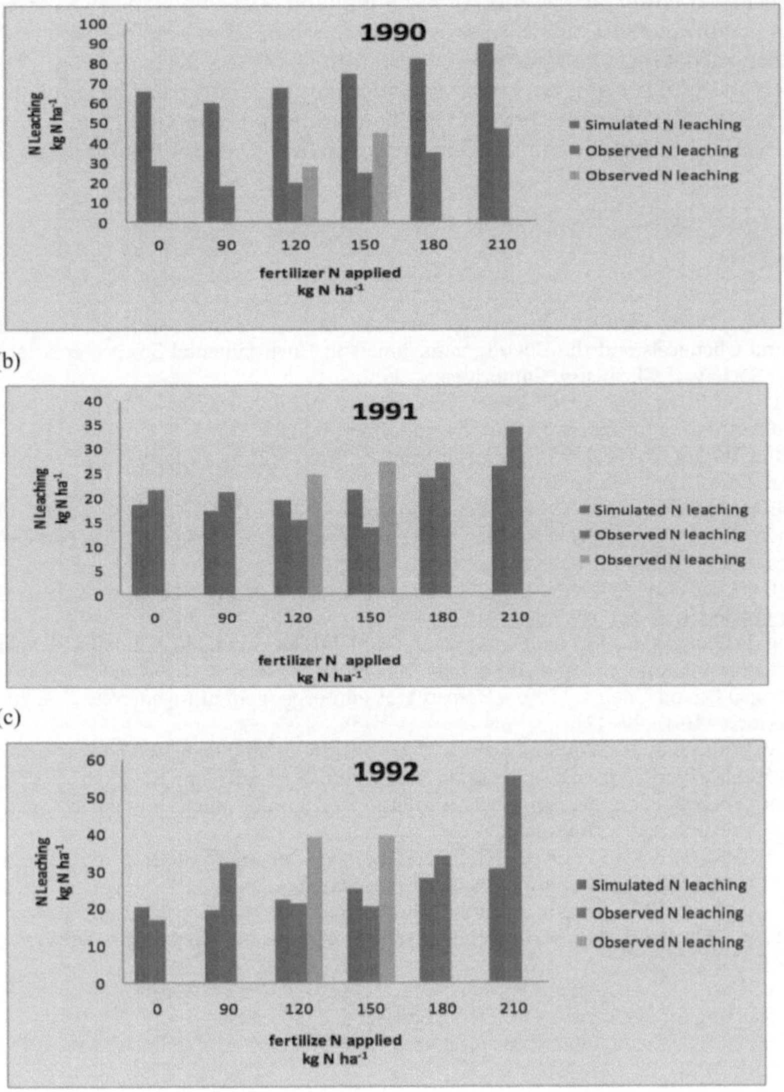

Figure 4. Summary of annual N leaching in spring barley fields with different amounts of N fertilizer.

tion of 90–150 kg N ha^{-1} of N fertilizer, followed by a sudden rise. The difference between simulated and observed data increases with increase of the amount of the fertilizer. One important reason for this difference is that the DNDC model underestimates the effect of the amount of fertilizer on N leaching. When the amount of fertilizer applied exceeds 180 kg N ha^{-1}, the inaccuracy of DNDC is obvious.

4 CONCLUSION

1. DNDC is an appropriate model for predicting nitrate leaching in fields. The total simulation result of No. 3 field is broadly ideal. The prediction of the model fits fairly well with the validation data, and the seasonal differences in N leaching are represented very well.

2. Water input including precipitation and irrigation significantly contribute to N leaching, but the model may overestimate the effect of water input on N leaching. The high correlation also indicates that the model is highly dependent on water input. For example, prediction is much higher than validation at both the beginning and end of 1990 with excess precipitation.
3. The effect of fertilizer amount on N leaching may underestimate the simulation by the DNDC model. The difference between the simulation and observation data increases with the increase in the fertilizer amount. The appropriate amount for fertilization is 90–150 kg N ha^{-1}.

REFERENCES

Addiscott, T.M., (1996). Fertilizers and nitrate leaching. In: R.E. Hester and R.M. Harrison, Editors, Agricultural Chemicals and the Environment. Issues in Environmental Science and Technology 5, The Royal Society of Chemistry, Cambridge: 1–26.

Bergström, L. and Brink, N., (1986). Effects of differentiated applications of fertilizer N on leaching losses and distribution of inorganic N in the soil. Plant Soil 93: 333–345.

Dise, N.B. and Wright, R.F., (1995). Nitrogen leaching from European forest in relation to nitrogen deposition, Forest Ecol. Manage. 71: 53–161.

Li, C.S., (2004). Modeling terrestrial ecosystems. Complex Systems and Complexity Science1 (1): 49–57.

Li C. et al., (2006). Modeling nitrate leaching with a biogeochemical model modified based on observations in a row-crop field in Iowa. Ecological Modelling 196: 116–130.

Li C.S., (2007), User's Guide for the DNDC Model, University of New Hampshire http://www.dndc.sr.unh.edu/papers/Li_biogeochemistry.pdf.

Li, H., Wang, L.G. and Qiu, J.J., (2009). Application of DNDC model in estimating cropland nitrate leaching. Chinese Journal of Applied Ecology, 20(7):1591–1596.

Li, Z.B., Wang Q.C. and Chen J., (2002). Research of agroecosystem nitrogen cycle. Soil and Environmental Science, 11(4): 417–421.

O'Sullivan, M.F. and Ball, B.C., (1982). Spring barley grouth, grain quality and soil physical conditions in a cultivations experiment on a sandy loam in Scotland. Soil & Tillage Research, 2: 359–378.

Singh, B. and Sekhon, G.S., (1979). Nitrate pollution of groundwater from farm use of nitrogen fertilizers—a review. Agric. Environm., 4: 207–225.

Vinten, A.J.A., Vivian, B.J. and Howard, R.S., (1992). The effect of fertiliser on the nitrogen cycle of two upland arable soil of contrasting textures. Proc, Fert. Soc., 329.

Vinten, A.J.A., Vivian, B.J., Wright, F. and Howard, R.S., (1994). A comparative study of nitrate leaching from soils of differing textures under similar climatic and cropping conditions. Journal of Hydrology, 159: 197–213.

Vinten, A.J.A., Castle, K., and Arah, J.R.M., (1996): Field evaluation of models of denitrification linked to nitrate leaching for aggregated soil, Eur. J. Soil Sci., 47, 305–317.

Vinten, A.J.A., (1999). Predicting Nitrate Leaching from Drained Arable Soils Derived from Glacial Till. J. Environ. Qual. 28: 988–996.

Hydraulic Engineering V – Hong, Liu & Xie (Eds)

A spur-dike wetland hydrodynamic experiment with sediment discharge

H.X. Liu, M.C. Zhu, T.J. Huang & Y.J. Lu
State Key Laboratory of Hydrology-Water Resources and Hydraulic Engineering, Nanjing Hydraulic Research Institute, Nanjing, China

ABSTRACT: Recently, the ecological effects of hydraulic projects have attracted worldwide attention. Structures such as spur dikes are effective for river regulation as well as a threat to river ecology. Therefore, in this paper, we discuss the application of spur dikes' artificial wetland in river restoration projects. To study the effect of this ecological engineering on the emission of polluted sediments, laboratory experiments were carried out. The concentration field was measured under different conditions in the surrounding area of the project, including different vegetation and sediment inputs. The results indicated that the peak concentration decreased significantly due to the change in the flow velocity. The increase in the vegetation occupancy rate will lead to a transition of the concentration process from unimodal to multimodal. This will be very useful for the enhancement of water quality and pollutant decomposition.

1 INTRODUCTION

Hydraulic projects have caused several environmental and ecological problems. In general, the river ecology is threatened by the change of the dynamic structure of water and sediment and the boundary conditions of rivers. For example, the construction of hydraulic structures may increase the turbidity and directly affect the photosynthesis of aquatic vegetation. The riverbed erosion caused by the project will also destroy the habitat of the organism. In the 1970s and 1980s, more than 330 hydraulic structures, such as spur dikes, were constructed by the Pearl River Nanning-Guangzhou waterway project, enormously destroying aquatic habitats. Therefore, how to avoid the ecological threat of hydropower project and study the related environmental protection technology is the key to the sustainable development of economy. Therefore, research and engineering practice of the traditional hydraulic structures with ecological restoration technology are very important. For example, Lu & Li (2011) and Wang (2011) studied the application of artificial fish nests and other ecological methods in the management project. Artificial wetland is the most common and effective method in ecological engineering, which greatly helps the formation of aquatic communities. The flow velocity in the spur dam area is often lower, which is very suitable for the reconstruction of aquatic vegetation. Constructed wetlands and eco-banking areas will form. Studies have shown that lake wetland vegetation can reduce sediment and other input flow pollutants from 25% to 62% and enhance water quality. Similarly, the ecological dikes of constructed wetlands also affect land-based pollution flows. Suspended sediment particles absorb heavy metals and other pollutants (James & Martin 1997, Pan 1999), especially in vegetation interception. Therefore, the purpose of conducting laboratory experiments in this study is to quantitatively analyze the interception of pollutants in an ecological spur dike. Then, improvements to the project and related parameters will be further studied.

2 LABORATORY HYDRODYNAMIC EXPERIMENT

As shown in Figure 1, the dimension of the experiment flume is 65 m (length) × 4 m (width) × 0.5 m (height). Near both ends of the flume, there were brick grids for flow stabilization. A fence-type gate is set in the downstream end to control the water depth. Water flows in the flume can be recycled after passing through the gate. The water from upstream was supplied by the pumps (the largest discharge is 150 L/s).

The experimental setup is put in the middle of the flume; four spur-dike models with inter-distance of 2 m were placed here in order to simulate the real spur-dike group. In the setup, the distance between upstream and downstream is very long (~30 m); therefore, the influence of entrance and tail gate is negligible. The dam-field zone between the middle two spur-dikes is the wetland simulation zone, in to which polluted water is diverted. In the downstream of spur-dikes, a measuring bridge for the placement of turbidimeters is set in order to monitor the variation of sediment concentration in this cross section. At the same time, another bridge for velocity measurement is set to obtain the flow field. In this experiment, the turbidity processes in the downstream were recorded under different discharge, wetland area, and input polluted sediment concentration. Therefore, the interception effect of ecological spur-dikes can be tested under different conditions.

Electromagnetic flow velocimeters, which can detect 2D instantaneous velocity and fluctuation, were applied for the measurement of velocity and flow field. Sediment concentration in water was measured by laser turbidimeters with an accuracy of 0.001 g/L. The wooden solid spur-dike models were 100 cm long and 6 cm high, with side slope and head slope of 1:1.25 and 1:1.3, respectively. On the basis of several trial runs, we selected fly-ash for the simulation of polluted sediment particles. The vegetation in artificial wetland was simulated by plastic grasses, which are distributed in the interval of about 5 cm.

The prototype of this experiment is a spur-dike regulation section. According to the local hydrodynamic conditions, it was decided to use undistorted model and the horizontal and vertical scales were about 100, and the velocity scale is 10. Two typical input discharge values, regulation stage discharge and flood discharge, were selected. Then, the corresponding discharge in model was calculated by scales. Thus, in this experiment, there are two conditions, namely non-submerged (regulation stage discharge) and submerged (flood discharge).

"Polluted water" with six different pollutant concentrations, 1, 2, 4, 6, 8, and 10 g/L, were poured into the artificial wetland, in order to test the response of vegetation interception. For the convenience of comparison, the total volume of every experiment run was the same, that is, 50 L. In the aspect of artificial wetland, four schemes of different vegetation proportions were determined, namely no vegetation, 1/3 vegetation (from the bank to 1/3 spur-dike length), 2/3 vegetation (from the bank to 2/3 spur-dike length), and full vegetation.

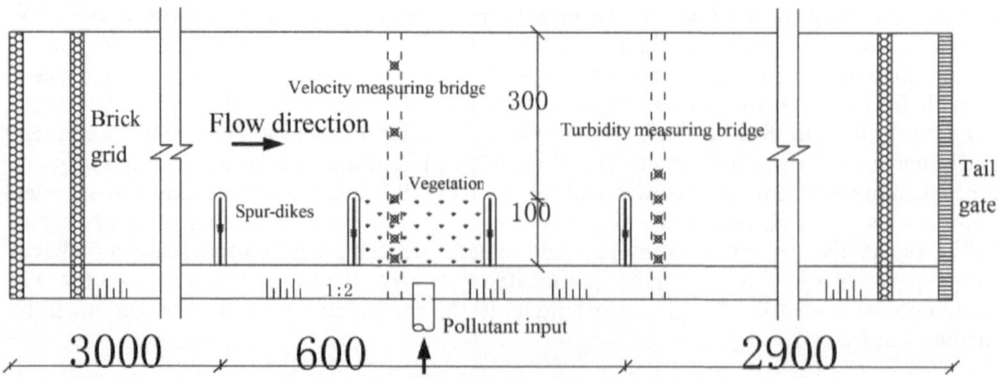

Figure 1. Layout of the laboratory hydrodynamic experiment (units: cm).

3 DOWNSTREAM CONCENTRATION

Figures 3 and 4 show downstream concentration process curves plotted under the non-submerged condition, and Figures 5 and 6 show similar curves plotted under the submerged condition. The positions of all measuring points are shown in Figure 2. In Figure 2b, the values obtained at points a and b are mostly zero, indicating that the diffusion of pollutant would not exceed the range of a–b when non-submerged (#1–a when submerged). The length of the ranges was approximately 130–160 cm.

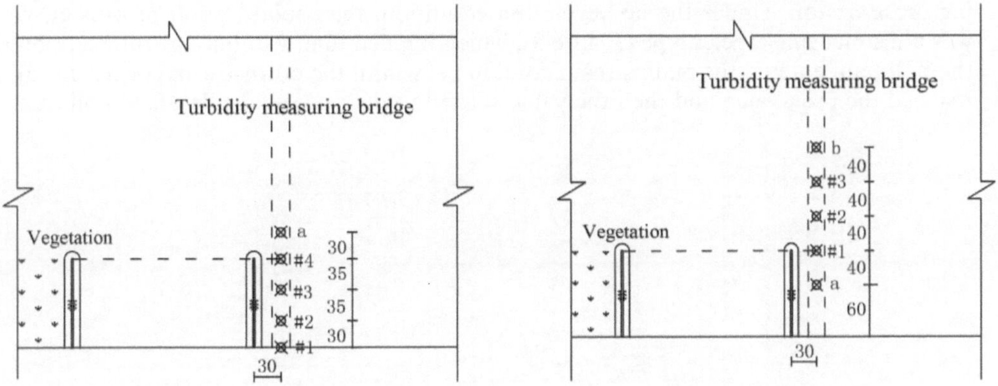

(a) Submerged condition (b) Non-submerged condition

Figure 2. Placement of concentration measurement points (units: cm).

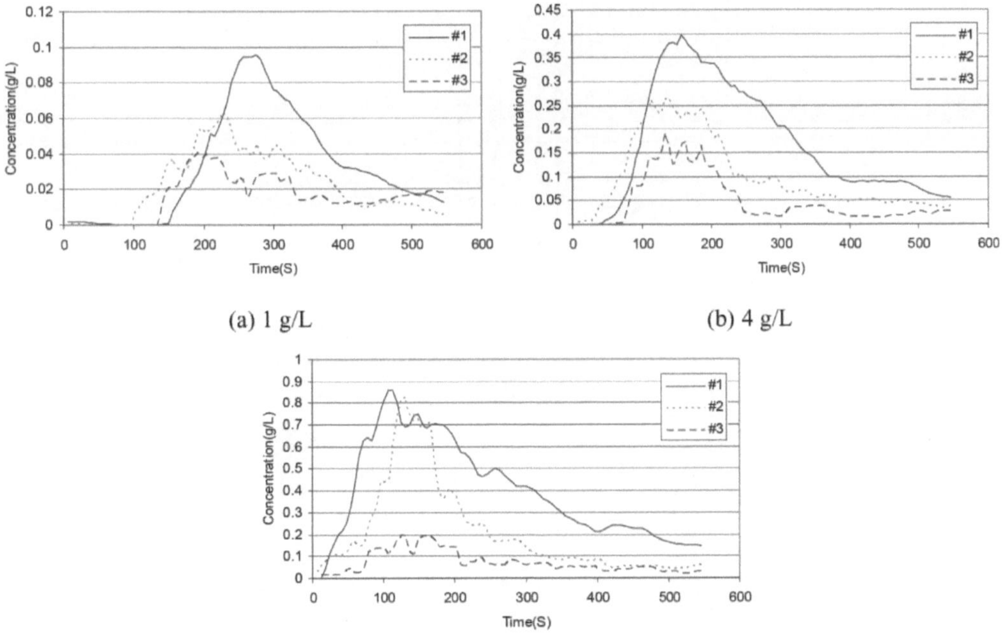

(a) 1 g/L (b) 4 g/L

(c) 10 g/L

Figure 3. Downstream concentration processes (non-submerged, no vegetation).

3.1 *Non-submerged condition*

1. The downstream concentration was much smaller than the upstream input concentration due to water dilution. When the input concentration was 1 g/L, the peak value on the downstream cross section was only approximately 0.1 g/L. And the two values would change proportionally. When the vegetation area became larger, the downstream peak value decreased to a certain extent. The peak value under the full vegetation condition was exactly two-thirds of the corresponding value under the no vegetation condition. That is, the vegetation showed the effect of peak clipping.
2. The shape of the concentration process curve shows that every point-value starts to rise almost at the same time, which is the time when polluted water reaches the measuring cross section. Under the no vegetation condition, the concentration process curve was a distinct single-peak type (Figure 3), which implied that, without any interception, the pollutant entered the main stream continuously until the downstream concentration reached the peak value and then the value started to decrease as the detained pollutant

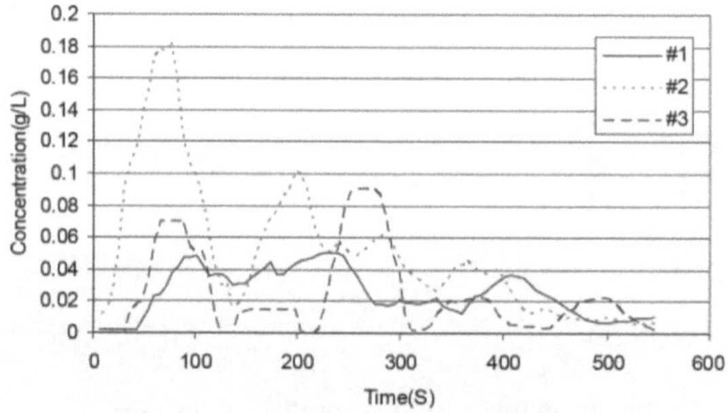

Figure 4. Downstream concentration processes (non-submerged, input concentration 2 g/L).

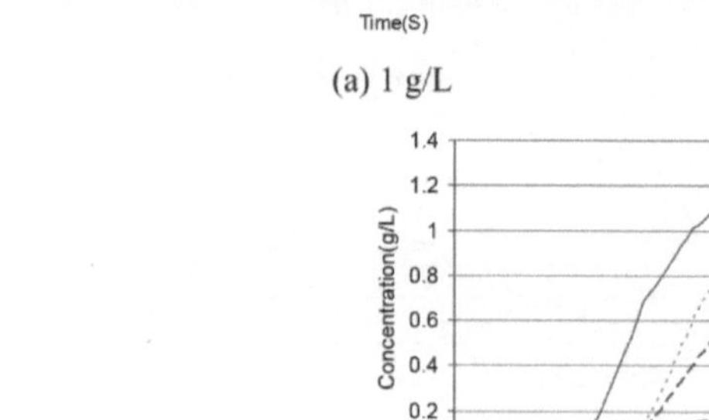

(a) 1 g/L

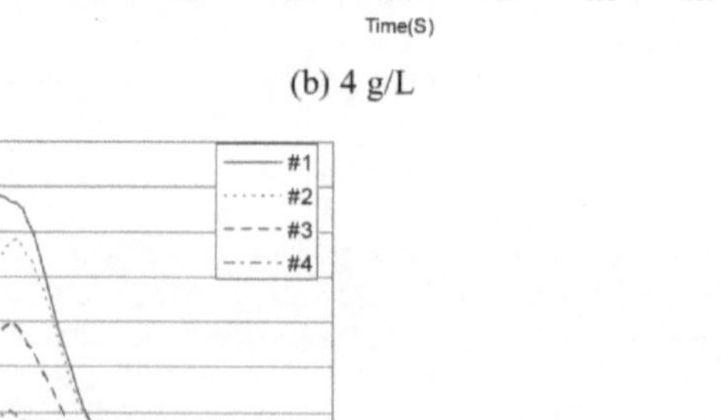

(b) 4 g/L

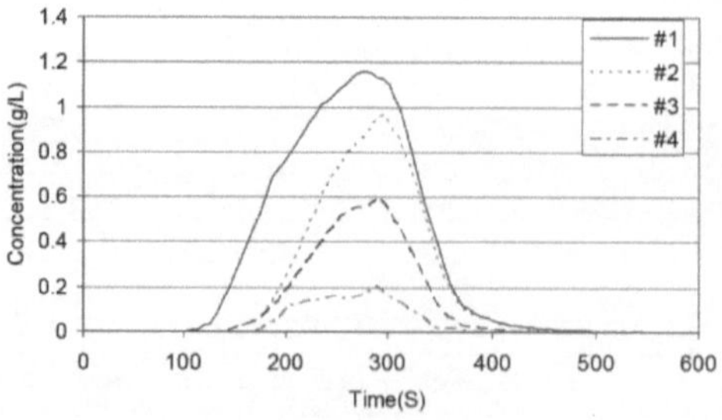

(c) 10 g/L

Figure 5. Downstream concentration processes (submerged, no vegetation).

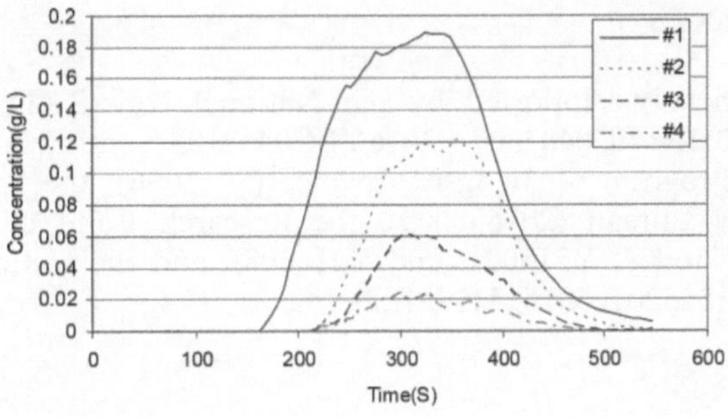

Figure 6. Downstream concentration processes (submerged, input concentration 2 g/L).

became gradually less after the input stopped. However, after adding more vegetation, a multi-peak feature appears in the concentration process curve (Figure 4). This means that the pollutant was intercepted and stored in the vegetation zone, so that the continuous supply of pollutant into the main stream was disturbed and multiple peaks were formed.

3.2 *Submerged condition*

1. The downstream concentration was also much lower than the input ones. Furthermore, there was a proportional variation feature. However, the peak clipping effect of vegetation was not as evident as in the non-submerged case.
2. Regardless of the vegetation scheme, the concentration of measuring points showed the relationship of #1>#2>#3>#4 (Figures 5 and 6). This result suggested that the polluted current flowed near the bank. Vegetation showed no effect on this phenomenon.
3. As for the curve shape, in all vegetation schemes, #1 value was the earliest one to increase compared with other measuring points. That is, part of the pollutant input from the bank was almost instantly carried to the downstream bank by the strong flow. Furthermore, compared with the non-submerged case, the transition from single peak to multi-peaks together with the increase in the vegetation area was not much impressive (Figures 5 and 6).

4 CONCLUSIONS

In this study, a laboratory hydrodynamic flume experiment was carried out to investigate the function of artificial wetland in an ecological spur-dike system. The following conclusions were drawn from this study:

1. A transition from single peak to multi-peaks together with the increase in the vegetation area was shown by the downstream concentration process curves. The peak clipping function of artificial wetland can reduce the downstream concentration, indicating that the polluted sediments were intercepted and stored in the ecological spur-dike zone. Furthermore, water quality can be enhanced and the detained pollutant can be biologically treated in the real wetland.
2. It can be concluded that the pollutant-interception effect of ecological spur-dikes on artificial wetlands is good. The ecological function is evident under the normal discharge condition (non-submerged). If the wetland area is restricted by the requirement of flood discharge or navigation, then the near-bank part of wetland can still play a similar role. For frequently flooded or highly bank-source-polluted river sections, the wetland area should be expanded as much as possible. During the floods, some pollutants can directly overflow the spur-dikes. Thus, improvements such as ecological design of the dike body (e.g. surface vegetation) should also be considered.

ACKNOWLEDGMENTS

This study was financially supported by the National Key R&D Program of China (2017YFC0405903, 2016YFC0402307, 2016YFC0402103, and 2016YFC0402108), the Hydraulic Scientific Program of Jiangsu Province (2016006), the Scientific Program of Changjiang Waterway Bureau (201630009), the Research Fund of Nanjing Hydraulic Research Institute (Y216006, Y215021, and Y214008), and the National Natural Science Foundation of China (51520105014, 51379127).

REFERENCES

Hohmann, J. & Konold, W. 1992. Flussb aum as snahmen an der wutach und ihre bew er-tung au s oekologischer sicht. *Deutsche Wasserw irtschaft* 82(9): 434.440.

James, R.T. & Martin, J. 1997. A sediment resuspension and water quality model of lake okeechobee. *Journal of the American Water Resources Association* 33: 661–680.

Lu, Y. & Li, S.Q. 2011. Ecological effects and stability of an artificial reef spur-dike. *Proceedings of the 7th IAHR Symposium on River, Coastal and Estuarine morphodynamics. Beijing, 2011*: 512–522.

Pan, G. 1999. Adsorption kinetics in natural waters: a generalized ion-exchange model, in adsorption and its application in industry and environmental protection. *Studies in Surface Science and Catalysis* 120: 745–761.

Wang, P.F. & Yang, C.Q. 2011. Study on effect of ecological groin on velocity of flow and sediment deposition. *Proceedings of the 23rd national Conference on hydrodynamics, Xi'an, 2011*: 697–701.

Hydraulic Engineering V – Hong, Liu & Xie (Eds)

 ISBN 978-0-8153-7471-8

Physicochemical properties and carbon cycle of soil on the intertidal mud flat to the north of the Yangtze Estuary

X. Shen
Nanjing Hydraulic Research Institute, Nanjing, Jiangsu Province, China
State Key Laboratory of Hydrology-Water Resources and Hydraulic Engineering, Nanjing, Jiangsu Province, China

P. Wang
Hohai University, Nanjing, Jiangsu Province, China
National Engineering Research Center of Water Resources Efficient Utilization and Engineering Safety, Nanjing, Jiangsu Province, China

Y.D. Wang
Hohai University, Nanjing, Jiangsu Province, China

ABSTRACT: Soil samplings and field monitoring were conducted to investigate CO_2 fluxes properties on the interface of soil-atmosphere on intertidal zones to the north of the Yangtze Estuary. The net CO_2 flux values indicated that the mud flat performed generally as a carbon sink. The low tidal flat (point III) was observed having the greatest CO_2 flux where the higher Chl-a content brought stronger CO_2 sequestration, and the lower TOC content led to less CO_2 release in the processes of soil respiration and mineralization. In addition, groundwater level was studied as another critical factor of carbon cycle which had negatively correlated with CO_2 net fluxes. The results provide a better understanding on restoring and protecting coastal wetlands and present an important opportunity for mitigating greenhouse gas emissions.

1 INTRODUCTION

Climate change has emerged as the most prominent of the global environment issues. The atmospheric concentrations of the greenhouse gases CO_2, CH_4, and N_2O have increased since 1750 due to human activity. In 2011, the concentrations of these greenhouse gases were 391 ppm, 1803 ppb, and 324 ppb, exceeding the pre-industrial levels by about 40%, 150%, and 20%, respectively (IPCC, 2013). Coastal wetland, as blue carbon ecosystems, are global carbon hot spots where area-based carbon pools and fluxes far exceed those of other terrestrial and aquatic ecosystems (Moffett et al. 2010, Kirwan & Megonigal 2013, Hopkinson 2012, Donato 2011). Vegetated coastal systems, including seagrass beds, mangroves, and intertidal marshes currently sequester 67–215 Tg C/a, globally (Hopkinson 2012), which presents an important opportunity for mitigating greenhouse gas emissions and global warming. There have been increasing amount of researches focusing on carbon flux of coastal wetland in recent years. Comparing ecosystem CO_2 fluxes among 22 inland wetland and 21 coastal wetland sites showed that coastal wetlands had higher annual gross primary productivity, ecosystem respiration, and net ecosystem productivity than inland wetlands (Lu et al. 2017). In a reclaimed coastal wetland in the Yangtze Estuary, the net ecosystem exchange displayed marked diurnal and seasonal variations, and the system performed as an important CO_2 sink. (Zhong et al. 2016).

Intertidal soil acts an important zones for organic matter deposition within coastal areas. As such, high rates of organic matter mineralization may be observed within these zones.

An availability of light and nutrients means that intertidal soil may also be important habitat for microphytobenthos. Microalgae have been shown to contribute significantly to total primary production in a number of coasts (Underwood & Kromkamp 1999). Intertidal sediments are, therefore, likely to play an important role in both the primary production and respiration on mudflats. Environmental factors controlling these processes include light, sediment TOC, aquatic organism, exposure and temperature (Perran et al. 2004). Soil respiration rates had significant relation with soil temperature variation (Freeman et al. 1993). The period of waterlogging on intertidal mudflat influenced carbon cycle process via impacting soil respiration and algae growing (Ewanchuk & Bertness 2004, Barr et al. 2010). Also, it was found CO_2 fluxes negatively correlated with groundwater level, and the value increased from 646 $mg{\cdot}m^{-2}{\cdot}d^{-1}$ to 1590 $mg{\cdot}m^{-2}{\cdot}d^{-1}$ when water lever descended (Freeman et al. 1993). For the Chongming east intertidal flat at the Yangtze estuary, the low tidal flat was the sink of carbon with the average carbon flux of −13.23 $mg{\cdot}m^{-2}{\cdot}h^{-1}$, and temperature and sunshine intensity were major impact factors which control the carbon fluxes and their variation (Yang et al. 2006).

The length of coastal line of Jiangsu Province is nearly 1000 km, among which the vast majority is silty mud coastal zone. The area of the coastal wetland accounts for 1/4 of the total amount in China. For the past 20 years, the landscape of this area has changed greatly due to the development of the intertidal wetland in Jiangsu province. The major transformation characteristics including salsa, reeds and spartina represented the natural wetland were changed into aquaculture ponds, salt pan and agricultural land represented artificial wetlands (Huang et al. 2014). Also, the mud flat area has been generally increasing as a consequence of reclamation and silting. The carbon flux on the soil-air interface of intertidal mud flat is rarely studied before the present research. Therefore, the intertidal zone to the north of Yangtze River Estuary was chosen to investigate carbon cycle by field measuring and laboratory experiments. The result can improve our understanding of the roles of coastal wetlands in the global carbon cycle. Our results also have implications for informing wetland management and climate change policymaking, for example, the efforts being made by international organizations and enterprises to restore coastal wetlands for enhancing blue carbon sinks.

2 RESEARCH AREA

The research area is an intertidal zone by the Yellow Sea which is located to the north of the Yangtze Estuary, southeast part of Jiangsu Province. The shore is a depositional mud flat whose deposition speed is about 25–30 m/year. The width of the coastal wetland is 4–6 km, and could be divided into several belts, such as grass flat, Artemisia mud flat, mud-silt flat and silt flat offshore. The sampling and monitoring area (31°47'19"N, 121°54'16"E) is a part of Artemisia mud flat. It is submerged periodically with the tide where rare plants grow on the flat and seldom influenced by human activities.

3 PHYSICOCHEMICAL PROPERTIES OF SOIL

3.1 *Soil grain composition*

Intact soil core samplings were conducted to reveal its physicochemical properties in April, 2017. Three sampling points were numbered as I, II and III offshore, and represented high tidal flat, middle tidal flat and low tidal flat. Each soil core was divided into five layers vertically with the thickness of 5 cm for each layer. Number each layer as 1 to 5 from top to bottom.

All the soil samples were taken to the laboratory where they were frozen and dried, then crushed and ground. Fully Automatically Laser Particle Size Analyzer was applied to analyze soil grain composition. The median diameter of soil particles of five layers at three points were presented in Figure 1. As shown, the diameter values of different samples ranged from 170 μm to 190 μm. No significant differences in soil grain composition were observed across

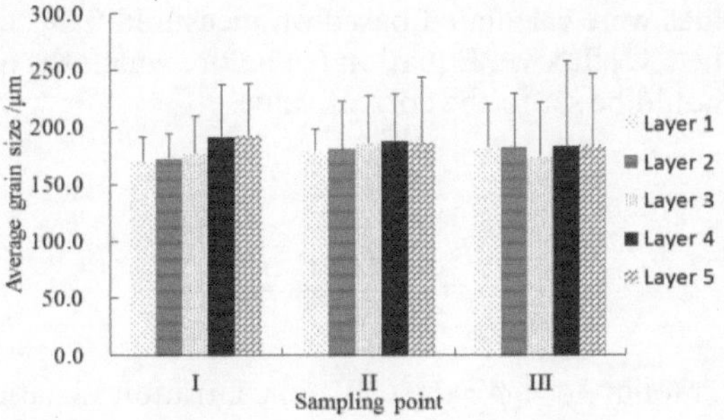

Figure 1. The average diameter of each soil sample.

the shore. The median diameter values of point I and II increased gradually from surface to bottom. It was suggested that the tidal current had more sizing power in the area where it's closer to the coast.

3.2 *TOC contents*

TOC contents of each soil sample were determined by Combustion Oxidation-Undispersed Infrared Method (Multi N/C2100) after they were frozen and impurity abandoned. The vertical distributions of TOC at each point were analyzed. In the surface layer (layer 1), the TOC content at point I (2.2%) was higher than both point II (1.79%) and III (1.56%). In vertical direction, the average TOC content of three points were 1.65%, 1.64% and 1.49% respectively. There was no obvious trend of TOC vertically because of cultivation activities, disturbance of aquatic organism and tidal current probably. Former results suggested that there were complex factors to affect TOC content (Yin et al. 2012), and the main environmental factor on TOC distribution of intertidal mud flat was tidal current and aquatic organism (Xu et al. 2014).

3.3 *Chl-a contents*

It's studied that Chl-a existed in the top layer mainly in this mud flat (Shang et al. 2009). As such, the surface soil was sampled to monitor Chl-a contents. The frozen samples were taken to the laboratory, and a small amount (2 g) of three soil samples were determined according to Specifications for Oceanographic survey (GB 12763–2007) after the sample pretreatment.

Chl-a content at three points (I, II, III) were 7.16 μg/g, 7.69 μg/g and 8.16 μg/g respectively. The Chl-a values were generally high since the environment were extremely suitable for microalgae growing (Shang et al. 2009). The small soil particles also had strong absorptivity and brought sufficient nutrient substances for benthic microalgae (Li & Ding 2009). Furthermore, the sampling activities were conducted in Spring when the sunlight and temperature were critical for algae (Yang 2006). It was suggested that Chl-a content increased as it's approaching seawater by comparing data of three points.

4 CO_2 FLUX PROPERTIES ON SOIL-AIR INTERFACE

4.1 *Measuring instrument and data processing*

WEST Systems portable soil flux meter (LICOR LI-820) for Carbon dioxide was applied to measure CO_2 concentration over time on soil-air interface. The systems include the double beam infrared sensor, the air pump, the accumulation chamber, and the palmtop. The CO_2 monitoring duration was 4 minutes for each time at every point, and there were 5 monitoring times totally at the interval of 30 minutes. The monitoring period of about 2 hours was a part of tide-falling period.

The CO_2 variation rates were calculated based on measured CO_2 concentration and time. Afterwards, calculate the CO_2 flux via Equation (1) before which the accumulation chamber factor (Equation (2)) should be set to the correct value.

$$\phi = \varphi \cdot k \tag{1}$$

$$k = \frac{86400 \cdot P}{10^6 \cdot R \cdot T} \cdot \frac{V}{A} \tag{2}$$

where ϕ = the CO_2 flux, $mol \cdot m^{-2} \cdot d^{-1}$; φ = the CO_2 concentration variation rate, ppm/s; k = the accumulation chamber factor; P = the barometric pressure, HPa; T = air temperature, K; V = the chamber net volume, m^3; A = the chamber inlet net area, m^2; R = the gas constant, $J \cdot mol^{-1} \cdot K^{-1}$).

4.2 *CO_2 concentration and CO_2 flux*

Intertidal mud flat is one of the most complicated reservoirs of the carbon cycle where the carbon cycle interacts with other biogeochemical cycles. Microalgae uses CO_2 to build their cell structures which is how carbon is absorbed or sequestered to the soil. Stored carbon can be released or emitted through natural respiration or when cell structures decompose, are burnt, or in the case of soil carbon, are disturbed. Therefore, CO_2 concentration in soil-air interface varies constantly.

The average values and the standard deviation of CO_2 concentrations on mudflat surface of three monitoring points were analyzed (Figure 2). The average CO_2 concentration at point I was 312–335 ppm, point II 300–310 ppm, and point III 295–315 ppm. The value of point I was the greatest since it was close to the coast where human activities could produce more CO_2 emission to atmosphere. The CO_2 concentration of point II and point III were generally lower as they were less influenced by human activities. The standard deviation statistics of three points indicated that the data of point I was steadiest and varied most slightly over time.

The CO_2 flux was calculated according to Equation (1). Here, we defined net CO_2 flux as carbon sequestration minus carbon emission. Figure 3 was a boxplot revealing CO_2 net flux on soil-air interface of three points. In the boxplot, the top line and the bottom line of box represented 75% and 25% of the flux values, and the inside line meant the median value. Besides, the small square inside expressed the average value, and the short lines above and below the box represented the maximum value and minimum value respectively. The positive net CO_2 fluxes meant that the carbon sequestration fluxes were larger than the emission fluxes. It was indicated that the intertidal wetland had carbon sequestration function in total, and point II and point III were stronger that point I. In addition, the average values of CO_2

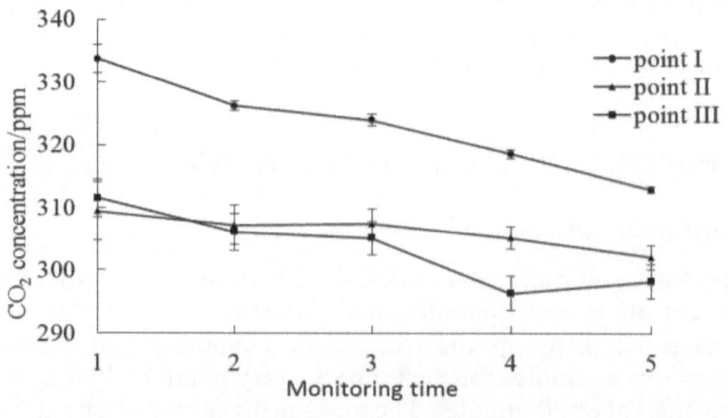

Figure 2. CO_2 concentrations on mudflat surface at different time.

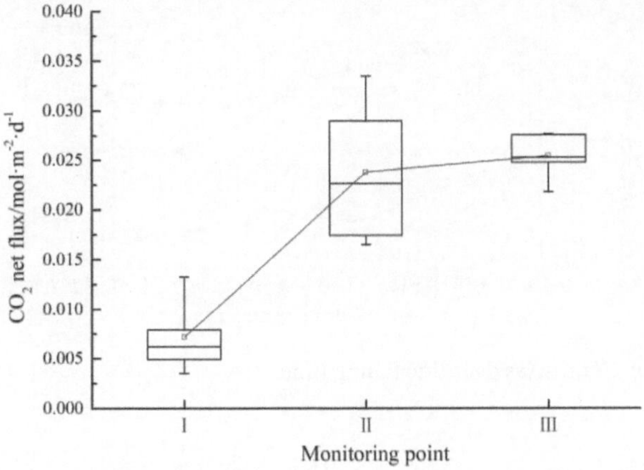

Figure 3. CO_2 net fluxes on soil-air interface.

fluxes were higher than the median ones which indicated that data appeared skew distribution and focused on larger values mainly. These results were in agreement with former research (Heinsch 2006) which investigated CO_2 flux in 38 types of wetland of Northern American for four years and found that intertidal wetland had CO_2 capture and storage function in the soil condition of high water content and low salinity.

4.3 *Controlling factors of carbon cycle*

Previous research (Blanca et al. 2008) of wetland carbon fluxes have indicated that soil TOC content and plant distribution played a significant role in soil-air carbon flux, as well as soil Chl-a contents, sunlight condition and groundwater level. Here, TOC contents, Chl-a contents and tide-ebbing time were analyzed to reveal the controlling factors of carbon cycle in the intertidal mud flat.

1. CO_2 net flux related with soil TOC contents and Chl-a contents
Soil TOC was the main source for microbe metabolism and mineralization. Soil released more CO_2 due to faster respiration rate of soil in the condition of high TOC contents (Susumu et al. 2017). Generally, comparison of CO_2 release fluxes of three points suggested that point I was the greatest, and point III was the lowest.

Chl-a contents were used to indicate microphytobenthos biomass. The microalgaes that live in the soil mediate many of biogeochemical interactions in addition to driving important carbon cycle processes by which soil systems absorb CO_2. The CO_2 sequestration fluxes were impacted by Chl-a contents and sunlight distinctly. No marked differences among three sampling points were observed. CO_2 sequestration fluxes at point III was the highest because the largest Chl-a content of point III led to the highest photosynthesis rate.

In brief, both CO_2 release fluxes and CO_2 sequestration fluxes composed the net CO_2 fluxes on soil-air interface. The lowest release flux and the highest sequestration flux of point III resulted in the greatest net CO_2 flux among these three sampling points.

2. CO_2 net flux variation over time
CO_2 fluxes on soil-air interface were monitored for 5 times for the interval of 30 minutes. It was proposed that groundwater level changed over time, whereas other environmental factors, such as soil temperature, soil TOC content, Chl-a content maintained steady over time. Figure 4 described the variation curve of CO_2 net fluxes over tide-ebbing time. It was shown that the CO_2 net fluxes decreased when the groundwater level descended. The reason was that respiratory rates of microbes in the soil strengthened when they were less submersed with the groundwater level declining (Liveseley & Andrusiak 2012). That induced larger CO_2 release fluxes. However, CO_2 sequestration flux was supposed not to change over time since Chl-a content as the dominate factor nearly keep steady in this short monitoring period.

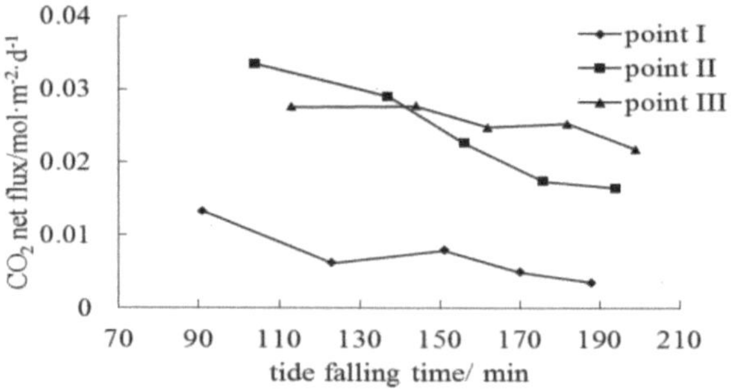

Figure 4. Variation of CO_2 fluxes over tide falling time.

5 CONCLUSIONS

The soil grain sizes were general fine, and the median diameter of particles increased vertically from top to bottom. The TOC content of high tidal flat was the greatest, and inclined gradually from coast to sea. In contrast, Chl-a content distribution appeared to be rising offshore.

CO_2 concentrations on the soil-air interface deceased over time which indicated that intertidal mud flat had great potential as carbon sinks. The net CO_2 fluxes of middle tidal flat and low tidal flat were higher comparing to high tidal flat which was supposed to have close connection with soil TOC content and Chl-a content, as well as groundwater level variation.

The net CO_2 flux depended on the balance of two processes—CO_2 release and CO_2 sequestration. CO_2 emission flux was higher due to higher TOC contents and consequent faster respiration rate of soil. CO_2 sequestration flux was influenced noticeably by photosynthesis rate of micro algae and had positively correlated to Chl-a content. The groundwater level played as another controlling factor of CO_2 net fluxes which decreased due to the increasing CO_2 release flux affected by soil respiration.

ACKNOWLEDGEMENTS

This research was financially supported by National Natural Science Foundation of China (51609151) and Natural Science Foundation of Jiangsu Province (BK20150078).

REFERENCES

Barr, J. G. et al. 2010. Controls on mangrove forest-atmosphere carbon dioxide exchanges in western Everglades National Park. *Journal of Geophysical Research* 115(G2): G02020.

Blanca, B. et al. 2008. A comparison of soil carbon pools and profiles in wetlands in Costa Rica and Ohio. *Ecological Engineering* 34(4): 311–323.

Donato, D. C. et al. 2011. Mangroves among the most carbon-rich forests in the tropics. *Nature Geoscience* 4(5): 292–297.

Ewanchuk, P. J. & Bertness, M. D. The role of waterlogging in maintaining forb pannes in northern New England salt marshes. *Ecology* 85(6): 1568–1574.

Freeman, C. et al. 1993. Fluxes of CO_2, CH_4 and N_2O from a welsh peatland following simulation of water table draw-down: potential feedback to climate change. *Biogeochemistry* 19(1): 51–60.

Heinsch, F. A. et al. 2006. Evaluation of remote sensing based terrestrial productivity from MODIS using regional tower eddy flux network observations. *IEEE Transaction on Geoscience & Remote Sensing* 44(7): 1908–1925.

Hopkinson, C. S. et al. 2012. Carbon sequestration in wetland dominated coastal systems—a global sink of rapidly diminishing magnitude. *Current Opinion in Environmental Sustainability* 4(2): 186–194.

Huang, H. et al. 2014. The landscape changes of intertidal wetland in Jiangsu Province and its environmental effects (in Chinese). *Hubei Agricultural Sciences* 53(4): 821–825.
IPCC 2013. Climate change 2013. The Physical Science Basis: 486–487.
Kirwan, M. L. & Megonigal, J. P. 2013. Tidal wetland stability in the face of human impact and sea-level rise. *Nature* 504(7478): 53–60.
Li, W. H. & Ding, P. X. 2007. Study on relation between variation of chlorophyll a concentration and sediment grain size on an intertidal flat. *Journal of East China Normal University (Natural Science)* (4): 26–33.
Liveseley, S. J. & Andrusiak, S. M. 2012. Temperate mangrove and salt marsh sediments are a small methane and nitrous oxide source but important carbon store. *Estuarine, Coastal and Shelf Science* 97: 19–27.
Lu, W. Z. et al. 2017. Contrasting ecosystem CO_2 fluxes of inland and coastal wetlands: a meta-analysis of eddy covariance data. *Global Change Biology* 23, 1180–1198.
Moffett, K. B. et al. 2010. Salt marsh atmosphere exchange of energy, water vapor, and carbon dioxide: effects of tidal flooding and biophysical controls. *Water Resources Research* 46(10): W10525.
Perran, L. M. et al. 2004. Carbon and nitrogen cycling on intertidal nudflats of a temperate Australian estuary. I. Benthic metabolism. *Marine Ecology Progress Series* 280: 25–38.
Shang, X. et al. 2009. Distribution characteristics and contribution to total primary production of microphotobenthos in the salt marshes of the Changjiang Estuary (in Chinese). *Acta Oceanologica Sinica* 31(5): 40–47.
Susumu, Y. et al. 2017. Effects of light, temperature and groundwater level on the CO_2 flux of the sediment in the high water temperature seasons at the artificial north salt marsh of Osaka Nanko bird sanctuary, Japan. *Ecological Engineering* 98:330–338.
Underwood, J.C. & Kromkamp, J. 1999. Primary production by phytoplankton and microphytobenthos in estuaries. In: Nedwell, D.B. & Raffaelli, D.G. (eds), *Advances in ecological research—estuaries*: 93–153. San Diego: Academic Press.
Xu, X. et al. 2014. The characteristics of surficial sediments organic carbon in Yancheng coastal wetland. (in Chinese). *Journal of Natural Resources* 29(11): 1957–1967.
Yang, H. X. et al. 2006. Characteristics of carbon fluxes through intertidal flat wetland-atmosphere interface of Yangtze estuary (in Chinese). *Acta Scientiae Circumstantiae* 26(4): 667–673.
Yin, S. L. et al. 2012. Seasonal changes of chlorophyll and organic matter contents in sediment in the intertidal zone of Dagu estuary, Jiaozhou bay (in Chinese). *Transactions of Oceanology and Limnology* 2:97–106.
Zhong, Q. C. et al. 2016. Carbon Dioxide Fluxes and Their Environmental Control in a Reclaimed Coastal Wetland in the Yangtze Estuary. *Estuaries and Coasts* 39, 344–362.

Hydraulic Engineering V – Hong, Liu & Xie (Eds)
© 2018 Taylor & Francis Group, London, ISBN 978-0-8153-7471-8

Experimental study into the blocking effectiveness of silt curtain on suspended materials

S.G. Chen, C. Peng & H.B. Chen
Tianjin Research Institute of Water Transport Engineering, Tianjin, China

X.W. Zhang & Q. Zhang
CCCC Second Harbor Consultants Co. Ltd., Wuhan, China

ABSTRACT: This paper presents an experimental study into the Blocking Effectiveness (BE) of silt curtain on suspended materials under different conditions. Two types of sand samples were chosen for representing grains generated from outlet and dredging work. Moreover, two gaps between the silt bottom and the seabed were adopted to simulate the MHHW (Mean Higher High Water) and MLLW (Mean Lower Low Water) conditions. The current speed range was 0.02–0.04 m/s. The results indicated that the average permeability coefficient of the silt curtain was 0.0006 m/s. On the one hand, the blocking effectiveness of the silt curtain near the outlet ranged from 38% to 8%. The maximum value was observed at a speed of 0.02 m/s and at the water level MLLW. On the other hand, the blocking effectiveness of the silt curtain around the dredging area ranged from 59% to 40%. The maximum value was observed at a speed of 0.02 m/s and at the water level MLLW. From the results of the study, it can therefore be suggested that silt curtain around the dredging area is necessary and that near the outlet can be optional.

1 INTRODUCTION

Hassyan Energy is currently constructing the Hassyan Clean Coal Independent Power Plant (HCCIPP) along the Arabian Gulf coastline adjacent to the Dubai–Abu Dhabi border. Land reclamation and port construction, both of which involve dredging, are required to meet coal import and ash output. However, the HCCIPP is located within marine conservation areas in the UAE. Increased turbidity and sedimentation from dredging will result in some adverse environmental effects (Mcclanahan et al. 1997, 2007, Wang et al. 2001). For example, some sensitive coral reef systems show mortality after exposure (weeks) to concentrations as low as 30 mg/L (Erftemeijer et al. 2006), and light attenuation by suspended materials affects the amount of light available to the sea grass plants and associated epiphytes, microphytobenthos, and macroalgae (Erftemeijer et al. 2012).

Several mitigation measures have been proposed to minimize the above-mentioned negative impacts. Wang (Wang et al. 2001) proposed three measures, namely setting settling ponds to contain dewatering effluent, changing location of the outlet to slow down the effluent velocity, and using DGPS to control overdredging. Jiang (Jiang 2015) suggested that adopting an eco-friendly dredger and optimizing the operational schedule could mitigate the pollution of suspended materials. Cooper (Cooper et al. 2007, 2008) and Vargas (Vargas et al. 2006) studied coral health indicators to get early warning of dredging work. Researchers (Doorngroen 2009, Koskela et al. 2002, Sofonia et al. 2010) in Singapore and Australia have made some significant efforts in establishing reactive (feedback) monitoring programs to minimize coral mortality during large-scale dredging operations.

In this project of HCCIPP, some measures, including controlling suspension concentration outflowing from the outlet, setting silt curtains around the dredging and reclamation area, and establishing monitoring stations near sensitive areas, are adopted to protect the marine environment. To the best of the authors' knowledge, this is the first time to use silt curtains to prevent the diffusion of suspended sediments. Therefore, in this study, physical modeling results are presented to demonstrate the blocking effectiveness (BE) in mitigating the impacts. This paper is organized as follows. MetOcean and silt curtain conditions are first reviewed to determine the test conditions. The modeling design is introduced in the second part. Third, testing results are discussed. Conclusions are drawn in the final section.

2 TEST CONDITIONS

2.1 *Water levels*

Water levels including tidal planes at the site are from Jebel Ali Port Tide Station presented in Table 1. At the MLLW, the gap between the silt curtain and the sea bottom is 0.2 m, while it becomes 1.5 m at the MHHW due to tidal rise. The water depths where silt curtains are configured are 5 m at the MLLW and 6.3 m at the MHHW.

2.2 *Current*

In general, the current conditions within the project area and immediate vicinity are dominated by tidally generated currents. There is a well-known southwest to northeast residual current due to the large-scale circulation within the Arabian Gulf, which is discernible at the regional scale; however, at 0.02–0.04 m/s, it is small. The current speeds along silt curtains configuration are extracted from the tidal numerical modeling (Chen & Guan 2016), and the current mean velocity is 0.004–0.039 m/s.

2.3 *Seabed conditions*

Seabed grab sampling data and borehole data are available for the site. In general, materials from borehole are coarser than those from seabed. Therefore, in order to obtain conservative results, particle sizes of grab sampling data, which are smaller than the borehole data, are used in the physical modeling. As shown in Figure 2, stations D2, D6, and D7 are within silt curtains. The following two particle sizes are chosen in the test: (1) 0.075 mm, whose passing percentage is about 15%, and (2) 0.125 mm, mean value of D_{50} of the above samples. Fine sand is the simulated material, which flows from the outlet of the reclamation area, whereas coarser sand is the simulated material produced during dredging work.

2.4 *Silt curtains*

The layout of silt curtains is shown in Figure 3. Most of the dredging and reclamation areas are surrounded by silt curtains, whose characteristics are listed in Table 2. The mesh size of

Table 1. Tidal planes at the HCCIPP site.

Tidal plane	Abbreviation	Level (m DMD)
Highest astronomical tide	HAT	+2.29
Mean higher high water	MHHW	+1.79
Mean lower high water	MLHW	+1.39
Mean sea level	MSL	+1.09
Mean higher low water	MHWL	+0.89
Mean lower low water	MLLW	+0.49
Dubai Municipality Datum	DMD	0
Lowest astronomical tide	LAT	–0.11

*m DMD refers to meters with respect to the Dubai Municipality Datum.

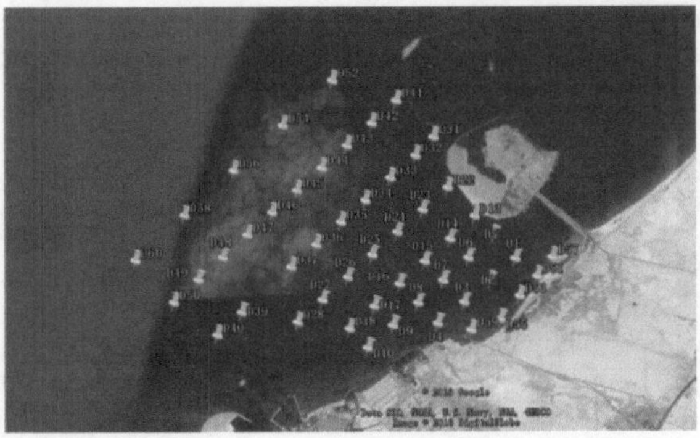

Figure 1. Grab sampling location.

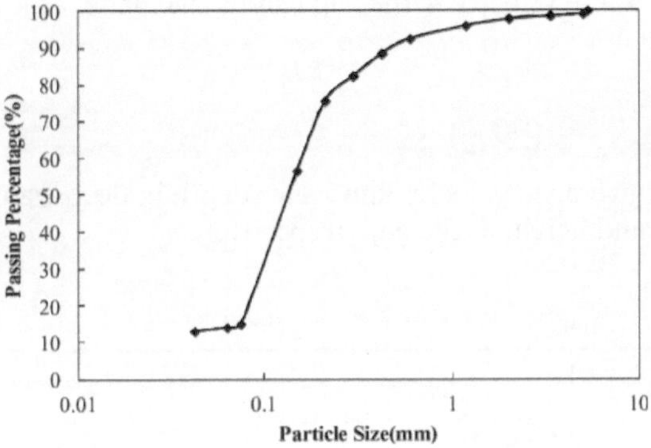

Figure 2. Grading curve at station D6.

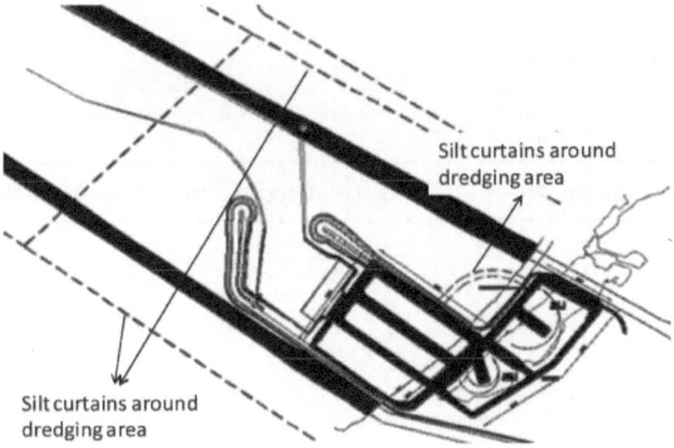

Figure 3. Layout of silt curtains.

Table 2. Characteristics of silt curtains.

Item	Unit	Value
Weight	g/m^2	301
Thickness	mm	2.40
Tensile strength	kN/m	9.51
Mesh size	mm	0.042

silt curtain is 0.042 mm, which is smaller than the sand size used in the test. Therefore, it can also be noted that sand particles cannot pass through the silt curtain and only leak from the gap between the seabed and the silt curtain bottom.

3 MODEL DESIGN

As mentioned in Section 2.4, leaking from gaps between the seabed and the silt curtain bottom is a key factor affecting the BE. Furthermore, start-up and transportation of sediments are determined by current velocity and particle size. Particle sizes are chosen to be equal to those in the prototype. Thus, physical modeling is designed to make current velocity at the gap in the modeling equal to that in the prototype. The following condition will be met:

$$v_{gm} = v_{gp} \tag{1}$$

As mass conservation, the velocity at the gap can be calculated as

$$v_g = \frac{VD}{h_g} \tag{2}$$

where V and D are the averaged velocity and water depth in deep water, respectively, and v_g and h_g are the velocity and height at the gap, respectively.

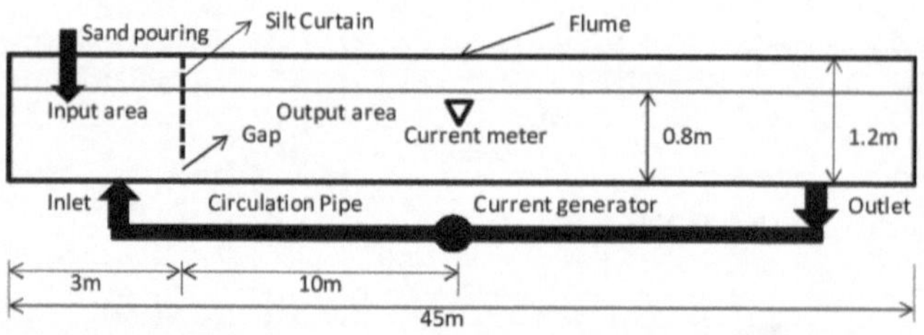

Figure 4. Physical modeling layout.

Table 3. Test conditions.

Test no.	Gap height (m)	Current speed (cm/s)	Particle size (mm)
Test 1	0	2	0.075
Test 2	0	2.5	0.075
Test 3	0.03	2	0.075
Test 4	0.03	2.5	0.075
Test 5	0.03	3	0.075
Test 6	0.03	4	0.075
Test 7	0.19	2	0.075
Test 8	0.19	2.5	0.075
Test 9	0.19	3	0.075
Test 10	0.19	4	0.125
Test 11	0.03	2	0.125
Test 12	0.03	2.5	0.125
Test 13	0.03	3	0.125
Test 14	0.03	4	0.125
Test 15	0.19	2	0.125
Test 16	0.19	2.5	0.125
Test 17	0.19	3	0.125
Test 18	0.19	4	0.125

Substituting eq. 2 into eq. 1 and considering V in model and prototype to be equal, the following equation can be obtained:

$$\frac{h_{gm}}{D_m} = \frac{h_{gp}}{D_p} \tag{3}$$

where h_{gm} and h_{gp} are the gap length in modeling and prototype, respectively, and D_m and D_p are the water depth in modeling and prototype, respectively.

Physical modeling is conducted at a current flume with length, height, and width of 45, 1.2, and 0.5 m, respectively. The silt curtain is fixed at one side of the flume, and the sand is poured into the input area of the flume, as shown in Figure 4. The currents at different depths are tested in order to obtain the target value. The test is conducted in the following steps:

1. Fixing the silt curtain and filling the flume with water to 0.8 m depth (D_m).
2. Starting current generator and adjusting the frequency converter to get the target speed V.
3. Pouring sand into the input area (Figure 4) and recording the input weight.
4. Making current generator work until no sediment leaks from the gap.
5. Draining out the water and measuring the remaining sand weight falling onto the bottom of the input area.

Two gap heights corresponding to the water levels MHHL and MLLW are tested. Current speeds in deep water are set at 2, 2.5, 3, and 4 cm/s. Two types of particle are chosen as input sediments. Moreover, the condition that no gap exists is used to determine the horizontal permeability coefficient. The test conditions are listed in Table 3.

4 RESULTS AND DISCUSSION

The horizontal permeability coefficient can be calculated by the following equation:

$$k = \frac{v}{h/L} \tag{4}$$

Table 4. Results of the horizontal permeability coefficient k.

Case	v (m/s)	h (m)	L (m)	k (m/s)
Test 1	0.02	0.072	0.0024	0.000667
Test 2	0.025	0.11	0.0024	0.000545

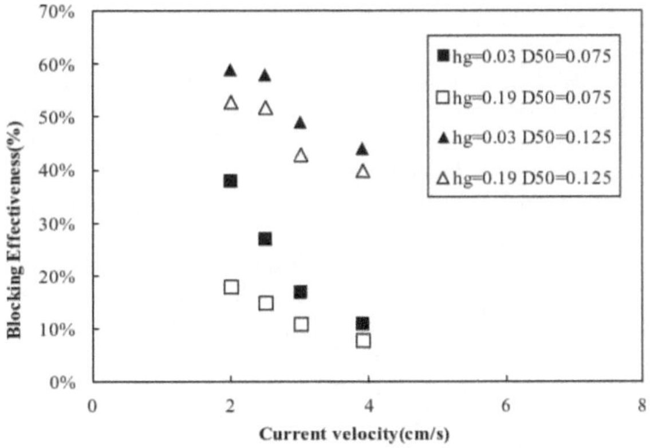

Figure 5. Results of BE under different conditions.

where k is the permeability coefficient, h is the water difference between two sides of silt curtain, and L is the thickness of the silt curtain. The results are given in Table 4. The average permeability coefficient is 0.0006 m/s.

The results of BE under different conditions are shown in Figure 5. It can be seen that the BE of silt curtain decreases with increasing current speed and gap length. For fine sand, the maximum BE is 38% when the current speed is 2 cm/s and the gap is 0.03 cm (which means that the water level is at MLLW). For coarse sand, the maximum BE is 59%, which is larger than that for fine sand. It is because the coarser the sand is, the larger the setting velocity is and the more the deposition will be. When the gap becomes larger, the BE decreases, which means that more sediment can escape from the gap at high water levels.

It can also be found the BE is more sensitive to the particle size than to the gap height. The mean BE value is 50% for coarse sand but 18% for fine sand. Therefore, it is suggested that silt curtains around dredging area are necessary and those near the outlet can be optional.

5 CONCLUSION

This paper presented an experimental study into the blocking effectiveness of silt curtain on suspended materials by conducting a wave flume test. Different current velocities, sand sizes, and bottom gaps were investigated. The results of the study indicated that the BE of silt curtain decreased with increasing current speed and gap length. Furthermore, the mean BE value was 50% for coarse sand (particle size, 0.125 mm) but 18% for fine sand (particle size, 0.075 mm). Therefore, it can be suggested that silt curtain around the dredging area is necessary and that near the outlet can be optional.

REFERENCES

Chen, S.G., Guan, N. 2016. Tidal and sediment numerical modelling report for dubai hassyan clean coal power plant. Tianjin Research Institute of Water Transport Engineering. Dubai.

Cooper, T., Fabricius, K. 2007. Coral-based indicators of changes in water quality on nearshore coral reefs of the Great Barrier Reef. Reef and Rainforest Research Centre Limited, Cairns., 31.

Cooper, T.F., Ridd, P.V., Ulstrup, K.E., Humphrey, C., Slivkoff, M., Fabricius, K.E. 2008 Temporal dynamics in coral bioindicators for water quality on coastal coral reefs of the Great Barrier Reef. *Marine and Freshwater Research.* 59:703–716.

Doorngroen S.M. 2009. Environmental monitoring and management of reclamation works close to sensitive habitats. *Asian and Pacific Coasts.* 15:62–81.

Erftemeijer, Paul L.A. and Roy R. Robin Lewis. 2006. Environmental impacts of dredging on seagrasses: a review. *Marine pollution bulletin.* 52(12): 1553–1572.

Erftemeijer, Paul L.A., et al. 2012. Environmental impacts of dredging and other sediment disturbances on corals: a review. *Marine pollution bulletin.* 64 (9): 1737–1765.

Jiang J.M. 2015. The measures for mitigating environmental pollution during dredging work. *Water transport in the Pearl River.* 7:60–61.

Koskela, R.W., Ringeltaube, P., 2002. Using predictive monitoring to mitigate construction impacts in sensitive marine environments. *Recent Advances in Marine Science and Technology.* Hawai.

McClanahan, T., Obura, D. 1997. Sediment effects on shallow coral reef communities in Kenya. *Journal of Experimental Marine Biology and Ecology.* 209:103–122.

McClanahan, T., Polunin, N., Done, T. 2002. Ecological states and the resilience of coral reefs. *Conservation Ecology.* 62(2):18–23.

Sofonia, J.J., Unsworth, R.K.F. 2010. Development of water quality thresholds during dredging for the protection of benthic primary producer habitats. *Journal of Environmental Monitoring.* 12:159–163.

Vargas-Angel, B., Riegl, B., Gilliam, D., Dodge, R.E. 2006. An experimental histopathological rating scale of sediment stress in the Caribbean coral Montastraea cavernosa. *Proceedings Tenth International Coral Reef Symposium*:1168–1173, Okinawa.

Wang C., Zhang L. 2001. Influence of dredging and deepening of the outward channel on ocean ecology and environment in the vicinity of the Pearl River Mouth and the prevetative measures. *Marine Environmental Science.* 20(4):58–66.

Hydraulic Engineering V – Hong, Liu & Xie (Eds)
© 2018 Taylor & Francis Group, London, ISBN 978-0-8153-7471-8

Numerical simulation of high-pressure submerged waterjet cutting hard clay using the SPH method

Shaogang Shi, Guojun Hong, Jifu Yin & Wei Shu
Key Laboratory of Dredging Technology of CCCC, National Engineering Center of Dredging Technology and Equipment, Shanghai, China

ABSTRACT: This paper presents the study of the waterjet cutting mechanism using the Smoothed Particle Hydrodynamics (SPH) method to simulate the process of high-pressure submerged waterjet cutting hard clay with numerical software ANSYS/LS-DYNA. The characteristics of the cutting hole and the propagation of stress wave on clay during the cutting process are discussed. Furthermore, the effect of the waterjet velocity and the impact distance on the cutting depth was analyzed. The results indicate that waterjet cutting forms a V-shaped hole of radius about 10 times the width of the waterjet. When the waterjet velocity becomes lower, the submerged environment has a more significant effect on cutting. The results of the study can provide a reference for improving the high-pressure waterjet cutting mechanism of a trailing suction dredger with preferable engineering research values.

1 INTRODUCTION

High-pressure waterjet uses water as the working substance, squirts from nozzle, and obtains extremely high-level energy density by supercharging equipment (Yang et al. 2006). Because of its high efficiency, safety, and convenience, it is widely used in the fields of material processing, mining, oil drilling, and other engineering fields. In recent years, rake tooth cutting supplemented with high-pressure waterjet has been applied in dredging engineering. This improvement can effectively reduce the cutting force of rake tooth, mitigate the problem of rake block, and improve dredging efficiency, which has become an important factor influencing the economic benefits of dredging.

High-pressure jet cutting is a complex nonlinear dynamic problem involving many factors, such as waterjet impact load, large deformation of materials, and fluid–solid coupling, which make the relative mechanism and experimental research more difficult. However, with the development of numerical software, it is possible to speed up the research progress. At present, finite-element method (FEM) and smoothed particle dynamics (SPH) method are the commonly used numerical methods. The SPH method is a meshless particle method, which can effectively avoid the mesh reconstruction and distortion occurring in finite-element method in dealing with large deformation. The SPH method is advantageous over the finite-element method in simulating the problem in this study.

At present, many scholars have applied the SPH solver in large commercial software or written SPH program to simulate jet cutting problems. Wang et al. (2016) presented a two-phase SPH mixture model to simulate water–soil interactions, in which water is considered as a Newtonian fluid and soil is modeled as an elastic-perfectly plastic material. Bui H et al. (2007) simulated the process of waterjet cutting of soil using the SPH method and proved that the SPH method has several advantages in dealing with large deformation and soil failure problems. Ma et al. (2008) coupled the SPH and FEM methods to simulate the process of water jet affecting the plastic surface and obtained a relationship between jet velocity and maximum depth. Zhang et al. (2015) simulated the process of jet cutting of sands by ALE and SPH methods individually and analyzed the shapes of hole and change of stress under each method. Song et al. (2009) used the SPH method to simulate the process of high-pressure jet cutting of rock and the energy conversation relationship and obtain the time history curve of impact. Si et al. (2016) used the SPH method to simulate the process of pulse jet breaking rock and analyzed the regulation of the stress wave propagation inside rock.

In this study, the SPH method is used to simulate the process of high-pressure submerged waterjet cutting of hard clay. For the research, a two-dimensional numerical model of jet cutting is established, which significantly reduces the computational time. In order to achieve the effect of submerged jets, rigid plates are added between the jet and submerged water to prevent initial penetration between particles. Meanwhile, the boundary force method is applied to deal with the solid wall boundary, so that the boundary defect of the SPH method is well solved.

2 THEORETICAL MODEL

The SPH method is a meshless method based on interpolation theory, which uses integral approximation in the form of MWS formula and transforms the partial differential equation into an integral form by kernel function interpolation. Therefore, its complete system consists of kernel approximation, conservation equation discretization, kernel function and time integration. In order to obtain the basic equations of SPH, the integral approximation of any function or smooth function should be expressed first. We assume the following expression of a continuous function:

$$f(x) = \int_{\Omega} f(x')\delta(x - x')dx' \tag{1}$$

where x are the spatial coordinates of calculated particles; x' are the spatial coordinates of particles that contribute to calculation; dx' is the unit volume in space; and $\delta(x - x')$ is the Kronecker delta function, which can be expressed as

$$\delta(x - x') = \begin{cases} 1 & x = x' \\ 0 & x \neq x' \end{cases} \tag{2}$$

Because of the strong spike of δ function, it cannot be applied in numerical simulation. Here, we substitute smoothing function $W(x - x')$ for the function $\delta(x - x')$, and then, Eq. (1) is transformed into

$$f(x) \approx \int_{\Omega} f(x')W(x - x', h)dx' \tag{3}$$

where $W(x - x')/h$ is the smoothing function, also called kernel function, and h is the smoothing length of the kernel function.

We assume that kernel function is used to show the relative distance between particles, which can be expressed in polynomial form as

$$W(S) = \mathrm{a}_0 + a_1 S + a_2 S^2 + \ldots + a_n S^n \tag{4}$$

where $S = |x - x'|/h$.

In order to make it easy to change the shape of kernel function, Eq. (4) can be expressed as a piecewise function:

$$W(S) = \begin{cases} b_1(S_1 - S)^n + b_2(S_2 - S)^n & 0 \leq S \leq S_1 \\ b_2(S_2 - S)^n & S_1 \leq S < S_2 \\ 0 & S_2 \leq S \end{cases} \tag{5}$$

Eq. (5) is mostly written as B-Spline function:

$$W(S)=\frac{1}{N}\begin{cases}1-\frac{3}{2}S^2+\frac{3}{4}S^3 & 0\le|S|<1\\ \frac{1}{4}(2-S)^3 & 1\le|S|<2\\ 0 & |S|\ge 2\end{cases} \tag{6}$$

3 NUMERICAL MODEL

3.1 *Model building*

In this paper, the two-dimensional numerical model consists of hard clay particles, high-pressure waterjet particles, submerged water particles, rigid plate, and boundary particles. Figure 1 shows the model of submerged waterjet for cutting hard clay. The clay dimension is 60 mm × 150 mm; the width and height of submerged water are 60 and 150 mm, respectively; the length and width of waterjet are 150 and 2 mm, respectively; and the velocity is along the central normal direction. The rigid plates are 150 mm long and 3 mm thick, which are simulated by solid element without considering their deformation. The boundary particles are arranged at the left and right boundaries of the computational domain, whose particle density is the same as that of neighboring particles, which can effectively increase the computational accuracy and prevent nonphysical penetration of particles. Because the form of moving waterjet is frequently adopted in dredging engineering, the calculation time of model is only set to 1000 μs.

3.2 *Material model*

The material model of hard clay adopted in this study is MAT147 in LS-DYNA, which can define not only the general parameters, such as density, bulk modulus, cohesion, and internal fiction angle, but also the rate of water content, viscoplasticity, porosity, softening, hardening,

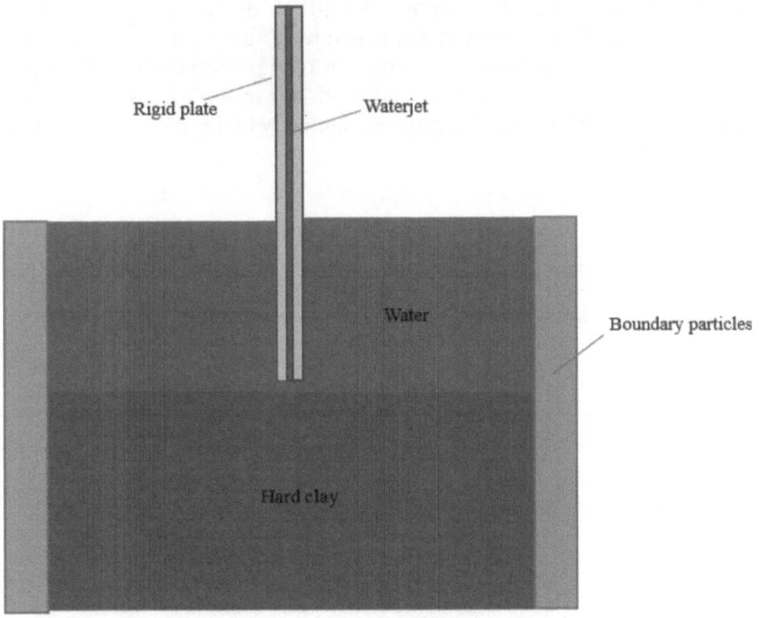

Figure 1. The model of submerged waterjet cutting hard clay.

and so on. This constitutive model is based on the work of ABBO et al. (2006), who modified the Mohr–Coulomb criterion; the modified yield surface is a hyperbola fitting that of the Mohr–Coulomb criterion. The modified criterion can be expressed as

$$Q = -p\sin\varphi + \sqrt{J_2 K(\theta)^2 + ahyp^2 \sin^2\varphi} - F_n \cos\varphi = 0 \tag{7}$$

where p is the hydrostatic pressure; φ is the internal friction angle; J_2 are the second invariants of deviatory stress tensor; $K(\theta)$ is the polar angle of the deviatory stress surface; F_n is the cohesion; and ahyp is the fitting precision between the modified yield surface and the standard surface. The material properties of hard clay are listed in Table 1.

Water can be considered as completely plastic material, the MAT_NULL model is used, and the state equation is given by

$$p = \frac{\rho_0 C^2 \mu \left[1 + \left(1 - \frac{\gamma_0}{2}\right)\mu - \frac{a}{2}\mu^2\right]}{\left[1 - (S_1 - 1)\mu - S_2 \frac{\mu^2}{\mu + 1} - S_3 \frac{\mu^3}{(\mu + 1)^2}\right]} + (\gamma_0 + a\mu)Ea \tag{8}$$

where ρ_0 is the initial density of water; C is the intercept of the curve $vs - vp$; μ is the viscosity coefficient; γ_0 is the coefficient of the state equation; α is the modifying factor of the state equation; S_1, S_2, and S_3 are the fitting coefficients of the curve $vs - vp$; and E is the internal energy per unit volume. The main parameters of state equation of water are listed in Table 2.

3.3 *Boundary treatment*

The SPH–FEM coupling algorithm is adopted between rigid plates, elements, and particles, and the keyword chosen is *CONTACT-ERODING-NODES-TO-SURFACE. In order to prevent the water particles from penetrating, the contact stiffness factor should not be too large, which is set to 0.1 in this study.

Because of the asymmetry of SPH boundary particles, instability may occur in boundary calculations, leading to abnormal calculation results; therefore, it is necessary to deal with the rigid wall boundary with special treatment. In this study, boundary force method is used to discretize the rigid wall boundary and appropriate center repulsion is applied to the calculated particles to prevent nonphysical penetration of particles close to the rigid boundary. In this study, the jet velocities are 50, 100, 150, and 300 m/s. The keyword in LS-DYNA is *VELOCITY-GENERATION, and the plus–minus of velocity depends on the direction of waterjet.

Table 1. Main material properties of hard clay.

Density (kg/m^3)	Bulk modulus (MPa)	Shear modulus (MPa)	Cohesion (kPa)	Friction angle (rad)	Specific gravity	Ahyp	Water content
2000	30	20	18	0.628	2.65	100	0.3

Table 2. Main parameters of the state equation of water.

Density (kg/m^3)	C (m/s)	GAMA	S1	S2	S3	α
1000	1467	0.35	1.921	−0.096	0	0

4 RESULTS AND DISCUSSION

4.1 *Analysis of the cutting process*

According to the modeling method mentioned above, high-pressure submerged waterjet cutting of hard clay was simulated. The impact distance was expressed in a dimensionless form as *D*/*d*. Figure 2 shows the cutting process at different times, with velocity of 200 m/s and *D*/*d* of 3.5. At the beginning of the cutting process, elastic deformation of water occurs together with the flow out of nozzle, leading to the streamline of waterjet deflecting and forming vortex near the nozzle. When waterjet contacts clay, an initial hole is formed in short time under compression of waterjet. The vortex formed by impact rotates in the hole, and the energy of the vortex increases as well, so that the clay particles at both ends of the hole pile up under the action of streamline lifting. The remaining jet particles continue to deepen the hole due to the hindrance of the submerged water environment and clay to the waterjet cutting process; the velocity of the waterjet that is away from the nozzle becomes more and more low; and the cutting energy is also reduced, leading to the gradual decrease of the cutting diameter in the vertical direction and the formation of a "V"-shaped erosion hole.

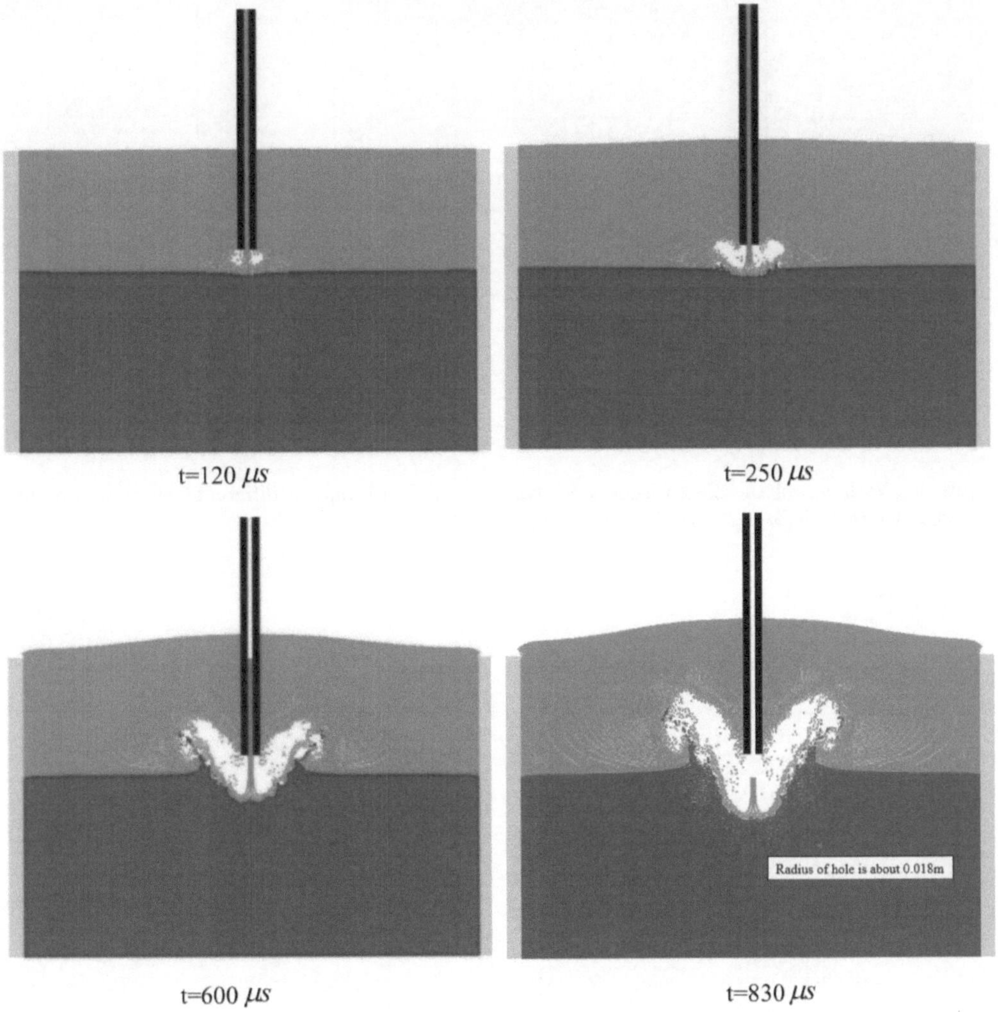

Figure 2. Images of the waterjet cutting hard clay at different times ($V = 200$ m/s, $D/d = 3.5$).

Figure 3 shows the curves of dimensionless vertical displacement (*H*/*d*) of the observation points selected on the edge of the cutting hole versus time. The observation points 2 and 3 are located at the center of the cutting hole, whereas points 1 and 4 are at both ends. For points 2 and 3, the cutting depth increases with time. Before 250 μs, the submerged waterjet is away from the clay, and a tiny hole is formed under the squeeze of water environment. During this period, the depth of the hole increases slowly, indicating that the submerged environment has a certain hindering effect on cutting. After the jet contacts clay, the deepening of the hole increases quickly. However, after 800 μs, the growth rate gradually decreases and tends to be stable due to the dissipation of waterjet particles. For observation points 1 and 4, the cutting depth increases first and then decreases. Initially, the change trend is similar to observation points 2 and 3, but then, due to the uplift of the waterjet particles, the particle displacement is relatively upward.

Figure 4 shows the equivalent stress on clay under waterjet cutting at 200 m/s velocity and 460 μs. According to the figure, the surface stress of clay mainly lies in the direct contact area between waterjet and hard clay. Because of the deflection of the streamline of water, the contact surface between waterjet and clay is increased, so the stress influence area is much larger than the width of the waterjet. The stress wave on the clay surface is propagated to the

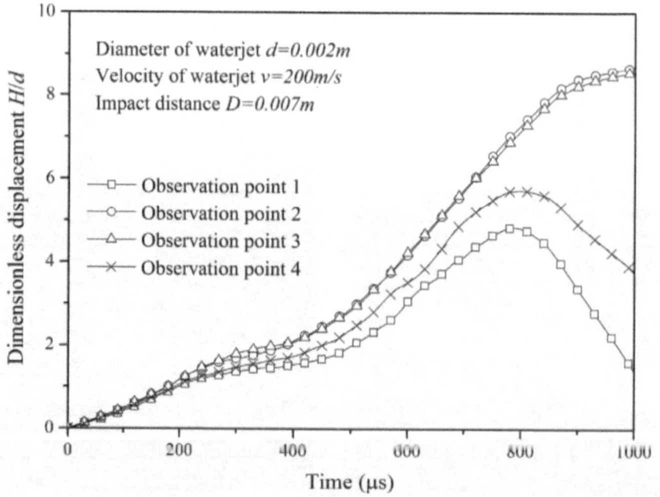

Figure 3. Variation of the dimensionless displacement *H*/*d* with time at different observation points ($V = 200$ m/s, $D/d = 3.5$).

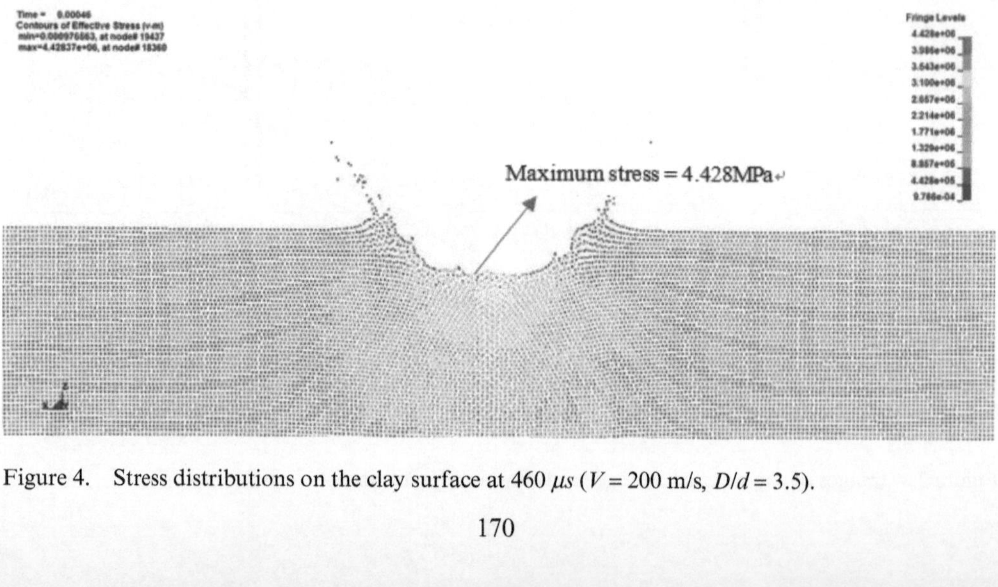

Figure 4. Stress distributions on the clay surface at 460 μs ($V = 200$ m/s, $D/d = 3.5$).

circumference in a spherical form around the contact point. There is an obvious attenuation of stress wave during propagation, which is caused by energy dissipation. Meanwhile, the stress on the clay in the vertical direction is significantly greater than that in the horizontal direction, because the clay particles below the jet are subjected to direct compression, making the deformation rate of the clay greater in the vertical direction than in the horizontal direction. Figure 5 shows the stress–time curves of particles at dimensionless distances (*x*/*d*) of 0.25, 2.5, and 7.5 away from the clay center in x direction. The curves show that the stress value of clay surface decreases with time and that the stress peak at distance 2.5 appears later than that at 0.25. The stress peak at distance 0.25 is approximately 6.5 MPa and that at distance 2.5 is 4.8 MPa. There is almost no peak value at distance 7.5, which shows that the stress wave has indeed undergone attenuation during propagation.

4.2 *Influence of waterjet parameters*

Figure 6 shows the variation of dimensionless cutting depth (*h*/*d*) with time at different velocities. The curves show that the variation trends under different velocities are similar in general. The first stage is the interaction of jet and submerged water environment, during which the

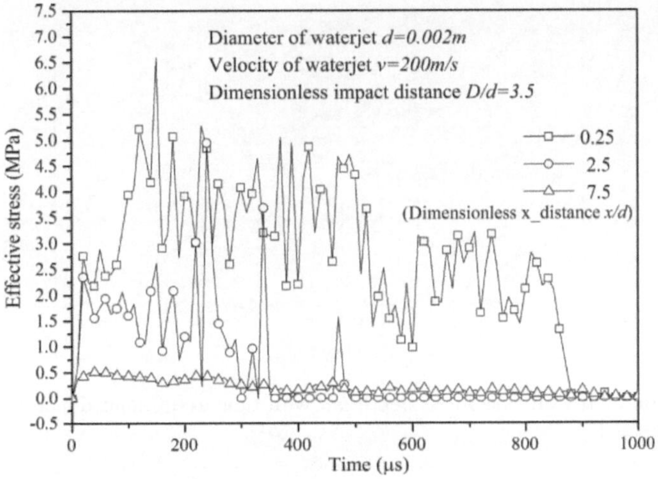

Figure 5. Variation of the effective stress with time at different distances ($V = 200$ m/s, $D/d = 3.5$).

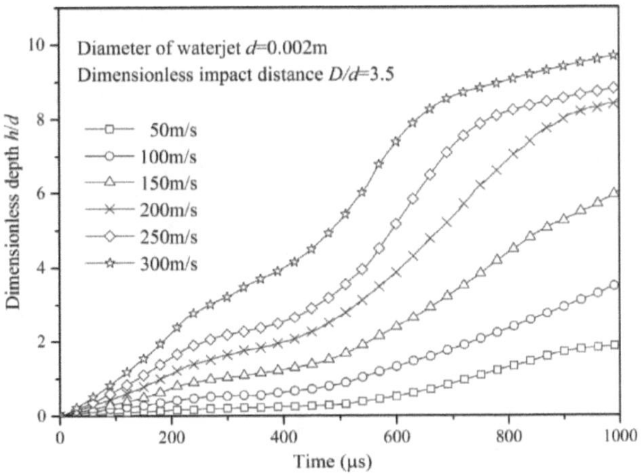

Figure 6. Variation of the dimensionless depth h/d with time at different velocities ($D/d = 3.5$).

upper surface of clay begins to sink under the squeezing of upper particles. At low velocities, the effect of submerged water on cutting is evident, whereas at high velocities, the cutting time is so short that the influence is negligible. The second stage is direct contact between waterjet and clay, during which a cutting hole is formed under the stress loading. In this stage, the cutting velocity is high, and at the same cutting time, with the increase of jet velocity, the cutting depth increases. In the third stage, the increasing trend of cutting depth gradually slows down. This is because the bottom of the hole formed by waterjet cutting will play the role of water cushion. Meanwhile, the backflow of waterjet occurs and the energy loss in the process of cutting reduces the impact pressure, which decreases the cutting efficiency gradually. Figure 6 also shows that, when the flow rate exceeds 200 m/s, the increasing range of cutting depth with velocity decreases. Therefore, in dredging engineering, we can choose appropriate jet pressure according to actual situation in order to reduce energy consumption and increase economic efficiency.

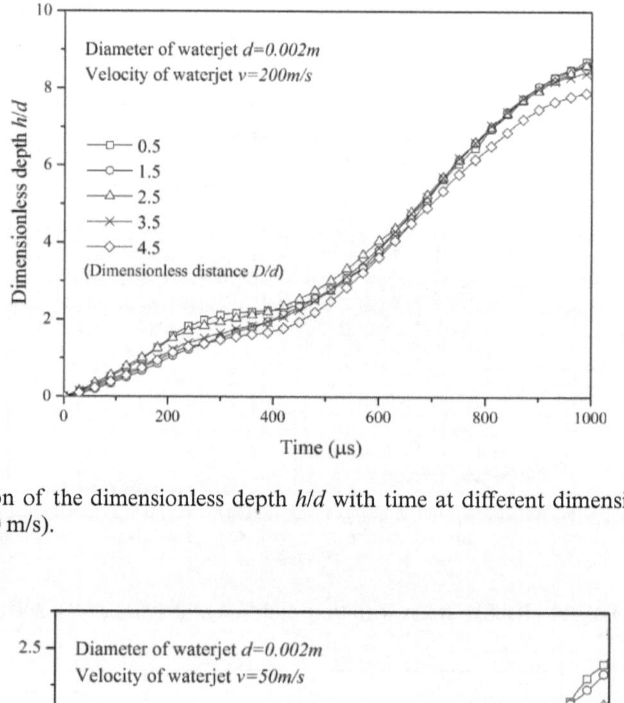

Figure 7. Variation of the dimensionless depth *h*/*d* with time at different dimensionless impact distances *D*/*d* (*V* = 200 m/s).

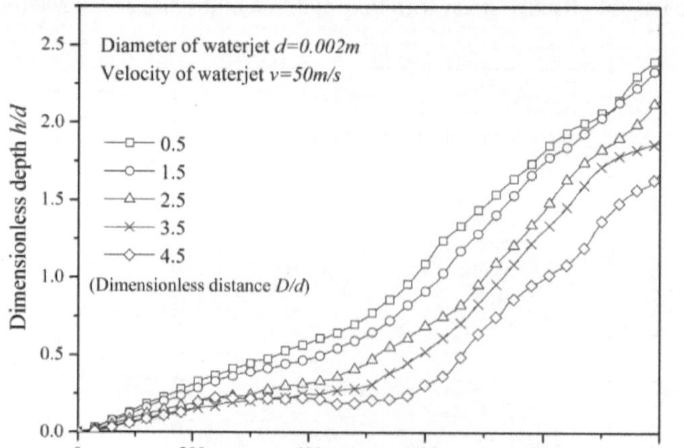

Figure 8. Variation of the dimensionless depth *h*/*d* with time at different dimensionless impact distances *D*/*d* (*V* = 50 m/s).

Figures 7 and 8 show the variation of dimensionless depth (*h*/*d*) with time under different dimensionless distances (*D*/*d*). The cutting velocity is 200 and 50 m/s in the figures, respectively. From the figures, the dimensionless depth *h*/*d* decreases with increasing impact distance, but the decreasing trend is smaller. This is because the waterjet needs to reach the nozzle through the nozzle and pass through the water environment before reaching the water–clay interface. The velocity of the waterjet decreases, and the time it takes to reach the interface is longer. In addition, the streamline of waterjet will spread during cutting, and the angle between waterjet and clay surface becomes larger, so that the impact pressure of waterjet decreases and the cutting depth is gradually reduced. Comparison of the two figures shows that, with the increase of jet velocity, the effect of cutting distance on cutting depth decreases gradually, which shows that the hindrance of the submerged water environment to waterjet cutting is negligible, whereas the waterjet velocity is sufficiently large.

5 CONCLUSION

This study used the SPH method for simulating the hard clay cutting mechanism of high-pressure waterjet. The simulation results were analyzed from three aspects, namely the cutting hole, the stress wave effect, and waterjet parameters. The numerical results of this study were compared with the present experimental results. From these results, the following conclusions can be drawn:

1. Submerged water environment and clay are a major hindrance to the waterjet cutting process, which leads to the gradual decrease in the cutting ability of the waterjet and in the vertical erosion radius. This will eventually form a "V"-shaped erosion hole with a maximum cutting radius of about 10 times the width of the waterjet.
2. During the cutting process, the stress on the clay surface is localized and mainly concentrated in the direct contact area between the waterjet and the clay, and the stress wave propagates from the center of the clay surface in the form of spherical wave. The stress wave is attenuated because of the energy dissipation during the cutting process. As the particles beneath the clay are directly squeezed by the waterjet, the deformation rate of the clay is greater in the vertical direction than in the horizontal direction.
3. Meanwhile, the higher the jet velocity, the higher the cutting depth. However, when the cutting velocity is large, the growth rate of the cutting depth decreases with increasing velocity. Therefore, to reduce energy consumption, more suitable waterjet pressure should be chosen according to the actual construction process.

REFERENCES

Bui H.H., Sako K., Fukagawa R. 2007. Numerical simulation of soil-water interaction using smoothed Particle Hydrodynamics (SPH) method. *Journal of Terramechanics*, 44:339–346.

LEWISBA. 2004. Manual for LS-DYNA Soil Material Model 147 [R]. *Department of Transportation: Federal High Way Administration*, 2004:10–13.

Ma L., Bao R.H., Guo Y.M. 2008. Water jet penetration simulation by hybrid code of SPH and FEA. *International Journal of Impact Engineering*, 35:1035–1042.

Si H., Xue Y.Z. 2016. Numerical analysis for stress wave effects of rock broken under pulse jets. *Journal of Vibration and Shock*, 35(5):146–152.

Song Z.G., Cheng J.M., Liu F. 2009. Numerical simulation for high-pressure water jet breaking rock mechanism based on SPH algorithm. *Oil Field Equipment*, 38:39–43.

Wang C., Wang Y.Q., Peng C., Meng X.N. 2016. Smoothed particle hydrodynamics simulation of water-soil mixture flows. *Journal of Hydraulic Engineering*, 142(10):04016032.

Yang N.H., Ni F.S., Sun D.D. 2006. Development of high pressure water jet technology and its application in Dredging Engineering. *China Plant Engineering*, 6(10):67–68.

Zhang H., Ni F.S., Gu L., Zhang D.Y. 2015. Comparison of ALE and SPH method for simulation of process of breaking soil with High-speed jet scouring. *Water Resources and Power*, 2015(11):75–78.

Hydraulic Engineering V – Hong, Liu & Xie (Eds)

Analysis of output and power for the dredger's cutter head under different axis angle conditions using the distinct element method

Jingzhao Hu & Ting Lin
CCCC National Engineering Research Center of Dredging Technology and Equipment Co. Ltd., Shanghai, China

Peng Jin & Liquan Xie
College of Civil Engineering, Tongji University, Shanghai, China

ABSTRACT: Cutter head is the main part of the dredger. The axis angle between the cutter head and the earth plays an important role in power consumption and production during the soil cutting process. Therefore, research on the relationship between the driving power and production of the dredger at different axis angles of the cutter head is important for the design and optimization of dredging equipment. This paper presents the analysis of the driving power and production for the general-type cutter head of a 3500 m^3/h dredger. For the numerical simulation of the dredger, a three-dimensional distinct element method is used to determine the efficiency of the rotating cutter head at axis angles of 15° and 45°. In addition, the productivity at these angles is discussed and compared.

1 INTRODUCTION

Cutter suction dredger is the key equipment in the dredging field, which is widely used for maintaining inland river navigation channels and port facilities. According to China Dredging Association, there were 93 cutter suction dredgers interiorly in 2008, with a total installed power of 232.6 thousand kilowatts. The dredging system of the cutter suction dredger consisted of an excavating subsystem, a transport subsystem, and a walking mechanism, which includes a cutter head, a ladder, a pump (underwater pump and inboard pump), pipeline, and a spud carrier, as shown in Figure 1, all of which support transporting soil from the seabed to the land.

As the main part of the dredger, the cutter head cuts the soil with teeth by self-rotating and swinging of the ladder, whose capability is directly associated with the productivity of the dredger. Therefore, studies on improving the cutting capability of cutter head are always inevitable for a dredging engineer. Because of the three-dimensional structural complexity of the dredger cutter head, most of the numerical studies were carried out on the basis of the Evans cutting model (Verhoef, 1997) and two-dimensional cutting theory (Van Os, 1997), in which the movement of the cutter is of two types, namely translational motion and rotational

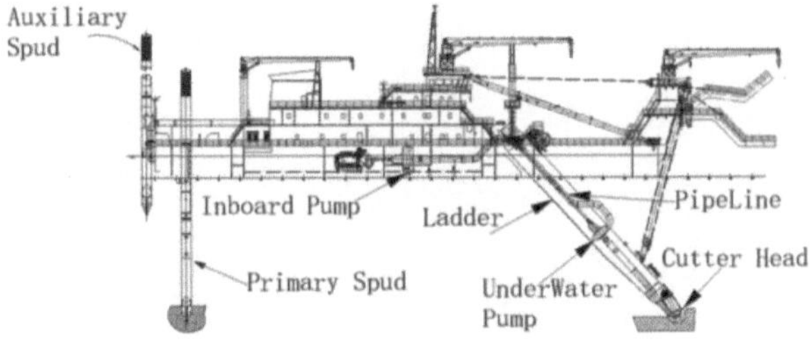

Figure 1. Profile of the 3500 m^3/h cutter suction dredger.

motion. This makes it easier for calculating the force exerted on one cutter tooth (Miedema, 1984) and considering the effects of the soil cutting thickness and pore water pressure (Van Os and Leussen, 1987). Moreover, many experiments have been conducted on saturated soil (He and Vlasblom, 1998) and unsaturated soil (Aluko and Chandler, 2004). However, it is still difficult to calculate the power and output of the whole cutter head.

In this study, numerical simulation based on distinct element method of YADE (Particle Flow Code) was conducted to study the power, production, and cutting efficiency of cutter head at axis angles of 15° and 45°.

2 DREDGING WORKING CONDITION

At present, most cutter suction dredgers are equipped with a fixed cutter head, so the axis angle cannot be adjusted between the cutter head and the earth. During dredging, the ladder is lowered to the seabed and the cutter head is used to remove soil from its original place and blend it with surrounding water. In this process, two important movements of dredger are swinging and advancing. Swing movement control has a remarkable impact on the output of dredging operations. At the stern of hull, there is a spud, one end of which is steadily plugged into the soil underwater. The dredger and the cutter head swing around the spud when the two winches on the ladder rotate in opposite directions. At the end of swing, the hull advances one step with the pushing from the spud carrier. When the carrier cylinder reaches the maximum stroke, the spud goes out, the other spud is plugged into the soil, and the carrier comes back to the original position. Then, a new process begins repeatedly.

Because of the difference in the dredging depth and the ladder angle, the axis angle between the cutter head and the earth is varied under different working conditions. Therefore, when the cutter head excavates the soil, the driving power in the teeth and arms of the cutter head is changed, indicating the influence on the efficiency and production of the cutter head.

3 VERIFICATION OF NUMERICAL SIMULATION

First, we compared numerical simulation of the cutter head and the actual conditions, which showed that simulation results were consistent with the actual situation.

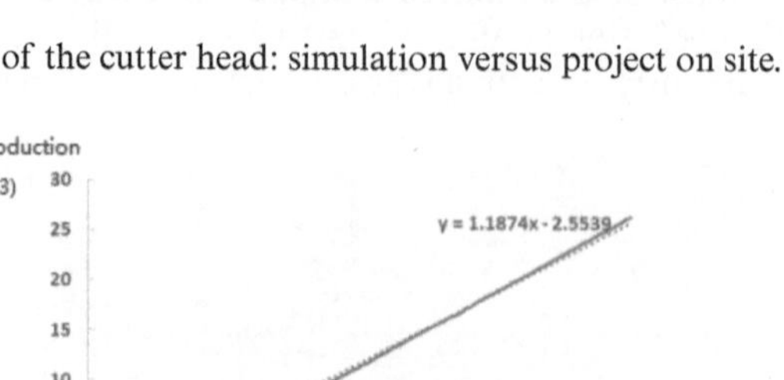

Figure 2. Driving power of the cutter head: simulation versus project on site.

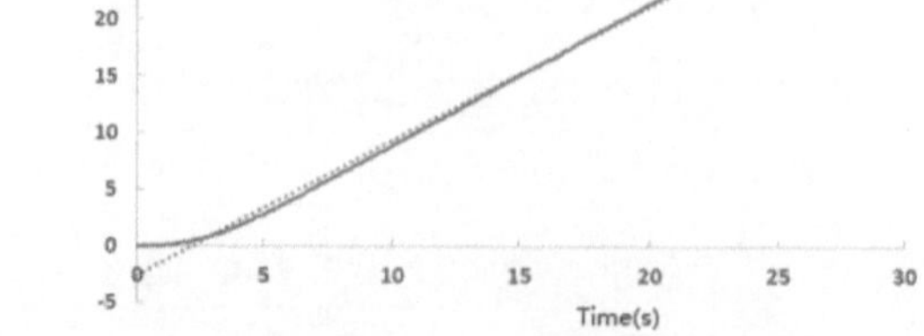

Figure 3. Output process of soil excavation for the simulation.

Figure 2 shows the power of the cutter head in the numerical simulation and the real project on site. Figure 3 shows the output process of soil excavation for the numerical simulation. For the cutter head to not touch soil spheres, when the rotation begins, the power of the numerical simulation slightly increases. After several seconds, the power of the cutter head in movement continuously increases and then gradually becomes stable.

From the results of Figure 2, we can find that the power of the cutter head in numerical simulation agrees well with the actual power in the project on site. Meanwhile, the fluctuation of the actual power is smaller than that of the numerical modeling. Calculation results showed that the average power of the actual condition is 275 kW, which is 1.04 times higher than the power of numerical simulation (263 kW). Figure 3 shows that the production of the simulation is 4274.66 m^3/h, which is 0.95 times higher than the production of rated power (4500 m^3/h).

Therefore, the numerical modeling can well simulate the actual situation. The result indicates that this numerical modeling method is effective.

4 CALCULATION MODEL

This study was carried out on the general-type cutter head of a 3500 m^3/h dredger, with a height of 1610 mm, an outer diameter of 2770 mm, and an inner diameter of 2300 mm. The cutter head has six arms, each containing eight or nine teeth (Figure 6).

On the basis of the software YADE, the numerical simulation analysis was conducted on the cutter head with a working condition of 26 rpm rotating velocity, 0.26 m/s swinging speed, and fine silt soil. The power and production were analyzed at cutter head axis angles of 15° and 45°, respectively, in the simulation experiment, as shown in Figures 4 and 5.

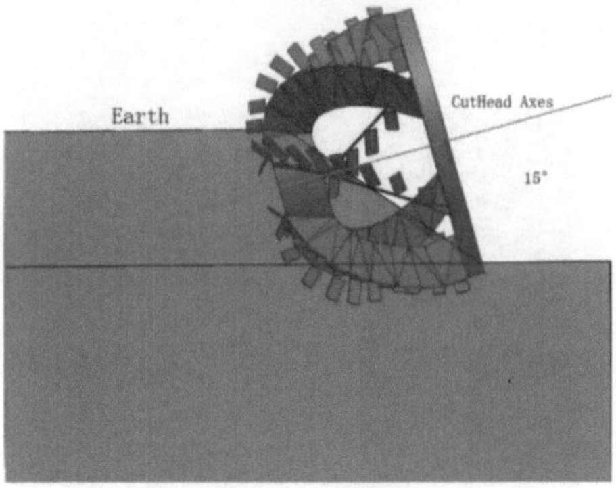

Figure 4. Cutter head under the 15° axis angle condition.

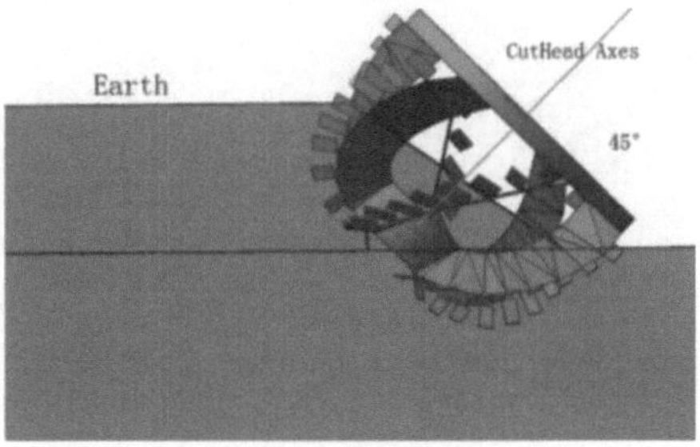

Figure 5. Cutter head under the 45° axis angle condition.

According to the drawing of the general-type cutter head, its numerical model was built with the software YADE, with a 1:1 scale of prototype. The model was painted with different colors to visually differentiate the parts of the cutter head. The cutter teeth were painted yellow and green, and the cutter arms were painted yellow and red. The YADE model of cutter head was built, which is shown in Figure 7.

The required driving power of every cutter tooth depends on many factors in the dredging engineering. In the analysis, the driving power of the ith cutter tooth (Pi) can be given by

$$P_i = F \times \frac{2\pi n}{60} \times d_i \tag{1}$$

where P_i is the power of the ith cutter tooth (kW); F is the equivalent tangential force exerted on the ith cutter tooth (kN); n is the rate of autorotation (rpm); and d_i is the distance from the ith cutter tooth to the axis wire of the cutter head (m). Accordingly, the total required power of the cutter head can be given by

Figure 6. Prototype of the dredger cutter head.

Figure 7. YADE model of the cutter head.

Figure 8. Complete model of the cutting simulation: the cutter head ran from one end to the other end of the soil bed and took away the loosened soil by the cutter teeth.

$$P = \sum_{i=1}^{m} P_i \tag{2}$$

The model of the cutter head was added into the seabed model at the axis angles of 15° and 45°. The complete model of the cutting simulation was assembled together after considering the interactions between the cutter teeth and the bed soil, as shown in Figure 8.

The simulation can be set under a specified condition. The initial location of the cutter head was at one end of the seabed model. When the cutter head swings to the other end at a specified speed, the soil within the range of the cutter will be cut, loosened, and taken away.

The time of each cutting process was 23 s, and the length of the cutting trail was 6 m. The output of the cutter head V might be given by the volume of the soil spheres loosened and taken away:

$$V = \frac{\sum_{i=1}^{n} v_i}{1-p} \tag{3}$$

where v_i is the volume of a single ball and p is the porosity of soil model.

5 CALCULATION RESULTS

5.1 *Power of the cutter head under two axis angle conditions*

The results of driving power of the cutter head under two axis angle conditions were plotted, as shown in Figure 9. It is evident from the figure that the driving power curves grow continuously and gradually become stable after several seconds of movement of the cutter head. Overall, the driving power of the cutter head at the 15° angle is lower than that at the 45° angle. In the steady state, the average driving power of the cutter head at the 15° angle is 272.9 kW and at the 45° angle is 374.8 kW. The result indicates that the cutter head under the 45° axis angle condition needs more driving power to cut the soil.

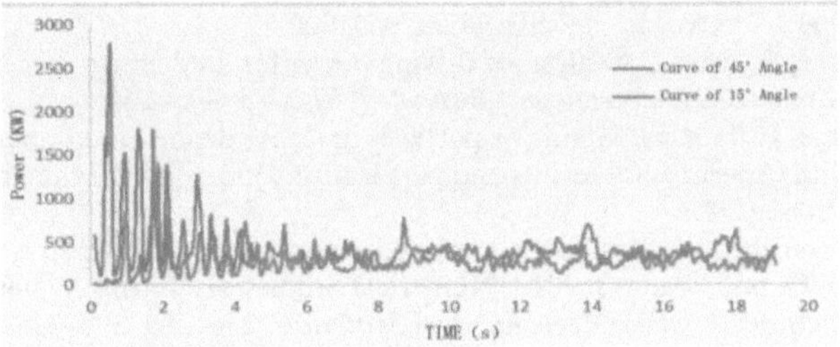

Figure 9. Driving power of the cutter head: 15° axis angle versus 45° axis angle.

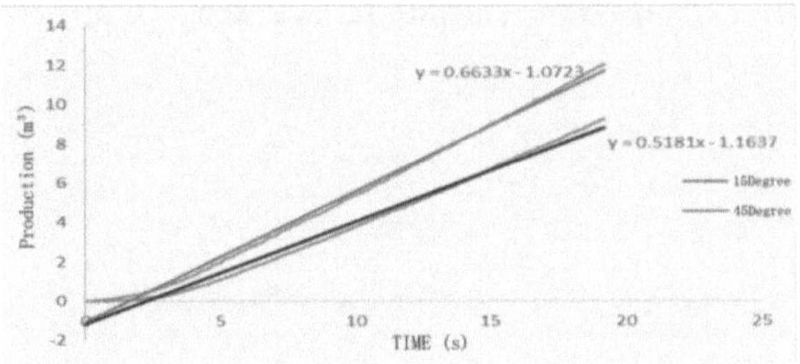

Figure 10. Output process of soil excavation in the simulation.

5.2 *Output of soil excavation under two axis angle conditions*

The output–time curves of soil excavation under two axis angle conditions are plotted in Figure 10. The trend lines of the curves are also depicted in the figure. The slope of the trend line can be considered as soil excavation volume per unit time. According to the output process of soil excavation in Figure 10, the outputs of the cutter head at the 15° and 45° axis angles are 1865 and 2387 m^3/h, respectively.

6 CONCLUSION

The numerical simulation of driving power and output for the cutter head under two axis angle conditions produced an interesting result, indicating that the ratio of output to power is different for two axis angle conditions. According to the computational result, the ratio of output to power under the 15° axis angle condition is 6.834, which indicates that 1 kW power contributes 6.834 m^3/h output, and that under the 45° axis angle condition is 6.369, which indicates that 1 kW power contributes 6.369 m^3/h output.

Therefore, from the numerical simulation results, some important conclusions can be summarized:

For the general-type cutter head, both the average driving power and the average output are lower at the 15° axis angle than at the 45° axis angle.

The reason for this can be explained from Figures 4 and Figure 5, which show that the number of cutter teeth participating in cutting soil is less at the 15° axis angle than at the 45° axis angle.

Therefore, the productivity of the general-type cutter head is higher at the 15° axis angle than at the 45° axis angle.

REFERENCES

Aluko, O.B., Chandler, H.W. 2004. Characterisation and Modelling of Brittle Fracture in Two-dimensional Soil Cutting. *Biosystems Engineering*, 88(3): 369–381.

He, J. & Vlasblom, W. 1998. Modelling of saturated sand cutting with large rake angle. *15th world dredging congress*, June 1998, Las Vegas, Nevada, USA: 905–920.

Liang, X., L. Xie, J. Hu & L. Zheng, Analysis of driving power for dredger cutter head based on distinct element method. *Advanced Engineering and Technology II 2015: 375–380.*

M. Kwapinska, G. Saage, E. Tsotsas. Mixing of particles in rotary drums: A comparison of discrete element simulations with experimental results and penetration models for thermal processes [J]. *Powder Technology*, 2006(161):69–78.

Miedema, S.A. 1984. The cutting of densely compacted sand under water. *Terra et Aqua*, (28), 4–10.

Qin, S., Xie, L., Hong, G. and Wang, J. 2013. Numerical Studies of Cutting Water Saturated Sand by Discrete Element Method. *Applied Mechanics and Materials*, 256–259, 306–310.

Van Os, A.G. 1997. Behavior of soil when excavated underwater. International Course Modern Dredging, Hague, The Netherlands, 1997, 135–141.

Van Os, A.G., Leussen, W. 1987. Basic jesearsh on cutting forces in saturated sand. *Journal of Geotechnical Engineering*, 113(12), 151–159.

Verhoef, P.N.W. 1997. *Wear of rock cutting tools* [M]. Taylor & Francis Ltd.

Hydraulic Engineering V – Hong, Liu & Xie (Eds)
 ISBN 978-0-8153-7471-8

Study on prestressed grid anchorage technology for slope and landslide treatment

Ronglu Wang, Yan Zhao, Yanxia Liu & Zhibin Liu
China Institute of Water Resources and Hydropower Research, Beijing, P.R. China

ABSTRACT: Prestressed grid anchorage is a new and reliable engineering technology that has a broad application in slope and landslide treatment. This technology uses pressure-dispersion tensile reinforcing bars, in which the wires extending from anchor holes are combined with the relative wires from all directions in the middle position and then tensioned together. Under this condition, external anchor piers are not necessary, and it can be tensioned repeatedly. Both the grid and the node show a plum-type layout. The construction of this composite slope anchor structure is simple, thereby overcoming the disadvantages of cast-reinforced concrete frame beams, such as easy cracking, easy deformation, and complex construction, and achieving a safe anchoring effect.

1 INTRODUCTION

With the recent rapid development of infrastructure construction in China, slope treatment has become more and more important in roads, railways, dams, and other maintenance management projects. In recent years, with the occurrence of various types of slope slumping incidents, the demand on favorable slope reinforcement has increased rapidly. Overall, compared with other countries, slope reinforcement techniques, including survey, design, construction, safety monitoring, and forecasting are less in China. As to the comprehensive administration combining slope reinforcement with green environmental protection, there are still many urgent practical problems needing to be studied and solved.

For most rock slopes, discontinuous surfaces in rock mass, especially joints nearly parallel to slope surfaces, are their characteristics. The application of prestressed slope anchorage exerts stress on the sliding surface to increase its resistance forces. Sometimes, supplementary drainage technology is applied in order to ensure that excessive pore water pressure will not occur on the potential slip surface. Prestressed anchor segment should be set away from the sliding surface as far as possible, and the unstable slope should be anchored in deep, stable rock. Thus, prestressed anchorage reacts to the slope rock mass, increasing the effective stress on the sliding surface (Zheng YR, Chen ZY, Wang GX 2007). Typically, reinforced concrete frame beams are cast on slope surfaces, and prestressed anchorage is generally used for fixation at the nodes. Shotcrete sprayed on steel net (across the beam between the two rows of an anchor cable) can prevent rock falling from the original slope. The disadvantages of this approach are unreasonable stress system, construction difficulties, and high cost, which limit its scope of application.

Prestressed grid anchorage technology for slope reinforcement is based on pressure-dispersion reinforcing bars, which is recognized as the most scientific and rational technology worldwide (Liu ZB 1997). This anchorage technology has been widely used in the slope-supporting engineering. It increases the strength and self-stabilizing ability of the surrounding rock and reduces the load of supporting structures, as well as saves cost. Pressure-type cable is a new anchorage system. Compared with the traditional pull-type anchorage, its force transmission mechanism is more reasonable and the distribution of bond stress is more uniform (Fu W 1996). Meanwhile, the anti-corrosion performance and durability are significantly improved.

The extended chain of the anchor hole is combined with the relative chain that extends from all directions in the middle station. Thus, on the slope, it can form a criss-cross steel strand without bond-prestressed grid anchorage system. In this study, we will summarize the technology, characteristics, and prevention programs of prestressed grid anchorage.

2 THE PRESTRESSED GRID ANCHORAGE TECHNOLOGY

2.1 *Grid layout of the anchor structure*

Prestressed grid anchorage technology for slope reinforcement is based on the pressure-dispersion anchorage, a widely accepted approach. The wires extending from anchor holes are combined with the relative wires from all directions in the middle station and tensioned together. Therefore, a vertical, horizontal cross-strand without a bonded anchorage system can be formed on the slope. The cross-nodes of the grid appear as anchor holes of pressure-dispersion type or tension sides, showing a plum-type layout (Figure 1). According to the anchoring force of the slope stability, the number of strands of each anchor hole is 4 n (n is a positive integer, e.g., 1, 2, 3, ...). With the cross-symmetrical nodes, the tension applies prestressed load symmetrically. The number of steel plates can be increased or decreased, if needed, to adjust the load of strand at any time.

2.2 *Mechanical properties of the anchor grid structure*

Prestressed grid anchor structure can be simplified to a number of anchoring units, as shown in Figure 2. Two sets of orthogonal inverted U-shaped unbonded strands are anchored in deep and stable rock. Geotechnical cylinders that are located in the scope of four-hole anchors are firmly anchored to the deep rock with four beam strands. Prestressed grid anchor structure transmits anchored tensile load to the slope rock only in the anchor segment, anchor cable outlet corner, the places in which exposed strands get a cross-precast concrete bearing plate, and the tension side of the anchor plate. After bending 90°, four-beam vertical anchor cables turn in the direction of the docking level of ground dome parts and are then tensioned, which not only changes the direction of the force of anchor cables but also significantly increases the contact area between the anchor and the rock mass, so that the rock within the four-hole anchors can be restricted. The mechanical properties of the prestressed grid structures and the unbonded prestressed concrete structures are similar. Under the load case, the elastic technology can be used for design. Under the limit state, analysis and design can be carried out on the basis of the unbonded prestressed concrete structure.

From the effect of structure force, the effect of anchor tension balance sliding force is direct and visible. From the displacement effects of the rock, prestressed force can limit the deformation of the slope, and this effect is difficult to replace by other structures. It should be noted that the prestressed grid anchor structure of slope reinforcement is usually anchored

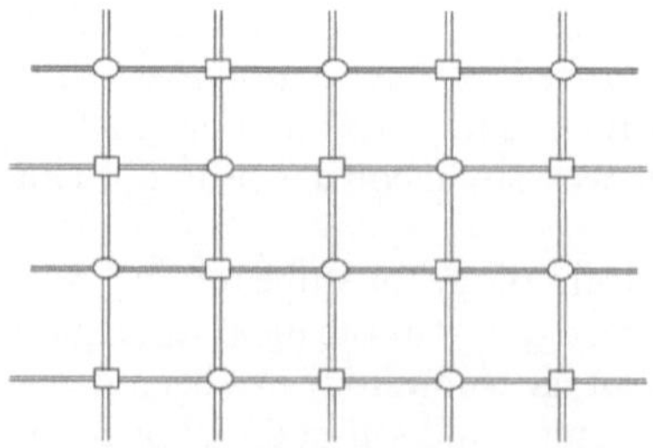

Explanation of symbols:○—Pressure - dispersion holes (of eight non-bonded strand)
□—repeated the tension of the tension side

Figure 1. Prestressed grid layout.

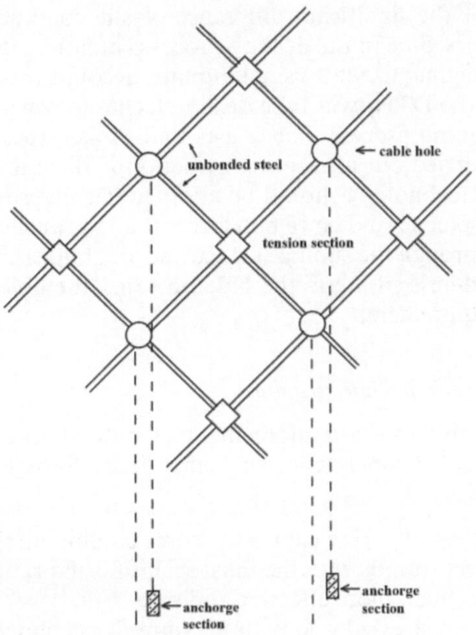

Figure 2. Anchor cell of the prestressed grid.

with a number of root anchors. Therefore, if little or random damages happen to the anchors, they will bring about only little effect on the slope stability. However, in a particular region (such as a fault zone or a weak interlayer area), when a few anchors lose effectiveness, the consequences would be severe. An anchor release load can be used to simulate the anchor cable failure on-site, and at the same time, the subsequent load increment of the adjacent anchor should be measured, to limit the initial stress in prestressed grid.

2.3 *Project of the grid anchor structure*

Prestressed girder structure involves the same anchorage construction processes as the prestressed anchor frame beam structure, including drilling, clear, knitting cable wear, and grouting, whereas it has no outside anchor pier, setting an additional tension end on the slope and adopting single symmetrical tensioning process. The required tensioning devices for this process have small size and lightweight, facilitating easy handling, movement, and operation on the high and steep slope. More importantly, the construction of poured reinforced concrete anchor pier and frame beam becomes unnecessary in this process, which can greatly shorten the construction period.

The construction of prestressed grid structures follows up–down order, which can guarantee safety. The construction procedures are as follows: clean-up of the slope according to the design requirements; drilling of plum-type anchor hole on tension side in vertical and horizontal intersection point; drilling holes and installing the anchor plates; placing the anchor into the hole and grout; spreading strands out anchor holes in all directions and putting them into the anchorage at the end of tension side; prestressing the anchors; straining circularly and symmetrically to the required tonnage; compensating tensioning and according to the feedback of slope stability warning systems; adjusting the load of the strand at any time; and finally, covering the exposed strands with polymer mortar or concrete to ensure anticorrosion performance and durability of stands.

It should be noted that drilling is a major factor affecting the cost and construction period of the anchorage process. In particular, there are many uncontrollable factors of the auxiliary work time, such as scaffolding, machine transfer, rig positioning, and the disassembly of the drill pipes. Because of the significant difference of site conditions (including workers), the length of auxiliary work time in the drilling process cannot be ignored. For most slopes, the rock mass is rather fragmentized; thus, taking into account the adverse effects of water on the mudding of rock, the DTH twin-tube drilling technology with normal air circulation should be used in the drilling process. For cracks and broken rock in the drilling process, pre-grouting should be carried out after pore formation if there is no collapse of the hole, whereas the tube-drilling technology should be adopted if collapse of holes and sticking is severe. The borehole diameter should be 10 mm larger than the anchor hole diameter, and the drilling depth should be through the sliding surface and 1 m longer than the anchor segment. Taking into account the debris silting in the hole bottoms, the hole depth should be 0.5 m deeper than the design requirements.

2.4 *Characteristic of the grid anchor structure*

This technology is applicable to the reinforcement of natural and excavated slopes. Compared with the past reinforced concrete lattice frame beam construction technology, it has the following characteristics:

1. The lightweight, flexible, pull string cable structures combining the prestressed grid consisting of unbonded steel strands with the most scientific and rational pressure dispersion make the slope reinforcement more effective and safe.
   - Overall—combining prestressed grid with pressure-dispersion anchorage as a whole.
   - Lightweight—the minimal additional deadweight load.
   - Flexible—adaptable to a variety of slope topographic and geologic conditions.
   - Tension—taking full advantage of high-strength, low-relaxation steel strands.
   - Small cross section of wiring—easy to protect due to the low thickness, and durability under double-layer protection.
   - Tension side—located in the center of the four-hole anchors, making the outside anchor pier unnecessary, and the tensioning repeatedly.
   - Simple and convenient construction—saving time, effort, and cost; matured technique and reliable quality.

   The advantage of unbonded strands is that any local load will be evenly distributed throughout the whole length, resulting in a smaller incremental load and allowing larger deformation.

   The working principles of prestressed grid anchor structure and the traditional frame beam anchorage are different, and their mechanical systems in the foundation are completely different, too. In order to play the full carrying capacity, the flexural framework beam should have a greater deformation, and thus, cracks are inevitable. If high-strength wires and ordinary reinforced concrete beams are arranged in series connection, the strength of the materials do not match, which will lead to material waste inevitably.
2. Prestressed grid and pressure-dispersion anchorage as a whole is more effective and has a safe anchoring effect. The combination of the pressure-dispersion anchorage cable strand and the unbonded prestressed grid makes the external anchor pier unnecessary, and it can be tensioned repeatedly because of the tension side located in the center.

3 TECHNICAL POINTS OF THE GRID STRUCTURE APPLICATION

3.1 *Selection of anchorage technology*

The effective anchor section of the prestressed anchor is the key forced section. It is critical to choose the anchorage appropriately according to geological conditions. On the basis of the transfer mode of loads acting on the anchor, the anchorage technology can be divided into tension type and pressure type. Dispersed pressure type is different from the general pressure type in that the load is not concentrated in one place when it passes to the cement paste, but passed to the surrounding cement paste dispersedly through unbonded strands and a few carrying bodies in the anchor section. Therefore, it is called pressure-dispersion type. In 1995, it was adopted in abutment reinforcement works of Sichuan Tongtou arch dam. Now, the technology of prestressed anchorage has been widely used in slope-strengthening projects.

3.2 *Curve design of anchor cable exports to the angle plate*

Overhang strand of prestressed anchorage is at the exit, so it must be through the turn angle plate to converse direction and thus form the strand curve segment. At the contact area between corner backing plate and the strand, prestressed steel strand is easily subjected to local hardening and excessive friction losses due to friction and lateral force squeezing action if corner plate curve has poor designed or inappropriate configuration.

The turning tensile strength test results of FIP standard strand are as follows: if the curvature radius of the strand satisfies $R_{tend} > \alpha\varphi_n \cdot N/n$, then the reduction of its tensile strength values should be 5%.

In the above equation, α is the coefficient; for smooth channel α = 20, for bellows pore, α = 40;

Φn is the strand nominal diameter (mm);
N is the total number of roots in a bunch of anchor strand;
n is the number of strand roots passing the radial component force.

Another function of the turn angle plate is passing concentrated load at the corner to the slope rock mass safely; therefore, cement mortar or concrete bond is required between the curve steel plate and rock.

3.3 *Cast concrete bearing plate*

A precast concrete pad is required to transfer the load at the interfaces between prestressed grid and the slope surface of the dome rock, and the strand in the same direction should be in the same horizontal position. In addition, exposed stranding should provide a smooth convex curve or a relatively smooth flat spherical curve, which is favorable for both load transfer and construction operation. On the basis of their shape, bearing plates are divided into square and rectangular plates. Outline dimension should be determined mainly on the basis of the bearing capacity of rock–soil layer of the slope and human carrying capacity, as well as the security, economy, and landscape.

3.4 *Design of cross-tensioning anchor plate*

The tensioning anchor plates should be set in the middle of the adjacent four-hole anchors, which are influenced by the load applied orthogonally and symmetrically in vertical and horizontal directions using single symmetrical tensioning technique. As a result, tensioning devices are small, lightweight, and easy to operate. The anchor load can be adjusted according to the change of prestress and the displacement of rock. The load can be changed with the increase or decrease of the number of steel pads in the process of application, which can help to avoid the effect of the clips on strand nicks.

3.5 *Durability of the structure*

Toward the end of the 20th century, two technologies of unbonded prestressed and cable-stayed bridge construction provided the conditions for the development of prestressed grid anchorage technology. With the generation and application of these technologies, rich experiences accumulated in the durability aspect of the exposed strand. First, the strand is filled with anticorrosion grease. Then, it is wrapped with high-density polyethylene pipe (HDPE), which is easily available in the market. Finally, polymer cement mortar (concrete) was covered on the unbonded strand. Thus, three protection layers were set on the surface of prestressing strand (grease, HDPE, and cement mortar), which provide reliable corrosion resistance and durability.

4 COMPREHENSIVE MANAGEMENT MEASURES

Because of its unique effect, simple construction technology, lightweight, and flexible and diverse structure types, prestressed anchorage technology is widely used in slope stability works. First, prestressed anchorage can provide sufficient active stabilizing force meeting the requirement of design to prevent the sliding of the slope. Second, prestressed grid plays the role of reinforcing shallow and deep rock mass, in addition to the general use of prestressed anchor friction pile to resist the sliding of the slope at the slope toe. Third, in accordance with the specific circumstances of the project, concrete spraying on steel can be applied to the closure of the exposed soil layer in the prestressed grid network to prevent the fall of shallow slope soil layer, rain erosion, and infiltration. Furthermore, hexagonal hollow bricks with quality evergreen vegetation planted in the center can be paved in the grid network. Fourth, drainage system is usually contained in the comprehensive management programs. Therefore, in the design of slope reinforcement, the supply of surface water and the change of groundwater must be clear. In general projects, comprehensive water drainage schemes, such as water interception on the slope top, water release along the slope, and water drainage at the toe, are commonly used and proved to be effective.

5 CONCLUSIONS

Prestressed anchorage, especially pressure-dispersion anchorage, has unique advantages, with the result that it has a broad application in slope reinforcement. However, anchor application alone cannot play a favorable role. In general, anchors are combined with other structures to form an integrated prestressed anchor anti-sliding structure, such as prestressed anchor pressure plate or pressure beam, prestressed anchor frame beam, prestressed cable retaining wall, and prestressed anchor sliding pile. The type of structure can be chosen according to the specific conditions of projects. The key point is that the designers should focus on the main design sectors and set up a holistic concept in the process of slope reinforcement. Only in this way can the slope treatment scheme be safe, reliable, technically feasible, economical, and reasonable.

Prestressed grid anchorage is a new type of composite anchor structure. It takes full advantage of the high-strength, low-relaxation, and unbonded strand, actively limiting or reducing the deformation of high and steep slopes. When evaluating in terms of its structural type, material saving, construction techniques, and treatment effect, it is innovative, advanced, and practical. It should be noted that a dynamic monitoring system should be established to monitor the stress and strain of prestressed anchorage and the dynamic characteristics of the reinforced slope, analyze the dynamic relationship between the grid anchor structure and the slope, evaluate the anchoring effect and consequently provide feedback for design, and make decisions about the adjustment of the anchor cable load or the application of other reinforcement measures. Finally, it is expected that this technology will be widely used in the treatment of slopes along the roads and railroads, the inlets and outlets of tunnels, as well as the banks of dams.

ACKNOWLEDGMENTS

The authors acknowledge the support of the National Key Research and Development Program of China (grant no. 2017YFC0404803) and the IWHR Basic Research Fund (No. SM0145B632017).

REFERENCES

Fu W. (1996), Techniques and Engineering Application of High-performance Prestressed Concrete, China Civil Aviation Press (in Chinese).

Liu Z.B. (1997), The study of the new anchorage method and system, Water Power (in Chinese).

Zheng Y.R., Chen Z.Y., Wang G.X. (2007), Engineering treatment of slope and landslide, China communications Press (in Chinese).

Hydraulic Engineering V – Hong, Liu & Xie (Eds)
 ISBN 978-0-8153-7471-8

Simulation and optimization of seepage control measures for gate dam

Q. Ren & G.Y. Li
Nanjing Hydraulic Research Institute, Nanjing, Jiang Su, China

ABSTRACT: The seepage characteristics of gate dam have become one of the important factors influencing its safety and normal operation. To reduce the leakage flow rate of dam abutment and improve the seepage stability of dam foundation, engineering design of impervious curtain and cutoff wall has been developed. Through establishing a deep site for a hydropower station by the overall three-dimensional seepage field analysis, the rationality of the seepage control measures for the dam body is evaluated and the possibility of optimization is demonstrated. The results of the study demonstrate that the seepage control effect of the impervious system is remarkable and the cutoff scheme is reasonable and effective. The cutoff wall extends to a certain length and depth, thereby having a significant effect on reducing the seepage stability of dam abutment and dam foundation as well as improving the seepage stability of the overburden layer.

1 INTRODUCTION

In Southwest China, the overburden layer is deep and the topography and geological conditions of the dam site area are complex. After the reservoir is filled, the seepage characteristics of the dam and the dam foundation become important factors that affect the safety and normal operation of the project (MAO X.Y., 2008; DENG C.C., 2011; HOU J.Q., 2012; QI L.J., 2012). Engineering measures including the design of impervious curtain and seepage cutoff wall are often used to reduce the leakage flow rate of dam foundation and dam abutment. In order to evaluate the reasonability of dam seepage control scheme of hydropower station, discuss the effect of different layout schemes of cutoff wall on the seepage field, and demonstrate the possibility of optimization, seepage control effects need to be finely simulated.

In this study, the overall three-dimensional finite-element model of the dam project on the deep overburden layer is established and the rationality of the seepage control measures and the possibility of further optimization are analyzed and proved, which provides reasonable suggestions for dam construction.

2 SEEPAGE ANALYSIS THEORY

For the seepage analysis of the gate dam hydropower station, evaluation of the seepage control effect is the main objective of long-term stable operation of the dam, thereby adopting steady seepage analysis.

According to Darcy's law, the seepage field in the dam area is regarded as a three-dimensional steady seepage field of nonhomogeneous anisotropic incompressible soil. The equation governing the seepage flow in the three-dimensional steady Darcy flow field is (MAO C.X. 2003).

$$-\frac{\partial}{\partial x_i}\left(k_{ij}\frac{\partial h}{\partial x_j}\right)+Q=0 \tag{1}$$

where x_i are coordinates, with $i = 1, 2, 3$; k_{ij} is the two-order symmetric Darcy permeability tensor, which describes the anisotropic permeability of rock mass; $h = x_3 + p/\gamma$ is the total head; x_3 is the position head; p/γ is the pressure head; and Q is the source–sink of seepage field.

The calculated boundary conditions are as follows:

$$h|_{\Gamma_1} = h_1 \tag{2}$$

$$-k_{ij}\frac{\partial h}{\partial x_j}n_i\Big|_{\Gamma_2} = q_n \tag{3}$$

$$-k_{ij}\frac{\partial h}{\partial x_j}n_i\Big|_{\Gamma_3} = 0 \quad \text{and} \quad h = x_3 \tag{4}$$

$$-k_{ij}\frac{\partial h}{\partial x_j}n_i\Big|_{\Gamma_4} \geq 0 \quad \text{and} \quad h = x_3 \tag{5}$$

where h_1, h_2 are head functions; n_i is the outward normal cosine of seepage boundary surface, with $i = 1, 2, 3$; Γ_1 = BG, CD is the first type of seepage boundary condition for a known water head; Γ_2 = GA, AF, FE, BC is the second type of seepage boundary condition for a known seepage flow; Γ_3 = GE is the seepage-free surface located between the seepage field and the imaginary area in the seepage basin, which is characteristic of nonlinear boundary; Γ_4 = ED is the seepage overflow boundary; because both the position of overflow point and the exact size of the entire escape are unknown at first, an iterative process is consequently required, which is a boundary nonlinear problem; and q_n is the boundary normal and the flow is positive.

3 CALCULATION MODEL

3.1 *General situation of engineering*

A hydropower station is located near the Tibet Autonomous Region Linzhi Zhong Da Upstream at a distance of approximately 3 km, which is a sluice and dam, with crest length of 505 m along the axis, crest elevation of 3154 m, dam height of 81 m, normal water level of 3152 m, and total storage capacity of 123 million 700 thousand cubic meters. The crest elevation is 3154.00 m, the maximum dam height of 81 m, crest width is 7 m, and crest length along the axis is 505 m.

The thickness of the layer covering the valley of the dam area is 5–127.6 m; the overall performance results in the left bank becoming the thinnest and the gradual increase in the thickness of the right bank. The overburden top-down hierarchy is as follows: ①drift gravel (Q_4^{pal}); ②-1 silty sand (Q_4^{eol}); ②coarse sand, fine sand (Q_3^{fgl}); ③water drift gravel (Q_3^{fgl}); ④sand gravel (Q_3^{fgl}); ④-1 medium coarse sand; ④-2 medium coarse sand; ⑤sand and gravel (Q_3^{fgl}).

The construction of the cutoff wall of depth 90 m on sand and gravel requires high technology, long construction period, and great efforts; furthermore, it is important to optimize the range and characteristic parameters of the cutoff wall. The following assumptions were

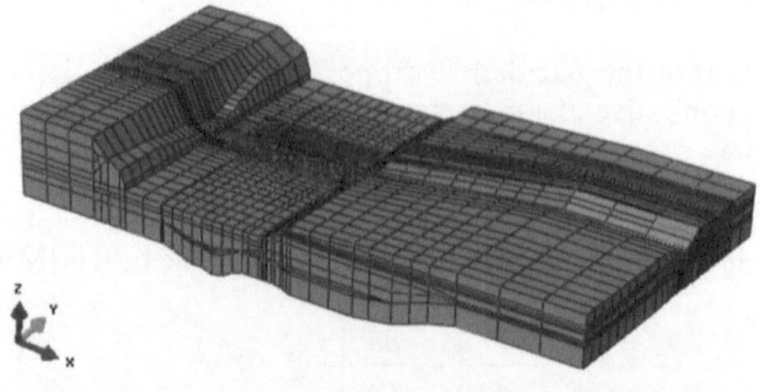

Figure 1. Entire three-dimensional finite-element grid model of the dam area.

made: sluice foundation, dam foundation, and dam abutment end with concrete anti-seepage wall; wall thickness, 1.0 m; and undetermined depth and extension length. In order to select the best cutoff wall length and depth, the characteristics of seepage field such as water head distribution and seepage flow under different schemes are analyzed.

3.2 *Finite-element model and boundary conditions*

The entire three-dimensional finite-element model of the dam area is shown in Figure 1. The whole model ranges from upstream boundary to the right guide wall extending upstream of 300 m, downstream boundary to tail water outlet extending downstream of 300 m, the left and right bank mountains along the dam axis extending outward from the 200 m model, and the lowest elevation covering the following 50 m.

The known water head boundary is the reservoir bank, the bottom of the reservoir, the river channel, and the intercepting boundary of the given groundwater level. The overflow seepage boundary is the slope of the left and right banks above the water level above the dam site. The impermeable boundary is the upper and lower sides and the left and right sides of the model. Two side intercept boundaries are partial boundary of a given groundwater level and the underside of the model (MAO C.X., 2003; LU L., 2006).

3.3 *Calculation parameters and calculation scheme*

According to geological generalization and field test results, the permeability coefficient of each covering layer and seepage control structure are shown in Table 1.

In order to understand the influence of different cutoff wall depths on the seepage field of dam foundation and dam abutment, the calculation scheme is formulated in Table 2. The finite-element mesh of cutoff wall and anti-seepage curtain is shown in Figure 2.

Table 1. Seepage calculation parameters.

Setting	Permeability coefficient cm/s	Slope range
Medium-coarse sand	1×10^{-2}	0.20–0.25
Gravel containing drift eggs	1×10^{-1}	0.07–0.10
Sand gravel	6×10^{-2}	0.10–0.15
Coarse sand containing floating eggs	2×10^{-2}	0.20–0.25
Gravelly medium-coarse sand	1.5×10^{-2}	0.25–0.35
Bedrock	7×10^{-5}	/
Concrete	1×10^{-7}	/

Table 2.

Scheme	Headwater level M	Downstream water level M	Extension of the cutoff wall length on the right bank	Cutoff wall depth m	Impervious wall form
1	3152.00	3121.3	150 m	Into the bedrock	Closed type
2			200 m	Into the bedrock	Closed type
3			300 m	Into the bedrock	Closed type
4			Into the bedrock	Into the bedrock	Closed type
5			150 m	50	Suspension type
6			200 m	50	Suspension type
7			300 m	50	Suspension type
8			Into the bedrock	50	Suspension type
9			200 m	30	Suspension type
10			/		

Figure 2. Finite-element mesh of the cutoff wall.

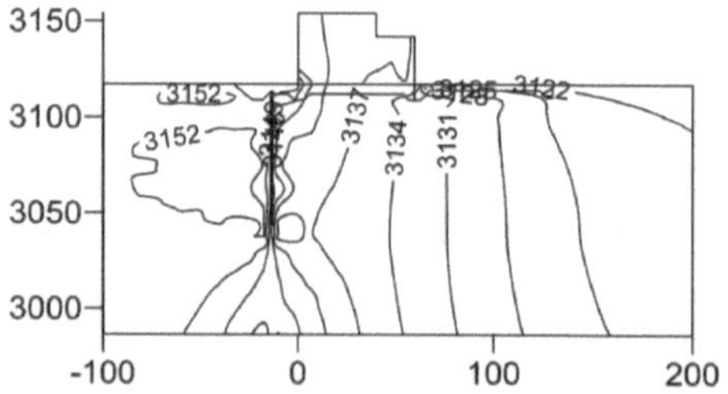

Figure 3. Head line contour of the sluice section.

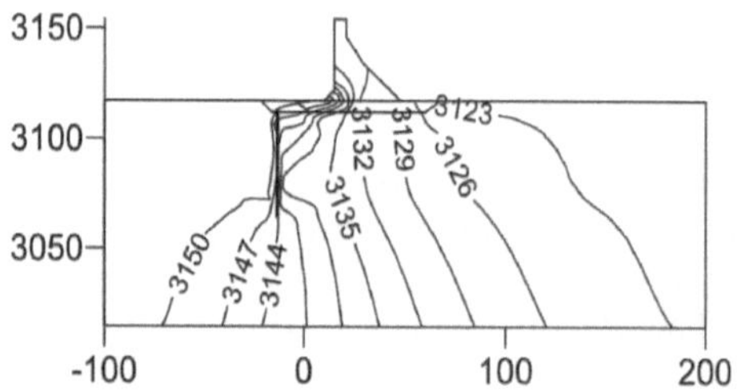

Figure 4. Head line contour of the gravity dam section.

4 CALCULATION RESULTS AND ANALYSIS

4.1 *Seepage field analysis of the typical section*

The total head line distribution of the spillway section of scheme 4 and the cross section of the gravity dam section during the normal operation of the reservoir are shown in Figures 3 and 4, respectively.

The contours of the total water head are densely distributed at the cutoff wall, which indicates that the impervious wall has a significant impact on water head dissipation. At the upstream side, the free surface of the seepage decreases rapidly in the dam body due to the floor of the gate covering layer of small permeability coefficient close to 4 orders of magnitude, thereby limiting the leakage flow rate, and the flow in the overburden layer around the internal diaphragm wall forms a stable free surface in the wall, close to the level and extends to the dam.

The head line's plane seepage of scheme 4 at an elevation of 3112 m is shown in Figure 5. The seepage field of the impervious curtain of the left abutment and the cutoff wall of the right bank are shown in the figure. It also shows the reservoir water level from dam foundation, anti-seepage curtain and cutoff wall of the right bank to the downstream, obvious seepage, downstream anti-seepage curtain, and gentle impervious wall infiltration, showing that the middle of the river is low and around the abutment is high trend. The head line away from the anti-seepage walls, dense distribution, and surface infiltration in the lower reaches of the impervious curtain and cutoff wall are formed suddenly from a distinct low head area near the downstream of the right, showing that the impervious wall effectively cuts head, thereby reducing the leakage flow rate of dam abutment.

4.2 *Gradient analysis of the key parts of dam seepage*

Table 3 presents the maximum hydraulic gradient in the internal anti-seepage wall at the dam axis 0+0. The larger the length of the cutoff wall on the right bank, the larger the average hydraulic gradient increase in the cutoff wall; the greater the depth of the cutoff wall, the greater the average hydraulic gradient increase in the cutoff wall, achieving the maximum value near the bedrock.

4.3 *Escape point elevation analysis*

The elevation of the escape points of different schemes is shown in Table 4. The analysis shows that the length of the cutoff wall extending on the right bank is 150 m. In the bedrock

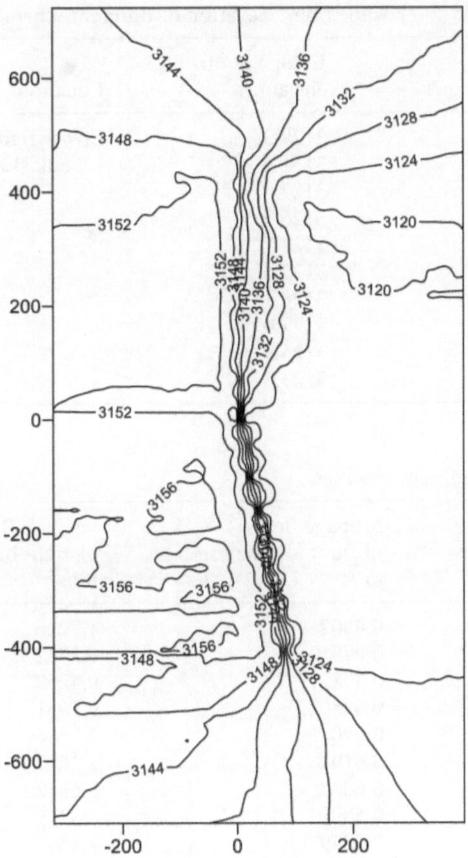

Figure 5. Plane seepage of the head line at an elevation of 3112 m.

Table 3. Key parameters of the seepage gradient of the dam cutoff wall.

Setting	Seepage slope	Location
1	19.30	Right dam end (dam axis 0 + 0)
2	22.21	
3	25.67	
4	29.13	
5	15.36	
6	17.06	
7	18.07	
8	18.98	
9	15.04	
10	/	

scheme (scheme 2), the escape point elevation is 3127.8 m, which is higher than the normal water level (3121.33 m) and the normal storage level (6.47 m). With the increase in the length of the cutoff wall on the right bank, the elevation of the escape point is reduced to different degrees. As for the scheme of reducing the cutoff wall depth to 50 and 30 m, the elevation of the escape point is increased to some extent compared with the impervious wall and the bedrock scheme.

Table 4. Escape point elevation of different schemes.

Schemes	Escape point elevation	Location
1	3129.3	Lower right bank slope
2	3131.6	
3	3127.8	
4	3126.9	
5	3124.7	
6	3122.6	
7	3128.1	
8	3127.4	
9	3125.9	
10	3123.8	

Table 5. Seepage flow of different schemes.

No.	Seepage flow of left bank m^3/s	Seepage flow of dam foundation m^3/s	Seepage flow of right bank m^3/s	Total permeability m^3/s
1	0.0029	0.4802	1.3226	1.8057
2	0.0029	0.4791	0.8742	1.3562
3	0.0029	0.4782	0.4258	0.9069
4	0.0029	0.4780	0.1051	0.5860
5	0.0030	0.6208	1.7516	2.3754
6	0.0030	0.6191	1.1577	1.7798
7	0.0030	0.6007	0.6842	1.2879
8	0.0030	0.5802	0.3319	0.9151
9	0.0034	0.7849	1.4563	2.2446
10	0.0051	0.9597	2.0066	2.9714

4.4 *Seepage flow analysis*

The calculation of seepage flow includes three parts, namely the downstream area of the left bank, the overburden area of the dam foundation, and the downstream area of the right bank. The seepage flow in each part of the working condition is shown in Table 5.

The infiltration rate in the left abutment is very small because of its good geological conditions, and the ratio of seepage flow to total seepage flow is larger for the dam foundation and the right bank with the existence of the deep overburden layer. It can be seen that the seepage flow around the right bank decreases as the depth of the cutoff wall increases on the right bank. Schemes 2, 3, and 4 reduce the range by 24.9%, 49.8%, and 67.6%, respectively, compared to scheme 1. As shown in Table 5, with decreasing cutoff wall depth, the seepage flow rate of the dam foundation and the right bank increases, and the rates of schemes 6, 9, and 10 are higher (31.2%, 65.5%, and 119.1%, respectively) than the rate of scheme 2.

Compared with all the working conditions, the depth of the cutoff wall on the right bank has a greater influence on the length of the cutoff wall. Despite the complex geological conditions of dam engineering, particularly the problems of thick covering layer and dam seepage control, the cutoff wall was extended to a certain length and depth on the right bank, blocking the entire river bed cover in two directions. Therefore, the truncated riverbed overburden layer could reduce the leakage flow rate of dam abutment and improve the seepage stability of dam foundation, thereby improving the covering layer. Overall, the length of the cutoff wall was extended to at least 200 m on the right bank and its depth reaching a maximum value near the bedrock. Therefore, the design of the cutoff wall met the requirements.

5 CONCLUSION

The results of the study demonstrated that the seepage control effect of the impervious system was remarkable during the operation period, and the seepage surface formed a depression at the downstream of anti-seepage curtain and impervious wall. The seepage pressure was effectively controlled, showing that the design scheme of the seepage control and the drainage system in the dam site area was reasonable and effective (ZHENG H.S., 2013).

Cutoff wall extended to a certain length and depth on the right bank, which blocked the entire river bed cover in two directions, thereby effectively reducing the leakage flow rate and improving the seepage stability of the overburden layer. From the economic and technological perspective, the length of the anti-seepage wall extended to at least 200 m on the right bank and its depth reached to the maximum value near the bedrock. Therefore, the design of the cutoff wall met the requirements.

REFERENCES

Deng C.C., Zhao J., Ren J. & Hong X. Seepage control measures for ancient river channel on deep overburden layer, *Engineering Journal of Wuhan University*, 2011, 44(2):161–165.

Hong X.P., Xu Q. & Yang Z.M. Seepage control measure optimization for Shanping ship lock foundation in Yellow River Engineering, *Journal of Wuhan University*, 2012, 45(5):584–589.

Lu L., Wang R.J. & Li Z.H. Study on Seepage of the Gate Dam on the Foundation of Deep and Overburden Layer with 3-D FEM. *Journal of Northwest Hydroelectric Power,* 2006, 22(5):9–12.

Mao C.X. Analysis and control of seepage calculation. Beijing: China Water power press. 2003.

Mao X.Y., Pan S.H. & Bai Z.X. Back analysis of initial seepage field of complex dam foundation at Shuangjiangkou Hydropower Station, *Rock and Soil Mechanics*, 2008, 29(z1):135–139.

Qi L.J., Liao H.S. & Li L.X. Influence analysis of seepage boundary conditions of anisotropy in rock-soil body of gate dam foundation. *Water Resources and Power*, 2012, 30(8):56–59.

Zheng H.S., Zhou S. & Liu W. Impact of cutoff wall depth on seepage characteristics of thick permeable layers of gate dam. *Water Resources and Power*, 2013, 31(9):76–79.

Hydraulic Engineering V – Hong, Liu & Xie (Eds)

Numerical simulation of the hydrodynamic response of regulation engineering in the Deepwater Navigation Channel of the Yangtze Estuary

T.L. Xu
School of Port and Transportation Engineering, Zhejiang Ocean University, Zhoushan, China

X.D. Mao
CCCC National Engineering Research Center of Dredging Technology and Equipment Co. Ltd., Shanghai, China

W. Chen
School of Port and Transportation Engineering, Zhejiang Ocean University, Zhoushan, China

C.P. Kuang
College of Civil Engineering, Tongji University, Shanghai, China

C.H. Mao, J.S. Zhang & S.J. Shi
School of Port and Transportation Engineering, Zhejiang Ocean University, Zhoushan, China

ABSTRACT: A depth-averaged 2D numerical model is set up to simulate the hydrodynamics of the Deepwater Navigation Channel of the Yangtze estuary due to the regulation engineering with MIKE21, which has been well calibrated and validated with the observed data of the tidal level, flow velocity and flow direction in the Yangtze estuary. The simulated results indicate that after the regulation engineering of the Deepwater Navigation Channel,1): The increase of tide amplitude at the lower reach, middle reach and fish mouth project part of the channel are about 0–0.2 m, 0.3–0.5 m and 0.2–0.3 m, respectively;2): The embankment and the spur dike are submerged during the flood tide. While at ebb tide, the embankment and the spur dike are exposed due to the decreasing of the tidal level.

1 INTRODUCTION

The Yangtze Estuary is the fourth largest river in the world in terms of sediment discharge, with an annual mean sediment discharge of approximately 0.486 billion tons before the 1990s and 0.2 billion tons after the 1990s. Nearly half of the fine-grained sediments from the Yangtze basin are deposited in the estuary and on the subaqueous delta, resulting in massive sand bars with a minimum depth of 6 m below the mean lowest low water level that seriously limit the feasibility of navigation. To improve the navigation conditions in the sand bar area, the engineering project Deepwater Navigation Channel (DNC) along the North Passage began in 1998 and was completed in 2011. Two dikes (south dike: 48 km long, north dike: 49.2 km long) and 19 groins (total length: 30 km) are built to increase the current speed and decrease the sediment deposition in the main navigational channel, which is designed to be 92.2 km long, 350–400 m wide, and 12.5 m deep (Li et al., 2016).

The DNC project is planned and implemented in three phases. However, since the completion of the second phase in 2004, a siltation problem in the middle segment of the navigational channel began to draw attention. Much greater amounts of deposited material needed to be dredged in routine maintenance to safeguard navigation than the original estimate of 30 million m^3/year. The annual amount of deposited material dredged from the DNC increased from 35.5 million m³ in 2005 to 60.0–80.0 million m³ after 2005. Silting along the DNC mainly occurs at the curve of the channel. Near this location, the amount of siltation has accounted for 70–80% of the entire DNC (Qi et al., 2005). Therefore, it is important to study the impact of regulation engineering in the Deep Navigation Channel of Yangtze Estuary.

The numerical simulation method has been widely applied to study the hydrodynamic response of regulation engineering in many coastal and estuarine areas. Ma et al (2013) applied the FVCOM model to study the physical mechanism of the migration of the sediment deposition in the Deepwater Navigation Channel of the Yangtze Estuary. Kuang et al (2014) applied Delft-3D model to simulate the hydrodynamics and sediment transport in the Yangtze Estuary. Xie et al (2010) based on the EFDC model to study the migration behavior of the fine silt particles of dredged sediment after dumping at different locations in the Deep Navigation Channel of Yangtze Estuary.

This study applies Mike21 model to set up a depth-averaged hydrodynamic model in the Yangtze Estuary in order to simulate the hydrodynamic change due to regulation engineering in the Deepwater Navigation Channel of Yangtze Estuary.

2 NUMERICAL MODEL

A depth-averaged two-dimensional (2D) model based on Mike21 Flow Model FM finite volume model, developed by the DHI, is established to study the hydrodynamic responses of regulation engineering in the Deepwater Navigation Channel of Yangtze Estuary. Mike21 Flow Model FM is based on a flexible unstructured mesh approach and has been used for applications in oceanographic, coastal and estuarine environments (DHI, 2011).

2.1 *Study area and computational mesh*

The flexible unstructured triangle mesh is applicable to the regime with three bifurcations and four river outlets in the Yangtze Estuary. The spatial extent of the model is from 29°N to 32.5°N in the N-S direction and from 120°E to 124°E in the E-W direction. The Jiangyin in the Yangtze Estuary and Cangqian in the Hangzhou Bay are specified as river boundaries. The computational domain is composed of an unstructured triangular mesh with 18,006 nodes and 33,656 elements with the largest mesh at the east ocean boundary and the smallest in the North Passage. The mesh size generally decreases from the open sea to the near shore, and the finest mesh appears in the river channels.

2.2 *Boundary conditions and parameter settings*

The open sea boundaries are specified as time-varying tidal levels derived from a calibrated model, covering all the China Sea and adjacent seas, whose open boundary is driven by the Global Tidal Model derived from 10 years of TOPEX/POSEIDON satellite altimeter data from 1992 to 2002. The wet and dry treatment for tidal flats is introduced into this model. The Courant Friedrich Levy (CFL) condition can be used as a guide for selecting the time step, and the self-adapting time step is in a range from 0.01 to 30 seconds to keep the CFL

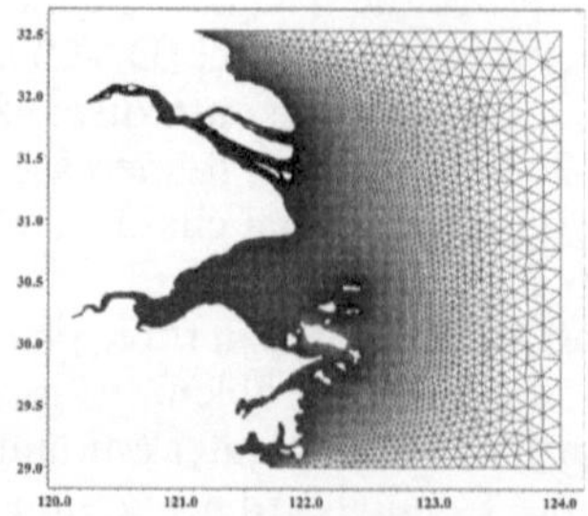

Figure 1. Computational mesh.

less than 1. The initial tidal level and velocity are defined as 1 m and 0 m/s respectively. The Manning number M in this study is in a range from 74 to 90.

2.3 *Boundary conditions and parameter settings*

The open sea boundaries are specified as time-varying tidal levels derived from a calibrated model, covering all the China Sea and adjacent seas, whose open boundary is driven by the Global Tidal Model derived from 10 years of TOPEX/POSEIDON satellite altimeter data from 1992 to 2002. The wet and dry treatment for tidal flats is introduced into this model. The Courant Friedrich Levy (CFL) condition can be used as a guide for selecting the time step, and the self-adapting time step is in a range from 0.01 to 30 seconds to keep the CFL less than 1. The initial tidal level and velocity are defined as 1 m and 0 m/s respectively. The Manning number M in this study is in a range from 74 to 90.

2.4 *Model verification*

The time series of hourly tidal level at Changxing, Hengsha and velocity at 4 stations (locations see the Fig. 2) from 06:00 May 5, 2004 to 06:00 May 6, 2009 are collected to validate the

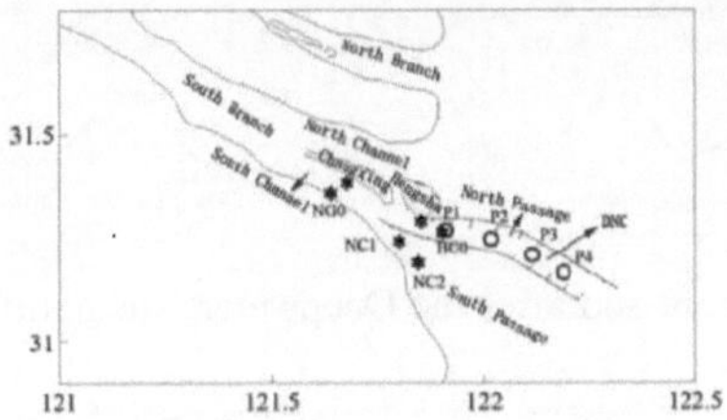

Figure 2. Locations of verification stations and selected points.

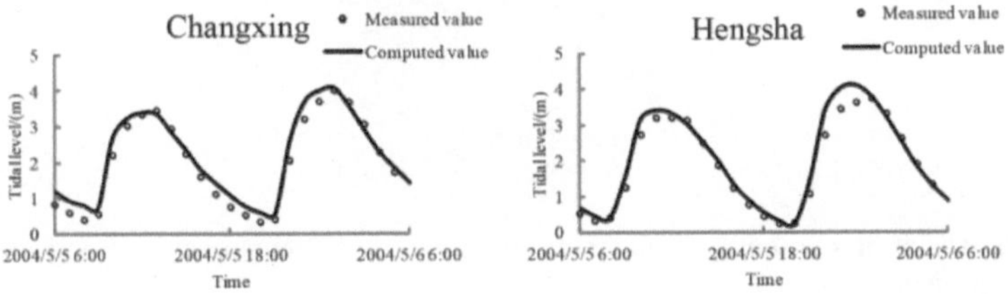

Figure 3. Verification of tide level.

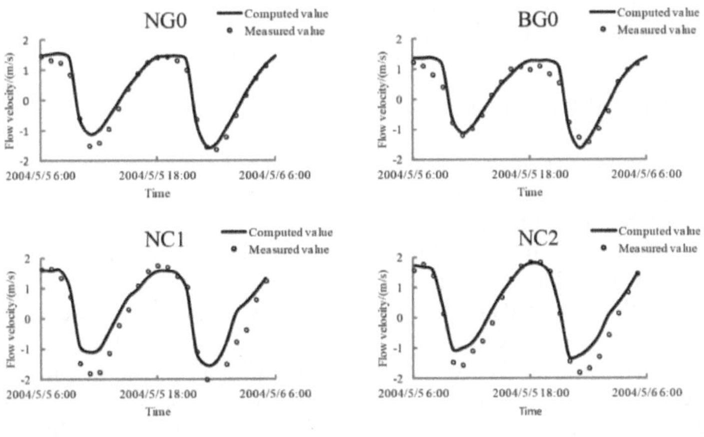

Figure 4. Verification of flow velocity.

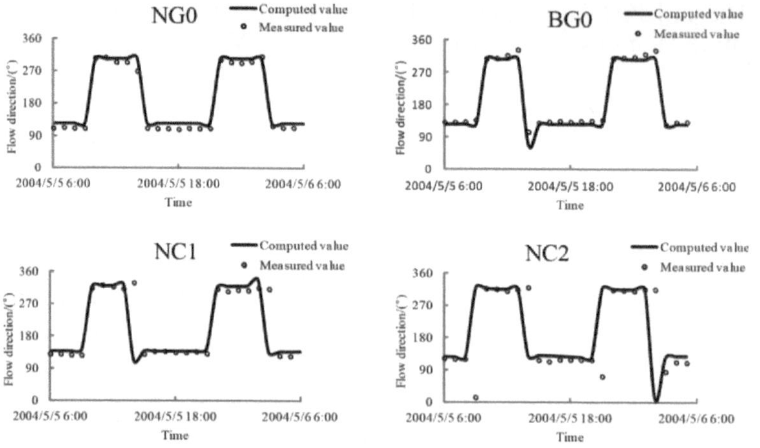

Figure 5. Verification of flow direction.

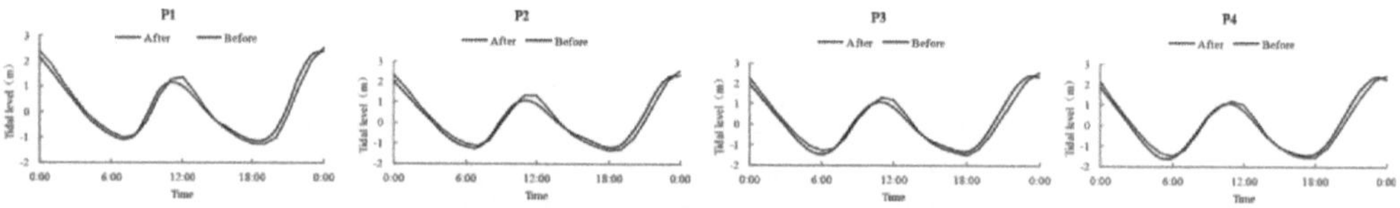

Figure 6. Tidal level process before and after the Deepwater Navigation Channel project.

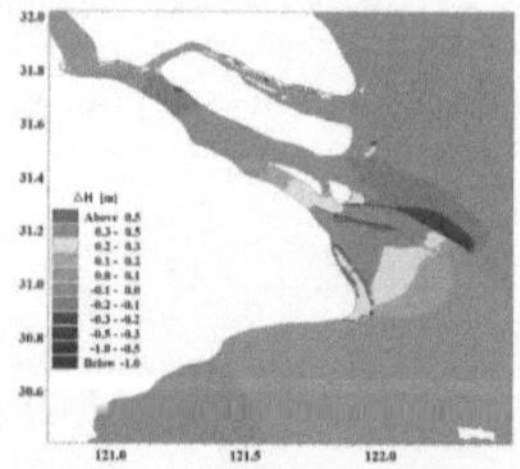

Figure 7. Tidal variation caused by deep-water channel renovation project in flood maximum (after project – before project).

simulations (Figs. 3–5). The computed values match well with the measured values. So this model can be used to simulate the hydrodynamic response due to regulation engineering in the Deepwater Navigation Channel of Yangtze Estuary.

3 RESULTS AND DISCUSSION

Four observation points are evenly arranged in the channel, the locations of which are shown in Figure 2 (represented by circle points). Figure 6 shows the tide curve of the four points within the channel before and after the deep-water channel regulation project. From the figure, the high tidal level and tidal amplitude in the channel increases obviously after the project, while the low tidal level decreased. Figure 7 shows the change of the tidal level after the project (at peak flood velocity): The increase of tide amplitude at the lower reach, middle reach and fish mouth project part of the channel are about 0–0.2 m, 0.3–0.5 m and 0.2–0.3 m, respectively. During the flood tide, the tidal level increases due to the constriction effect as the tide enters into the deep-water channel. The tidal level increases obviously until the vicinity area of Hengsha Island. However, the amplification at the upper reach and fish mouth project part decreases due to the connection of Hengsha channel. Near to the northern part of the northern embankment, especially for the area close to open sea, the tidal level decreases. The tide level decreases about 0.3–0.5 m. While the tide level increases at the southern part of the south embankment, the amplitude increases from the open sea area to the joint of north and south troughs.

Figure 8 shows the velocity variation of the four observation points before and after the deep-water waterway project. After the project, the average amplification of Pl to P4 are 10.45%, 4.32%, 27.72% and 20.25%, respectively. The maximum amplification appears at P3. Figure 9 shows the variation of the flow direction before and after deep-water channel regulation project. Obvious characteristics of rectilinear flow can be observed within the channel after the project. Figure 10 shows the flow field after the north and south dual breakwaters project. The embankment and the spur dike are submerged during the flood tide. While at ebb tide, the embankment and the spur dike are exposed due to the decreasing of the tidal level. The water flows backwards from beach to the trough, which would align and strengthen the ebb tidal current within the channel. Figure 11 shows the distribution of the velocity changes of flood and ebb tide caused by the projects (at peak flood and ebb velocities). The flood and

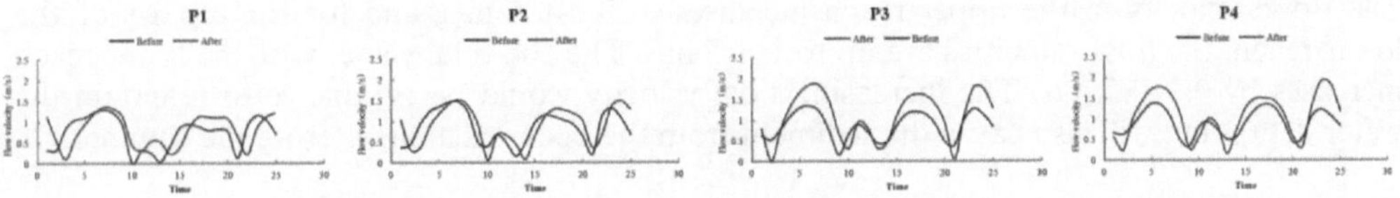

Figure 8. Velocity variation before and after the Deepwater Navigation Channel project.

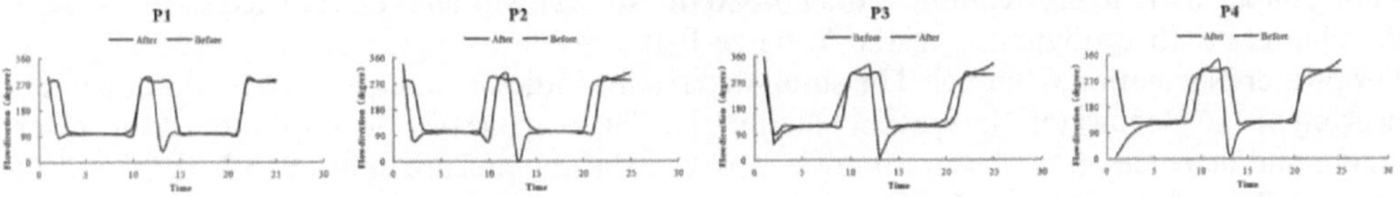

Figure 9. Variation of the flow direction before and after the Deepwater Navigation Channel project.

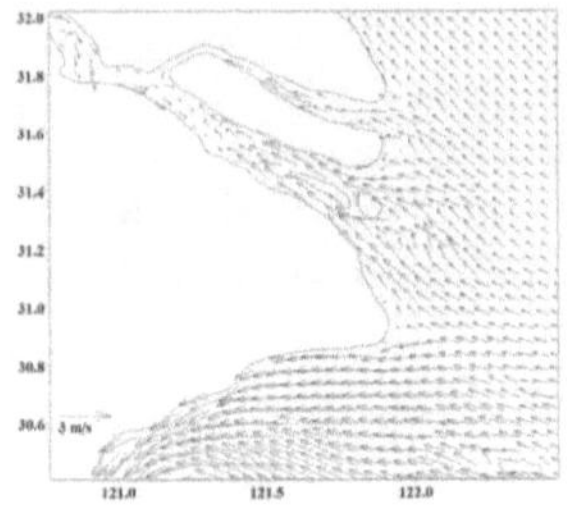

(a) Flood maximum before the project

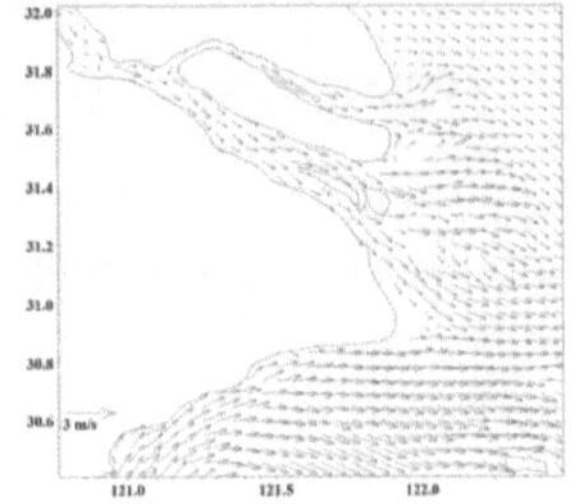

(b) Ebb maximum after the project

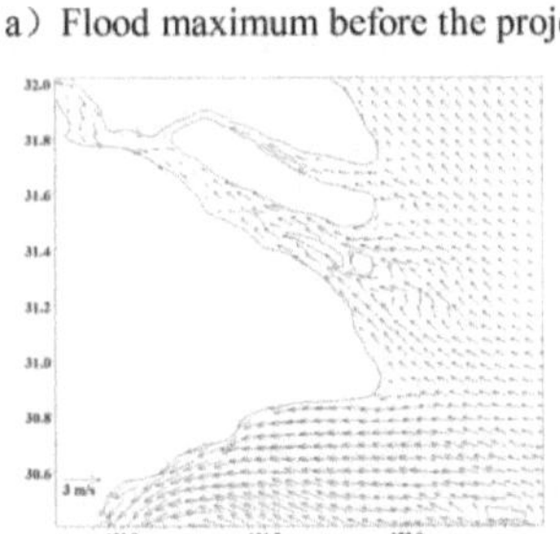

(c) Flood maximum after the project

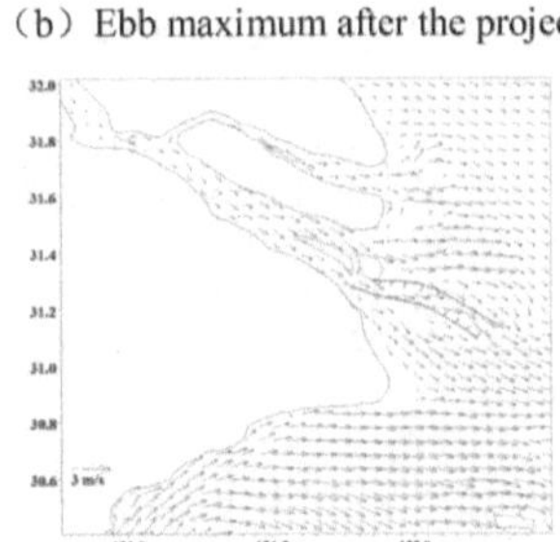

(d) Ebb maximum before the project

Figure 10. Flow field at flood and ebb maximum before and after the Deepwater Navigation Channel project.

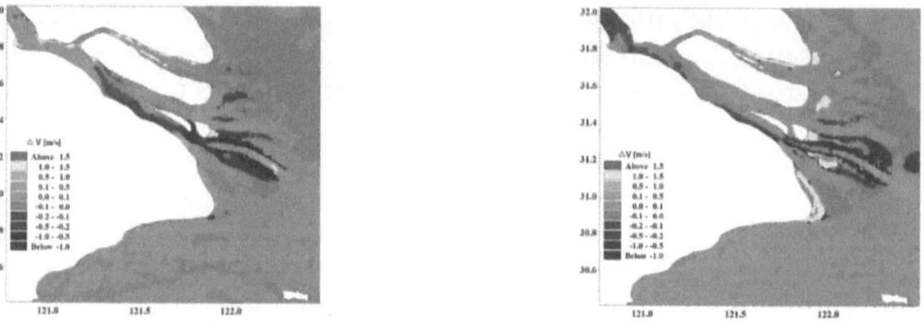

(a) Peak flood velocity (b) Peak ebb velocity

Figure 11. Distribution of the velocity changes of flood and ebb tide caused by the projects (at peak flood and ebb velocities).

ebb velocities dramatically decrease in both sides of the spur dike. The maximum decrement can reach to 1 m/s. The decreasing of the flow velocities will lead to sediment deposition. The flood velocity in the upper reach increases to 0.1–0.5 m/s, and for some parts of the lower reach, the tidal velocities are up to 1–1.5 m/s. The ebb tidal velocity of the lower reach increases by 0.1–0.5 m/s. The increasing tidal velocity would be conducive to maintain the water depth, but will also cause the sediment from the open sea area entering the channel.

4 CONCLUSIONS

In this paper, a 2D hydrodynamic model based on Mike21 Model is established to investigate the changes of hydrodynamic in the Yangtze Estuary due to regulation engineering in the Deepwater Navigation Channel. The simulated results indicate that after the regulation engineering of the Deepwater Navigation Channel,1): The increase of tide amplitude at the lower reach, middle reach and fish mouth project part of the channel are about 0–0.2 m, 0.3–0.5 m and 0.2–0.3 m, respectively; 2): The embankment and the spur dike are submerged during the flood tide. While at ebb tide, the embankment and the spur dike are exposed due to the decreasing of the tidal level.

ACKNOWLEDGEMENTS

The work was financially supported by the Innovation Foundation of Zhejiang Ocean University, Research starting of Zhejiang Ocean University and the Department of Education of Zhejiang Province.

REFERENCES

DHI. 2011. MIKE 21 & MIKE 3 FLOW MODEL FM Hydrodynamic and Transport Module Scientific Documentation.

Kuang, C.P., Chen., W., Gu, J., and He, L.L. 2014. Comprehensive analysis on the sediment siltation in the upper reach of the deepwater navigation channel in the Yangtze Estuary. Journal of Hydrodynamics, Ser. B, 26(2), 299–308.

Li, X., Zhu, J., Yuan, R., Qiu, C., and Wu, H. 2016. Sediment trapping in the Changjiang Estuary: Observations in the North Passage over a spring-neap tidal cycle. Estuarine, Coastal and Shelf Science, 177, 8–19.

Ma, G., Shi, F., Liu, S., and Qi, D. 2013. Migration of sediment deposition due to the construction of large-scale structures in Changjiang Estuary. Applied Ocean Research, 43, 148–156.

Qi, D., Gu, F., Wang, Y. 2005. The preliminary study on reason of back silting in the node W2 W3 in the North Passage. Shanghai Estuar. Coast. Sci. Res. Cent. 1–10 (in Chinese).

Rui, X., Wu, D.A., Yan, Y.X., and Hai, Z.H.O.U. 2010. Fine silt particle pathline of dredging sediment in the Yangtze River deepwater navigation channel based on EFDC model. Journal of Hydrodynamics, Ser. B, 22(6), 760–772.

Hydraulic Engineering V – Hong, Liu & Xie (Eds)
© 2018 Taylor & Francis Group, London, ISBN 978-0-8153-7471-8

Research on a ship bulkhead structure under the combined load

Qun Yin, Guangbo Cao, Renjie Xie & Shu Li
Jiangsu University of Science and Technology, Zhenjiang, Jiangsu, China

ABSTRACT: In this study, numerical simulation of the dynamic response of a typical ship bulkhead structure under the combined action of high-speed fragments and shock-wave loading was carried out. With the evaluation criteria based on the structural energy absorption and average residual velocity of fragments, we contrast and analyze the shock damage characteristics of the ship bulkhead structure under the shock-wave load, high-speed fragments, and the combined load. The results indicate that the combined action of shock-wave load and high-speed fragments can enhance the shock damage effect, differently from a simple superposition of the two load effects.

1 INTRODUCTION

Ship is the most important carrier of modern naval warfare; however, it is also the most vulnerable object to attack. Therefore, study on the improvement of the vitality of the ship is inevitable for any country. In the hull structure, the role of the bulkhead structure is very important, as it protects the important equipment in the cabin and ensures the safety of the crew. In view of the design requirements of the modern large-scale ship protection structure, the bulkhead structure will withstand the huge explosion load and the fragment penetration load when the explosion produces shock waves and high-speed fragments. Therefore, it is very important to study the shock damage characteristics.

There are some differences between the propagation speed and attenuation rate of shock waves and high-speed fragments, so are the researchers in the field. In recent years, researchers have realized that the combined action of shock waves and penetration is not a simple superposition of two loads but a superposed enhancement effect of structural failure. Therefore, the combined action is becoming a major research topic. Because the theory is mostly based on the assumption that the fragments are rigid and experimental methods are expensive and difficult to test, the numerical simulation method can be used to accurately describe the complex structure, material properties, load model, and can simulate the whole dynamic response process.

In this study, we investigate the dynamic response of a typical ship's single-deck bulkhead structure under the action of shock waves, penetration, and the combined action of two loads. We also analyze the shock damage characteristics of bulkhead structure under different loads to obtain the composite shock damage characteristics.

2 EVALUATION PARAMETERS OF ANTI-IMPACT DAMAGE PERFORMANCE

In order to study the anti-impact damage performance of various types of bulkheads, we need some parameters as the basis for evaluation. In this study, the average structural energy absorption and the average residual velocity are taken as the evaluation parameters.

3 NUMERICAL SIMULATION MODEL OF THE SHIP CABIN STRUCTURE

In order to ensure the accuracy of the numerical simulation results, we simplify the structure of the ship cabin into a regular three-compartment structure made of 945 steel. The material parameters are listed in Table 1. The length and width of the cabin are 9 and 8 m; the heights of the three compartments are 2.8, 2.45, and 2.35 m from the top to the bottom; and the middle compartment is the explosion chamber. Each layer of deck and side structure are designed with the corresponding rib. The model of the cabin structure is shown in Figure 1.

The explosive load in the model is simulated by a high-energy balloon using the same energy method, according to the gamma law equation. The gas specific heat ratio is $\gamma = 1.4$, density is $\rho = 1590$ kg/m^3, and unit mass internal energy is e = 4.2 MJ/g.

In order to accurately simulate the process of penetration, the region of the penetration of the elements has been refined in the model. The mesh size of the penetration region refinement is 10% of the length of the fragment, that is, 1 mm.

In this study, the warhead dosage is 150 kg, the shell thickness is 25 mm, and the fragment size is close to the thickness, burst away from 1.75 m. To simulate the scattering process of the warhead structure, it is necessary to set binding and detach strain on each pair of adjacent nodes of the warhead, while applying an initial pressure to the outer normal direction of the inner surface of the warhead unit. The pressure is the average pressure of the gas product when the explosive is detonated:

$$p_0 = \frac{0.5\rho_\omega D_\omega}{\gamma + 1} \tag{1}$$

where ρ_ω = explosive density; D_ω = explosive detonation speed; γ = adiabatic index of explosives; and $\gamma = 3$.

Table 1. Material parameters of the fragment and bulkhead.

Material	Density (kg/m^3)	Elastic modulus (105 MPa)	Poisson's ratio	Yield stress (MPa)	Hardening parameters (MPa)	Hardening index	Failure strain
Fragment	7830	2.11	0.3	835	320	0.28	0.3
Bulkhead	7770	2.05	0.32	440	220	0.26	0.3

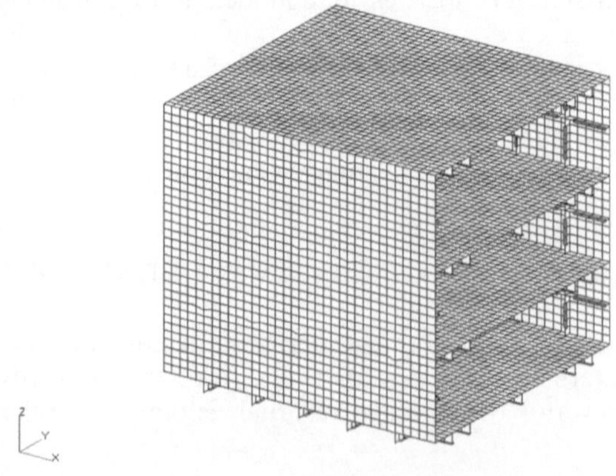

Figure 1. Model of the cabin structure.

The internal pressure of the shell is

$$p = p_0 \left(\frac{r_0}{r} \right)^{\gamma(\mu+1)} \tag{2}$$

where r_0 = initial radius of the shell, r = radius of a moment in the process of expansion of the shell (for spherical shell), and $\mu = 2$.

When the warhead explodes, the energy released from the explosive is partially consumed by the shell, causing the shell to break and drive the fragments, and the remaining energy is consumed by the expansion of the explosive product and the formation of shock waves. Therefore, the amount of explosives forming a shock wave is not the amount of charge in the real situation. According to the conservation of energy, the total energy released from the explosive charge of W is used for the increase of the energy and kinetic energy of the explosive product and the scattering of the fragments:

$$WQ_W = E_1 + E_2 + E_3 \tag{3}$$

energy of explosive product:

$$E_1 = WQ_W \left(\frac{r_0}{r}\right)^{b(\gamma-1)} \tag{4}$$

kinetic energy of explosive product:

$$E_2 = \frac{W}{2(a+1)} \mu_p^2 \tag{5}$$

kinetic energy of shell:

$$E_3 = \frac{m_p}{2} \mu_p^2 \tag{6}$$

where Q_W = explosive energy of explosives; m_p = mass of shell; a = 2/3 for the spherical charge as the shape factor; b = 3 for the spherical charge as the shape factor; and μ_p = the speed of expansion of the shell.

Substituting equations (4), (5), and (6) into formula (3):

$$\mu_p = \sqrt{\frac{2Q_W (1 - r_0 / r)^{b(\gamma+1)}}{1/\alpha - a/(a+1)}} \tag{7}$$

where α = loading factor; $\alpha = W/(W + m_q)$, from which rupture radius μ_{p0} and breaking speed μ_{p0} in the shell rupture can be obtained instantaneously. The energy left to the explosive product at this time is $WQ_W - m_q \mu_{p0}{}^2/2$. We substitute equation (7) to obtain the corresponding explosive equivalent. The initial velocity of the fragment is 1698 m/s, and the explosive equivalent is 54.26 kg.

4 STUDY ON THE DYNAMIC RESPONSE OF THE SHIP BULKHEAD STRUCTURE UNDER SHOCK WAVES

4.1 *Numerical simulation model under shock waves*

The cabin structure is described in Section 2, and the bulkhead structure under the shock-wave load is shown in Figure 2. The air field is established on the inside of the three compartments and the outside of the whole compartment, and the coupling surface adopts the multicoupling method with failure mode. For each coupling surface in the model, the Eulerian grids are parallel to the global coordinate system.

4.2 *Stress of the bulkhead structure under shock waves*

In this study, three representative positions are selected as the object of study, namely the junction of bulkheads and decks, the center of explosive bulkhead, and the junction of bulkheads and sides. Figure 3 shows the stress versus time curve for three aforementioned positions. The stress curves are found to fluctuate because the shock waves caused by the explosives reflected and refracted in the ship cabin. After the shock wave reaches the bulkhead for the first time and continues to gather the shock wave again on the bulkhead structure, again, there would be a convergence of shock-waves influencing the bulkhead structure continuously, which resulted in the alternative reduction and increment of the unit stress after reaching the peak. It can be found from the three points at the peak stress that the corner of the stress peak was higher than the bulk stress at the center of the bulkhead, because

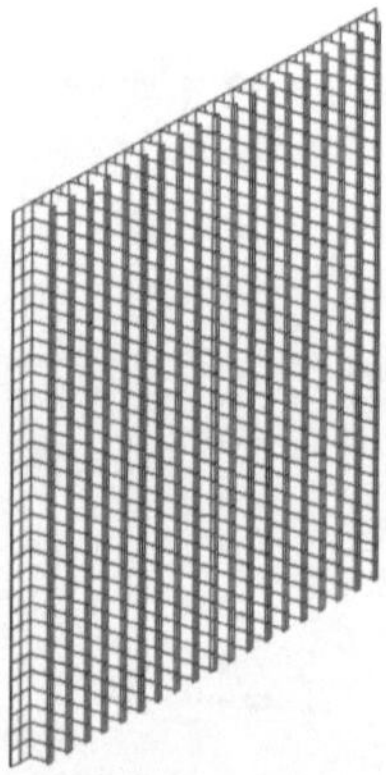

Figure 2. Bulkhead structure under shock waves.

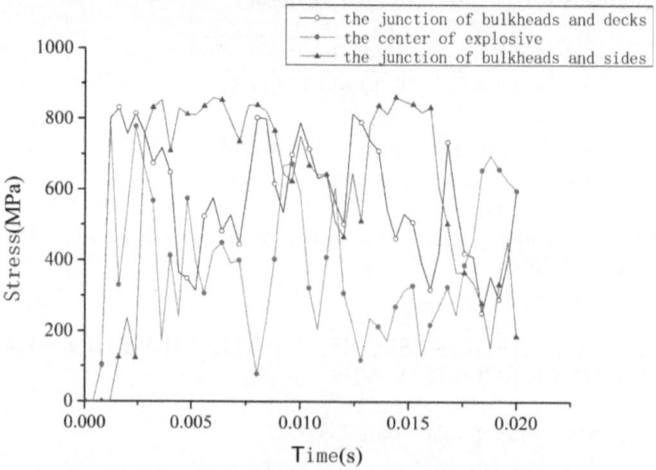

Figure 3. Stress–time curves of the bulkhead structure.

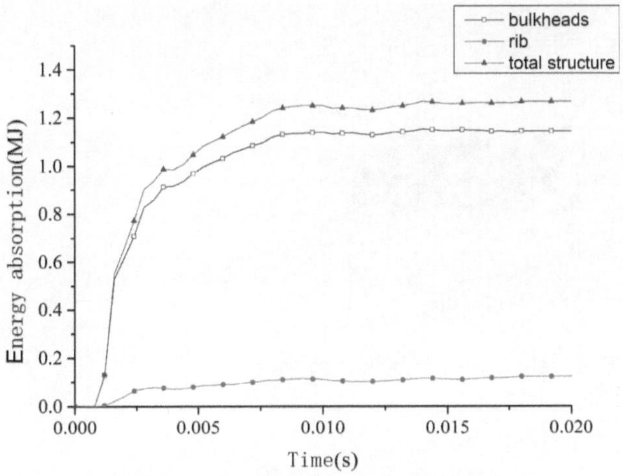

Figure 4. Energy absorption–time curves of the bulkhead structure.

Table 2. Energy absorption of bulkhead structure under shock waves.

Structure	Bulkheads	Rib	Total structure
Energy absorption (MJ)	1.144	0.123	1.266

the shock waves formed and converged more easily at the corners, resulting in greater stress at the corners. Therefore, the structure was more prone to tear at the corner.

4.3 *Energy absorption characteristics of the bulkhead structure under shock waves*

Figure 4 shows energy absorption of the components of the bulkhead structure versus time. It is evident from the figure that explosion occurred in the very short period of time, and the bulkhead and rib were free of energy because the shock wave has not yet reached the bulkhead structure. However, when the shock wave reaches the bulkhead structure, the energy absorbed by the bulkhead structure increased rapidly. Over time, the shock wave reflected and refracted in the cabin constantly and the formative shock wave affected the bulkhead again, thereby increasing the bulkhead structure energy absorption, which then stabilized gradually. Table 2 presents the energy absorption values of the bulkhead. It is evident from the table that the bulkhead structure under the shock wave absorbed more than 90.29% of the total energy, which was the main energy absorbing component.

5 STUDY ON THE DYNAMIC RESPONSE OF THE BULKHEAD STRUCTURE UNDER HIGH-SPEED FRAGMENT PENETRATION

5.1 *Numerical simulation model under high-speed fragment penetration*

In the study of the impact of the high-speed fragments on the bulkhead structure, the size of the compartments is consistent with the model size under the shock wave. Mesh refinement of the elements of the fragment penetration area and the bulkhead structure of the finite-element model are shown in Figure 5. In the numerical simulation process, an adaptive contact algorithm is used. In this section, Adaptive Master-Slave Surface Contact Adaptive Self Contact is used between the fragment and the bulkhead structure. Adaptive Self Contact is used between bulkheads and rib.

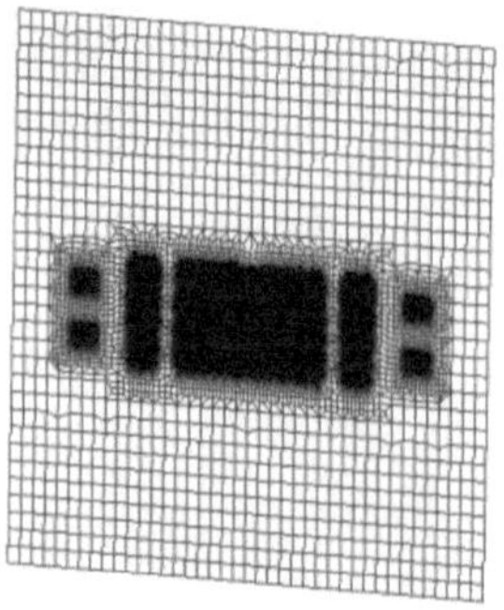

(a) Mesh refinement of bulkhead panels

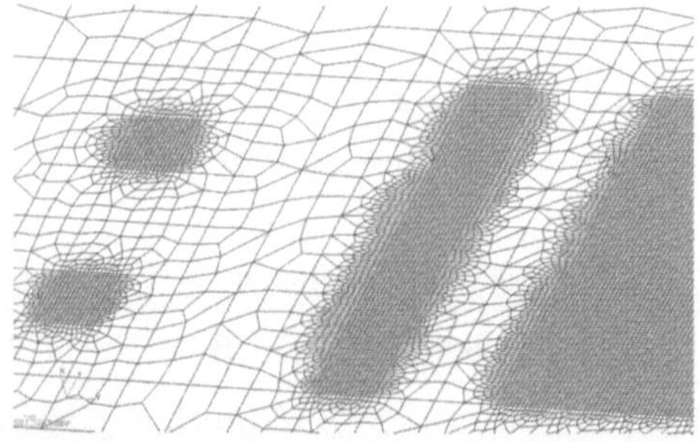

(b) Mesh magnified image of the penetrate area in the bulkhead

Figure 5. Model of the ship bulkhead structure under penetrating loads.

5.2 *Stress characteristics of the bulkhead structure under penetrating loads*

The time and angle of fragment penetration into the bulkhead structure showed a different trend, but the stress on the structure caused by the changing trend of fragments was consistent. The maximum stress was at the edge of the break, so we study the effect of one of the fragments on the structural stress.

Figure 6 shows the stress–time curve of three nodes from near to far at the edge of the break. It is evident from the figure that, when the fragments penetrate the bulkhead structure, the stress of measured point on the bulkhead structure increases sharply to a peak value and then decreases and fluctuates. The stress at the break edge was 514 MPa, which is greater than the material yield limit of 440 MPa. At 12 mm from the edge of the break, the maximum stress was 397 MPa. At 24 mm from the edge of the break, the maximum stress value was 151 MPa. This shows that the damage caused by the penetration of a single fragment has a local effect, the stress mainly concentrated around the bulkhead break, and the stress of the bulkhead structure away from the break zone was in the elastic range.

5.3 *Energy absorption characteristics of the bulkhead structure under penetrating loads*

As the number of fragments was more and the speed direction was not the same, the velocity of fragments in the direction perpendicular to the bulkhead plate was different, resulting in different times of fragments penetrating the bulkhead structure. Therefore, whenever a fragment penetrates the structure, the energy absorption of the bulkhead structure would increase and stabilize until the next penetrating fragment. The total energy absorption of the bulkhead structure would eventually stabilize when all fragments penetrated the structure.

Figure 7 shows the energy absorption–time curves of the bulkhead structure of the bulkheads under the penetration load. It is evident from the figure that the energy absorption

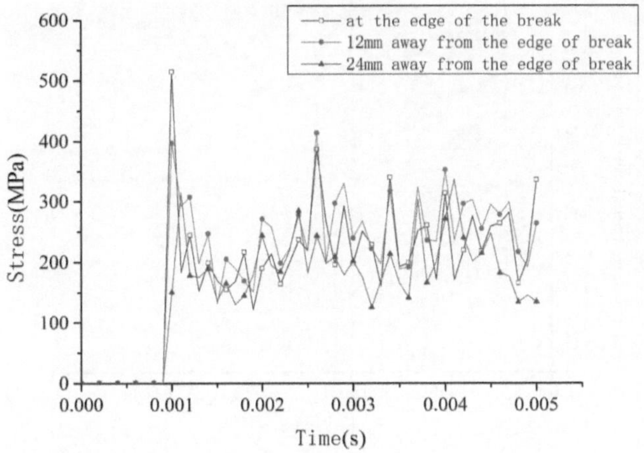

Figure 6. Stress–time curves of the bulkhead structure.

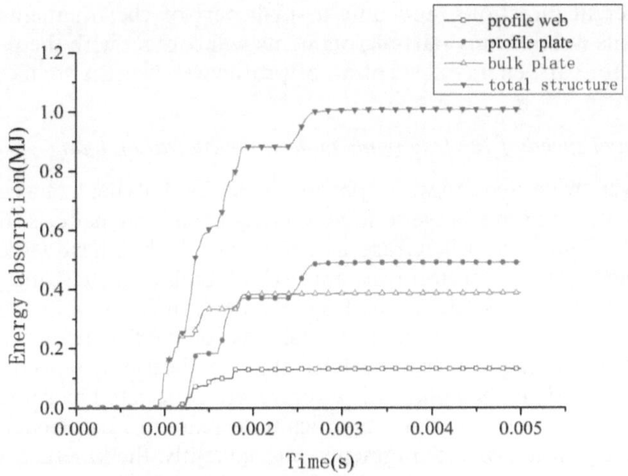

Figure 7. Energy absorption–time curves of the bulkhead structure.

of the bulkhead structure increased whenever fragments penetrated the structure, and the settling time between multiple increases was the time difference between the two adjacent fragments penetrating the bulkhead structure. The time interval between the two adjacent fragments penetrating the bulkhead structure increased because the fragments penetrating the bulkhead structure were almost perpendicular to the bulkhead structure, the spatial distribution of the fragments was spherical, and the velocity was along the normal direction of the fragmentation element. The longer the distance from the center of the bulkhead, the lower the normal velocity component perpendicular to the bulkhead structure, that is, the longer the fragmentation reaches the bulkhead structure. At the initial stages of penetration of the fragments, the time interval between fragment penetrations was short, so the energy absorption was stable for a short time after each fragment penetrates the structure. However, over time, the time interval of the follow-up fragment penetration increased and therefore the relative energy curve stabilization time increased. The total energy absorption of the bulkhead structure during the whole penetration process was 1.01 MJ. Among them, the energy absorptions of the bulkhead plate and profile plate were more than that of the profile web, which are 38.41% and 48.61%, respectively, of the total energy absorption, as the main energy absorption components. The energy absorption of the profile web was only 12.97% of that of the secondary energy absorption components. This was because the speed of fragmentation

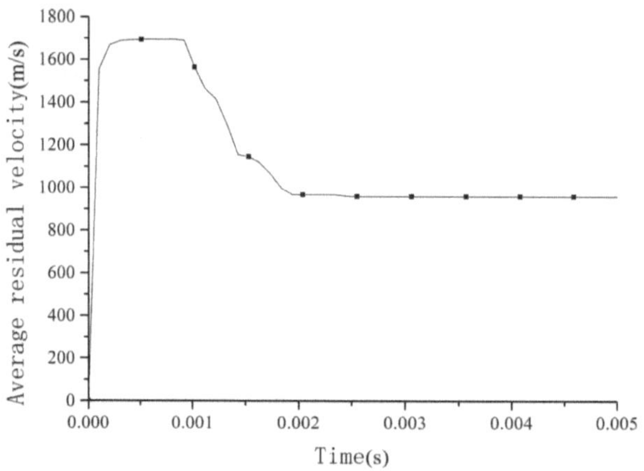

Figure 8. Average residual velocity–time curve of the fragments.

is different in different directions, and only a small part of the fragments would have an impact on the profile web, or only part of fragments will contact with the panel, and not the entire panel. Therefore, the energy absorption of profile web was the smallest.

5.4 *Average residual speed of the fragments under the penetration load*

The fragments driven by detonation waves penetrate the ship bulkhead at a certain speed after the explosives explode, and the average residual velocity of the fragments is shown in Figure 8. It is evident from the figure that, when fragments penetrate the bulkhead structure, as the fragments were subjected to the extrusion resistance of the bulkhead structure, the total kinetic energy of the fragments decreased. As the fragments had different velocities, their times of penetrating the bulkhead were different. The total mass of the fragments was certain, so the total kinetic energy of the fragments can be determined by the average speed of the fragments. At 0. 99 ms, for the first time, the bulkhead structure was penetrated by some fragments. This part of the fragmentation will undergo a significant decrease of kinetic energy, resulting in a decrease in the average speed of the fragments. Subsequently, the kinetic energy of the fragments which continue to penetrate the bulkhead structure subjected to compression resistance decreased several times, whereas the total mass of fragments remained unchanged, thereby decreasing the average residual rate of broken pieces several times. When all fragments penetrate the bulkhead structure, the average residual velocity of fragments was finally stabilized at 959 m/s.

6 STUDY ON THE DYNAMIC RESPONSE OF THE BULKHEAD STRUCTURE WITH THE COMBINATION OF SHOCK WAVE AND FRAGMENT PENETRATION

The order that shock waves and fragments reached the bulkhead was determined by the explosion distance. With the increase of the propagation distance, the shock-wave attenuation was relatively fast. According to the actual length of the cabin, more dangerous conditions are selected, 1.75 m burst. In this burst, the shock wave will reach the explosion chamber bulkhead structure before the break. When the shock load affects the bulkhead structure, the structure will be deformed. After that, the process of penetration was equivalent to the penetration of a target with a certain internal stress, which was easier to penetrate the bulkhead structure. After the break, a new shock wave formed by reflection and refraction inside the chamber will act again on the bulkhead structure, causing more severe damage.

6.1 *Numerical simulation model under the combined action*

The element size of the bulkhead structure under the combined action is consistent with the element size under the penetration load.

6.2 *Stress characteristics of the bulkhead structure under the combined action*

The corner of the bulkhead was far from the break, and the penetration process was strong locally; therefore, the stress at the corner of the bulkhead was mainly due to the shock wave. The point at the edge of the break was affected by the combined load, and the response of the point was the best way to reflect the response of the bulkhead structure. Figure 9 shows the stress versus time curve.

It is evident from Figure 9 that shock waves and fragments did not reach the bulkhead structure initially; then, the stress at the point was zero. With the propagation of the shock-wave load, the air in front of the bulkhead structure was squeezed, so that the stress was slowly increased. At 0.0007 s, the shock wave arrived at the bulkhead structure, and the stress at the measuring point was increased rapidly. At 0.001 s, the bulkhead structure was penetrated by the fragment, and the stress at the bulkhead structure increased again. The time interval between the two loads was short and hence there was no obvious difference. After that, the bulkhead structures were affected by the reflected and refracted new shock wave and reassembled in the cabin once again, which stretch on the break, and therefore the stress fluctuated at the measuring point.

6.3 *Energy absorption characteristics of bulkhead structures under combined action*

Figure 10 shows the time-dependent curves of the energy absorption of the bulkhead structure under combined action. At around 0.0007 s, the shock wave reached the bulkhead structure, and the energy absorption of the bulkhead structure increased rapidly. At 0.001 s, the bulkhead structure began to be penetrated by the fragments. From Figure 10, the penetration of different fragments makes the slope of each section of the bulkhead energy absorption curve different. The energy absorption of the bulkhead structure was no longer as obvious as the structural energy absorption under separate penetration. The combination of shock wave and penetration load propagates over the structure, and the energy absorption of the components of the bulkhead structure tends to be stable. The energy absorption of bulkhead plating was higher than that of the profiles, accounting for 65.20% of the total energy absorption, which was the main energy-absorbing component.

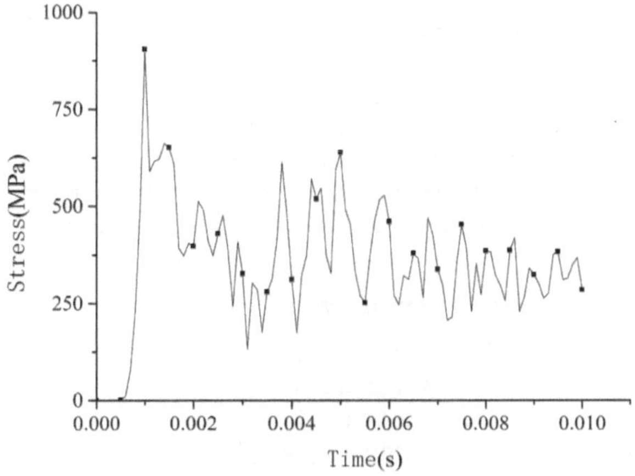

Figure 9. Stress–time curve of the bulkhead structure.

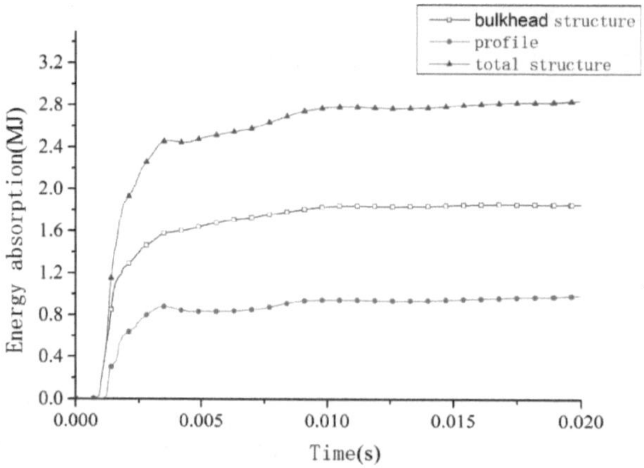

Figure 10. Energy absorption–time curves of the bulkhead structure.

Table 3. Energy absorption of bulkhead structure under different loads.

	Shock wave	Penetration	Simple stacking	Combined action
Energy absorption (MJ)	1.266	1.008	2.274	2.847

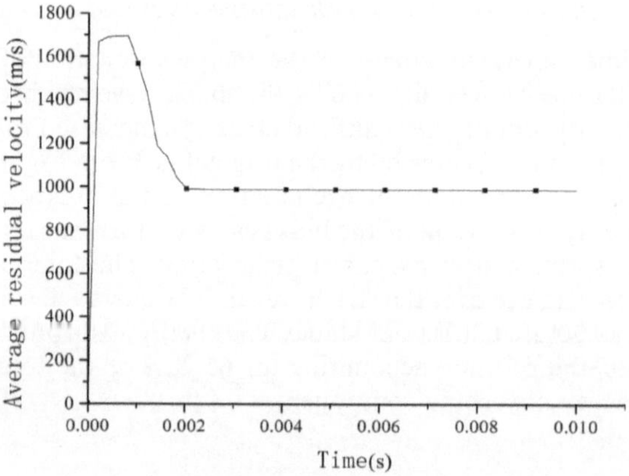

Figure 11. Average residual velocity–time curve of the fragments.

Table 3 compares the energy absorption under separate action and combined action. It is evident from the table that the combined effect was not a simple superposition of the two load effects, and that energy absorption under combined action was higher than that under simple superposition, that is, the combined action on the bulk structure damage has enhanced effect. Therefore, it is necessary to carry out the study of the combination of shock wave and penetration.

6.4 *Average residual speed of the fragments under combined action*

Figure 11 shows the average residual velocity–time curve of the fragments under combined action. Because the total mass of the fragments was constant, the total kinetic energy of the fragments was only related to the average velocity of the fragments. When all fragments penetrated the structure, the final average residual velocity of the fragment will remain at 987 m/s, higher than the average residual rate of fragments under separate penetration, because the explosion distance was 1.75 m. In this explosion distance, the fragments reached the bulkhead structure later than the shock wave. At this time, bulkhead structure effected by the shock wave produced a certain stress and the resistance of the fragments was reduced, so that the residual kinetic energy of the fragments penetrating the structure increased, that is, the average residual velocity increased.

7 CONCLUSIONS

In this study, the dynamic response process of a typical ship bulkhead structure under a separate load of shock wave or penetration as well as the combined action of shock wave and penetration were numerically simulated, compared, and analyzed, to draw the following conclusions:

1. When the shock-wave load acts alone, the entire bulkhead structure will respond, the maximum stress will appear at the corner of the bulkhead structure, and the corners in the structure of the bulkhead will get easily torn. The plate of the bulkhead structure is the main energy-absorbing component, and the T section is the secondary energy-absorbing component.
2. Numerical simulation of the process of high-speed fragmentation of the bulkhead was carried out. The damage effect of the fragment penetrating the bulkhead structure is local, the penetration of a single fragment only affects the area around the break, and there is little impact on the area far from the breach.
3. During the combined action of shock-wave and penetration effect, at an explosion distance of 1.75 m, the shock wave arrives at the bulkhead structure before the fragment. The bulkhead structure first produces stress and deformation under the action of shock waves. When the fragments penetrate the bulkhead structure, there is a certain stress inside the bulkhead structure, so that the resistance of the fragment penetrating the structure is reduced, leading to an increase in the average residual velocity of the fragments.
4. From the research on the dynamic response process and damage characteristics, it can be concluded that the combined effect is not a simple superposition of the two load effects. Therefore, the combined action on the bulk structure damage has an enhanced effect.

REFERENCES

Amdahl J. & Hopperstad S. 2009. On the resistance to penetration of stiffened plates, part II: numerical analysis [J]. International Journal of Impact Engineering, (36):875–887.

Awerbuch, J. & Bodner, S.R. 1974. Analysis of the mechanics of perforation of projectiles in metallic plates. International Journal of Solids and Structures, 10(1): 671–684.

Changqiang Huang & Hesong Zhu 1993. The establishment of the ultimate penetration formula of the target plate with spherical fragments. Journal of Projectiles Rockets Missiles and Guidance. (2): 58–61.

Di Wu & Shilian Zhang 2007. Numerical simulation of the dynamic response of ship under air explosion [J]. Ship Science and Technology. 29(6): 165–168.

Goldsmith W. 1985. In metal forming and impact mechanics [J]. Pergamon Press. 271–287.

Orphal D.L. & Anderson Jr. C.E. 2006. The dependence of penetration velocity on impact velocity. International Journal of Impact Engineering.

Taylor G. 1950. The Formation of a Blast Wave by a Very Intense Explosion. I. Theoretical Discussion, in Proceedings of the Royal Society of London. Series A, Mathematical and Physical Sciences. Royal Society. P:159–174.

Werbch A. & Bodner R. 1974. Analysis of the mechanics of perforation of projectiles in metallic plates [J]. Int. J. Solids Struct. 10(1):671–684.

Zhichun Zhang & Hongfu Qiang. 2008. Numerical simulation for fragments flight characteristics of high explosive projectile [J]. Journal of PLA University of Science and Technology. 9(6): 671–675.

Zhipeng Du. 2005. The anti-explosion mechanism and simulation study of the multi-cabin protection structure on the side of large surface ships [D]. Shanghai Jiaotong University.

Zhiyuan Mei & Xi Zhu. Analyze the penetration power of the cube fragment of missile warhead by MSC/DYTRAN [J]. Journal of Naval University of Engineering. 2002. 14(2): 39–42.

Zhou X.Q. & Hao H. 2007. Prediction of air blast loads on structures behind protective barrier [J]. International Journal of Solids and Structures. 35(5): 363–375.

Hydraulic Engineering V – Hong, Liu & Xie (Eds)
© 2018 Taylor & Francis Group, London, ISBN 978-0-8153-7471-8

Experimental study on the hydrodynamic characteristics of a geomembrane-enveloped concrete structure of ecological revetment

Ai-xing Ma
State Key Laboratory of Hydrology Water Resources and Hydraulic Engineering Science, Nanjing Hydraulic Research Institute, Nanjing, China

Min-xiong Cao
Nanjing Hydraulic Research Institute, Nanjing, China

Ya Deng
Hohai University, Nanjing, China

You-jun Yang
China Design Group Co. Ltd., Nanjing, China

Yu-ting Li
Nanjing Hydraulic Research Institute, Nanjing, China

Bin-rui Zhou
Nanjing R&D Tech Group Co. Ltd., Nanjing, China

ABSTRACT: Traditional revetment projects emphasize the durability and regulation effect but rarely consider river ecology and environment demand. Therefore, a new geomembrane-enveloped concrete structure of ecological revetment was put forward by waterway designers on the basis of dissipating the near bottom flow velocity, which could reduce the velocity through the block and shelter effects of the concrete structure and then form a moderate hydrodynamic environment benefitting sediment deposition and plant growth. An hydrodynamic experiment was carried out in a variable-slope flume to accurately measure the local flow conditions around the concrete structure by a three-dimensional velocity meter (Vectrino). The results indicate that due to the block effect, the longitudinal velocity dissipation rate of the designed structure reaches 68.4% and will further increase to 86.1% because of the block effect of the growing plants. Compared with the incoming flow, the turbulence intensity inside the lattice increases relatively, whereas the sharp dissipation of the flow velocity could form a moderate flow environment, which can prevent the erosion of nutrient soil on the slope. The near-bottom turbulence intensity will further reduce after plant growth, which is conducive to the protection of nutrient soil.

1 INTRODUCTION

In order to ensure the first target of durability and regulation effect, traditional revetment projects rarely consider the relationships between the projects and river ecology as well as the environment demand, resulting in negative impact on the surrounding original river ecological environment. In recent years, the idea of ecological revetment and slope protection was gradually recognized in waterway regulation engineering, while some new regulation structures are utilized and several improvements of ecological revetment technology are developed (Yang, 2004 and Yao, 2012). In the second phase of the 12.5 m deepwater channel construction in Yangtze River, downstream of Nanjing, the river dynamic conditions are complex, including the effects of the runoff, the tidal current, and the different routines of the ebb and flow, especially the intricate exchange of the cross-wise water volume on the shallow beach. Lately, a new geomembrane-enveloped concrete structure of ecological revetment was put forward by waterway designers (China Design Group *Co. Ltd.*, 2016), in order to weaken the flow of all directions and ensure slope stability, meanwhile providing breeding sites for surrounding aquatic creature and microorganism and improving the biology diversity.

The ecological geomembrane-enveloped slope protection structure of clathrate could adjust the local fluid flow conditions inside and around the structure, altering the flow velocity and turbulence characteristics, which would directly affect the protection ability of the slope surface, a similar result for planted structure. In this study, the local flow conditions inside and around the geomembrane-enveloped slope protection structure was measured accurately in a variable-slope flume, and the ability of the structure to reduce the near-bottom velocity and its mechanism were analyzed. Research results could be used for the design and application of clathrate's geomembrane-enveloped structure.

2 ECOLOGICAL GEOMEMBRANE-ENVELOPED SLOPE PROTECTION STRUCTURE AND FLUME EXPERIMENT

2.1 *Ecological geomembrane-enveloped concrete structure*

Ecological geomembrane-enveloped slope protection structure is mainly similar to a type of classical geomembrane-enveloped concrete structure, with a clathrate arrangement for geomembrane-enveloped concrete on the plane. After a series comparison experiments for relative height of geomembrane-enveloped concrete, net gap of clathrate, and arrangement pattern, the main shape scale of recommended slope protection structure was confirmed as follows: height of the geomembrane-enveloped concrete, 0.35 m; diameter of the similar semicircle cross-profile, 0.7 m; net gap of clathrate in the longitudinal direction, 1 m; and 1.5 m in the transverse direction, as shown in Figure 1.

2.2 *Setup of the flume experiment*

1. Experimental flume

The hydrodynamic experiments of the ecological revetment structure were carried out in a 44 m-long, 0.8 m-wide, and 0.8 m-deep variable-slope flume and a slope gradient of 1% at

(a) (b)

Figure 1. Experimental model of a geomembrane-enveloped concrete structure of ecological revetment: (a) designed structure and (b) planted structure.

Nanjing Hydraulic Science Research Institute. The flume adopts the self-circulation system of water flow, and the discharge of the inlet flow is controlled by the electric valve and the electromagnetic flowmeter with the PVC pipe group for energy dissipation. The water level at the outlet of the flume is controlled by a stepping motor by controlling the opening degree of the tailgate.

2. Model scale and experiment hydraulic conditions
In consideration of the scale of clathrate's geomembrane-enveloped concrete structure and the laboratory dynamic conditions, the flume experiment was carried out with a geometric scale ratio of 1:5.

The structure will be used on the bank of Kouanzhi waterway located in the lower reach of Yangtze River. The annual mean peak discharge (Q = 57,500 m^3/s) was taken as the flow condition (Nanjing Hydraulic Research Institute, 2016) in the laboratory experiment, with corresponding data of 2.81 m water depth and 1.14 m/s depth-averaged velocity. Assuming the logarithmic distribution of the vertical flow velocity, the flow velocity of the flow on the top of the tridimensional network structure is 0.95 m/s.

Because of the large scale of the model, the condition of the prototype depth can hardly be achieved directly according to the scale. Therefore, the model depth is reduced to 2 m, but the point velocity at the upper surface of the slope protection structure is guaranteed the same as that of the prototype. Finally, the experiment hydraulic conditions are an in-flow of 0.4 m depth and 0.43 m/s point velocity at the structure upper surface.

3. Layout of measuring instruments
There are 15 lattices in the geomembrane-enveloped concrete structure of clathrate along the flume, with a length of 5.24 m and the corresponding prototype length of 26.2 m. A total of 15 vertical measured lines were arranged streamwise (Figure 2), of which 1 line (0#) was set at the position of 10 m upstream from the structure front and 11 lines (3#–13#) were arranged in the lattices of the hydraulic stable region. A detailed measurement of perpendicular velocity was conducted, with eight measured points in the region of the structure height and six from the upper surface of the structure to water surface.

4. Measuring instrument of the velocity
The flow structure in the vicinity of the tridimensional network structure is measured by the Acoustic Doppler Point Velocity Meter (Vectrino). The instrument can measure the spatial distribution of complex hydraulic elements such as flow velocity and turbulence intensity near the structure by measuring the instantaneous three-dimensional velocity. In the experiments, the velocity-measuring frequency of vertical point is 90 Hz, the measuring time is 40 s, and the number of velocity sample data of each vertical point is 3600.

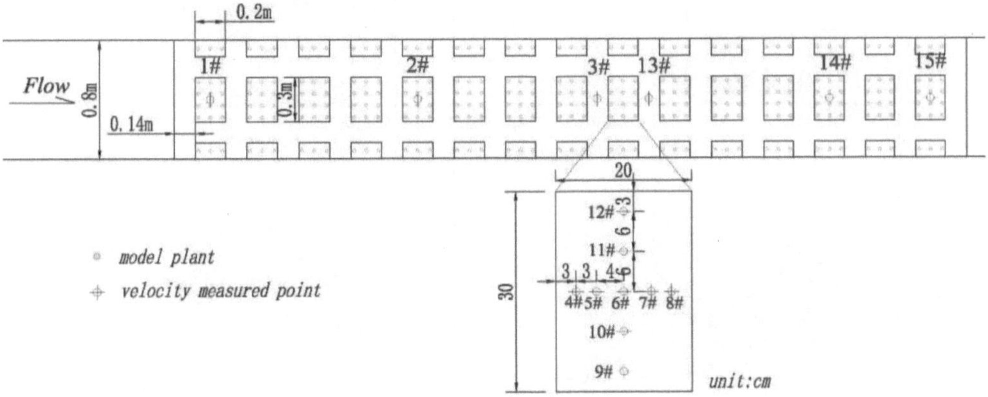

Figure 2. Layout of the vertical velocity measurement lines of clathrate's geomembrane-enveloped concrete structure.

3 ANALYSIS OF EXPERIMENTAL RESULTS

Flow conditions, such as velocity profile and turbulence intensity, around clathrate's geomembrane-enveloped structure changed obviously. The geometrical mean of the instantaneous flow velocity in the x direction (longitudinal direction, parallel to the direction of the flow) is taken as the time-averaged longitudinal flow velocity (u), and the root mean square of pulsating flow rate in the x direction is taken as the longitudinal turbulence intensity (σ_x). According to the model similarity law in an undistorted model, the model flow velocity (u_m) and the prototype flow (u_p), and the model turbulence intensity (σ_{xm}) and the prototype turbulence intensity (σ_{xp}), respectively, satisfy the following relations:

$$u_p = (\alpha_L)^{0.5} u_m,\ \sigma_{xp} = (\alpha_L)^{0.5} \sigma_{xm} \tag{1}$$

where α_L is the model geometric scale (equal to 5) for the model.

In order to facilitate the comparative analysis, velocity, turbulent intensity, and flow-measuring vertical position all adopt the values of prototype.

3.1 *Comparison of the flow conditions in the lattice between the designed structure and the planted structure*

The flow conditions and the contour cloud of u and σ_x of the designed structure and the planted structure in the vertical cross section are shown in Figure 3. The figure shows that because of the front geomembrane-enveloped concrete shelter, a larger trailing vertex was

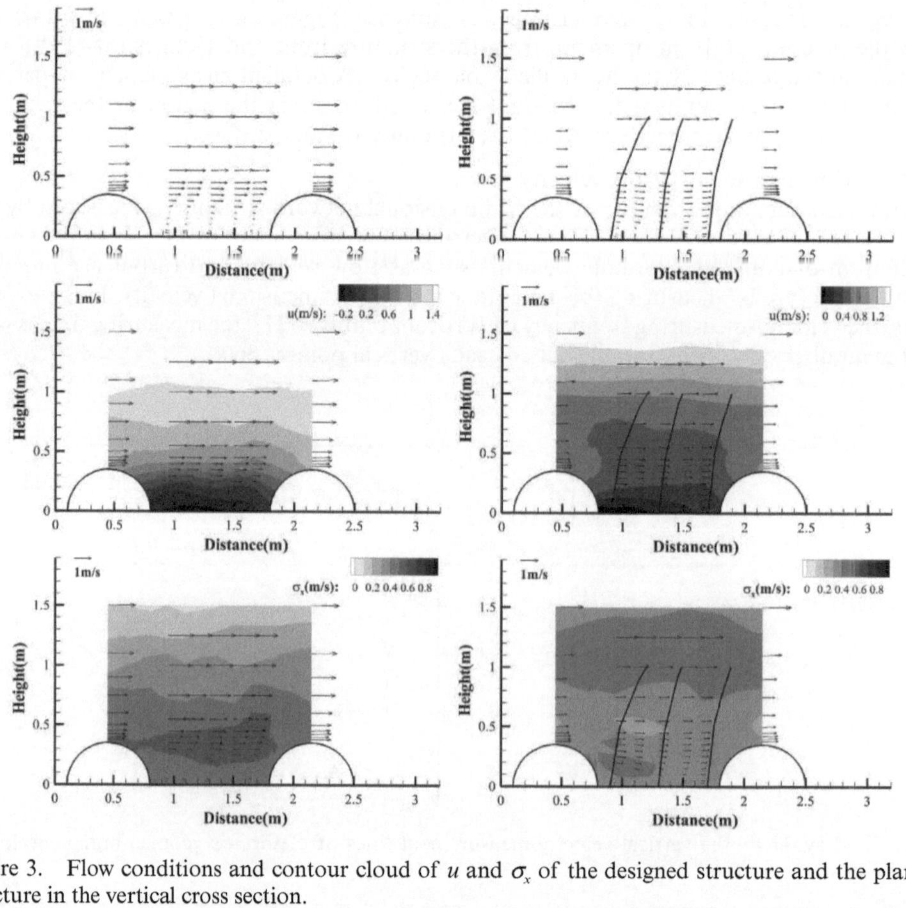

Figure 3. Flow conditions and contour cloud of u and σ_x of the designed structure and the planted structure in the vertical cross section.

fully developed in the lattice with two-thirds of the height of geomembrane-enveloped concrete. A smaller velocity occurred in the flow under the concrete shelter, $-0.2 < u < 0.2$ m/s. However, a stronger turbulence intensity appeared in the lattice region of the concrete height, $0.6 < \sigma_x < 0.7$ m/s.

The comparison results of the designed structure indicate that the flow velocity inside the lattice with plants further decreased, $0 < u < 0.2$ m/s, because of the effect of the energy dissipation of plants, without the obvious vertex. Except for the larger turbulence intensity ($0.4 < \sigma_x < 0.6$ m/s) appearing in the smaller rear region of the front concrete, other regions in the lattice were relatively weaker than in the designed structure ($0.1 < \sigma_x < 0.2$ m/s), and the turbulence intensity peak appeared in the plant upper region.

3.2 *Comparison of the velocity and turbulence intensity in the lattice between the designed structure and the planted structure*

Figure 4 compares the variation of u and σ_x along the relative depth z/h (z is the distance from the slope and h is the water depth) of the central vertical measured line (6#) in the hydraulic stable lattice between the designed structure and the planted structure. The figure shows that the flow velocity in the rear shelter region of the concrete sharply decreased compared to the incoming flow, whose bottom velocity was 0.72 m/s and velocity around the lattice bottom was –0.3 to 0.1 m/s, that is, the dissipation rate of bottom velocity was 68.4%. Meanwhile, the turbulence intensity obviously increased, the incoming bottom turbulence intensity was approximately 0.1 m/s, and the turbulence intensity in the bottom lattice was 0.2 m/s, while the peak along the whole vertical measured line occurred around the upper concrete with the value of 0.24 m/s.

For the planted structure of slope protection, the velocity profile in the vertical line was distinctly altered around the plant upper region, and the distribution profile was changed to "S" style from the logarithm style of the designed structure, whose overall velocity decreased obviously, and the velocity around the lattice bottom was 0.1 m/s, that is, with a velocity dissipation rate of 86.1%. Compared to the designed structure, there were two peaks of the turbulence intensity distribution in the vertical line, which appeared near the lattice bottom and around the plant upper region. In the lattice, the turbulence intensity was weaker around the upper concrete (0.14 m/s) and stronger around the bottom (0.2 m/s), but smaller than those of the designed structure in whole.

Figure 5 compares the variation of u and σ_x around bottom ($z = 0.03$ m) along the flow direction from the front concrete to the latter in the hydraulic stable lattice. The figure shows that because of the tailing vertex behind the front concrete structure in the lattice, the bottom velocities around the lattice from the front concrete structure to the latter were all in the reverse direction, $-0.2 < u < -0.05$ m/s. Meanwhile, the turbulence intensity gradually increased, $0.18 < \sigma_x < 0.21$ m/s. Because no obvious vertex appeared in the lattice after the

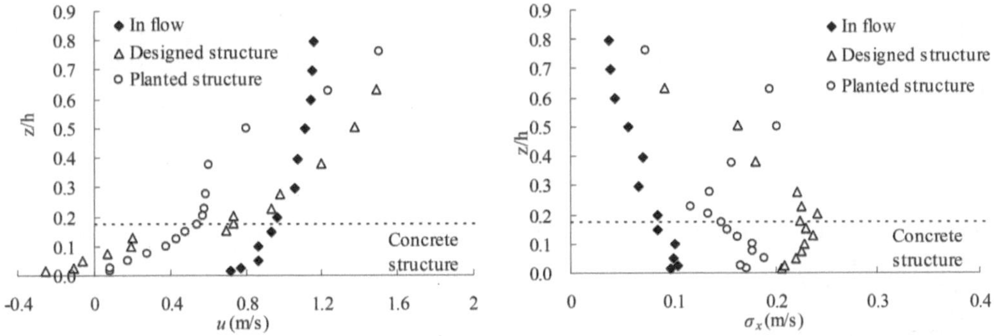

Figure 4. Comparison of u and σ_x along the central vertical measured line (6#) in the lattice between the designed structure and the planted structure.

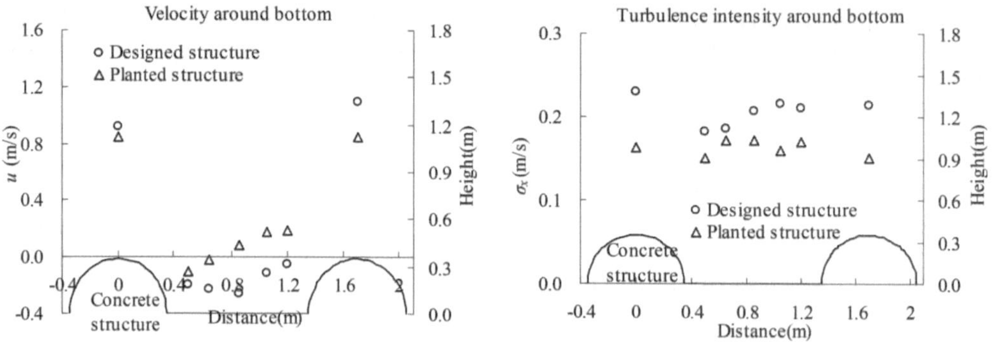

Figure 5. Comparison of u and σ_x along the flow direction in the lattice between the designed structure and the planted structure.

growth of plants, the bottom velocity was streamwise, except around the front concrete, $-0.1 < u < 0.19$ m/s, and the turbulence intensity was generally invariant, about 0.15 m/s.

4 CONCLUSION

Using the normal scaling (1:5) of the ecological geomembrane-enveloped slope protection structure, an experiment on the hydrodynamic characteristics of the designed structure and the planted structure was conducted in a variable slope flume. The main conclusions drawn from this study are as follows. (1) Through the block effect of the geomembrane-enveloped concrete on the incoming flow, a moderate hydrodynamic environment was produced in the shelter region of the lattice, which could prevent the nutrient soil from being washed away. (2) In comparison to the incoming flow, the flow velocity was sharply reduced in the lattice, and the velocity dissipation rates of the designed structure and the planted structure were 68.4% and 86.1%, respectively. However, although the turbulence intensity was relatively increased, the near-bottom turbulence intensity reduced in the planted structure compared with the designed structure. (3) There was little variation in the bottom velocity along the flow direction in the lattice. Under the condition of the incoming bottom velocity of 0.72 m/s, the lattice bottom velocities of the designed structure and the planted structure were small, accordingly being in the reverse direction (–0.2 to –0.05 m/s) and streamwise (–0.1 to 0.19 m/s), which could effectively protect the erosion of nutrient soil on the slope.

ACKNOWLEDGMENTS

This study was supported by the National Key Research and Development Program (No. 2016YFC0402100) and the National Natural Science Foundation of China (Grant Nos. 51309160 and 51579216).

REFERENCES

China Design Group *Co., Ltd.* and Nanjing Hydraulic Research Institute. 2016. Study on the ecological slope protection structure.

Nanjing Hydraulic Research Institute. 2016. Experiment study on the ecological slope protection structure model.

Yang H.J., Uchinda T. and Sheng L.X. *et al.* 2004. Advance in studies on the restoration of the damaged riparian ecosystem. *Journal of Northeast Normal University (Natural Science Edition)*, (1):95–100.

Yao S.M. and Yue H.Y. *et al.* 2012. Development trend of eco-revetment works in middle and lower reached of Yangtez River. *China Water Resources*, 6:18–21.

Hydraulic Engineering V – Hong, Liu & Xie (Eds)
© 2018 Taylor & Francis Group, London, ISBN 978-0-8153-7471-8

Numerical strategy for predicting erosion in a dredging pump impeller

T. Guo, L. Zheng, H. Zhuang & M. Liu
CCCC National Engineering Research Center of Dredging Technology and Equipment Co. Ltd., Shanghai, China

ABSTRACT: Numerically predicting erosive wear in dredging pumps is an important research aspect. Strong interactions between discrete particles and continuous water increase the complexity of the flow. The sand-water flow in a continuous elbow and the associated influencing factors, such as turbulence model, near-wall treatment, wall roughness, and volume fraction of sand, were simulated, and then the results were compared. Based on the comparison results, a problem-solving strategy is presented, in which water with small sand particles is assumed as a type of homogeneous Newtonian fluid and only the effect of large sand particles is considered. This strategy is used to predict the slurry flow and erosive wear in the impeller of a dredging pump DRC257-900. The result is compared with that of an actual dredging pump impeller. It is found that the strategy with two-way coupling can accurately predict the location of erosive wear in the impeller with low computational costs.

1 INTRODUCTION

Erosive wear refers to the thickness loss of the bounded wall caused by impacting, cutting, and grinding of particles in a particle-fluid flow. Erosion commonly occurs in oil and gas transmission, slurry transport, and other conveying systems (Bikbaev et al. 1972, Flemmer et al. 1988), which is considered as the main reason for the wearing out of pumps (Gandhi 1998). Previous studies on erosion in pumps have shown that the diameter of particle (Walker & Bodkin, 2000; Pagalthivarthi & Gupta, 2009; Li et al., 2011; Pagalthivarthi et al., 2013), the impact angle (Liu et al., 2004; Dong et al., 2009), and the volume fraction of sand (Ahmad et al., 1986; Addie et al., 1987; Peng et al., 2015) are the main factors influencing erosion. Figure 1 shows the erosive wear caused by silt and medium-coarse sand in the same type of

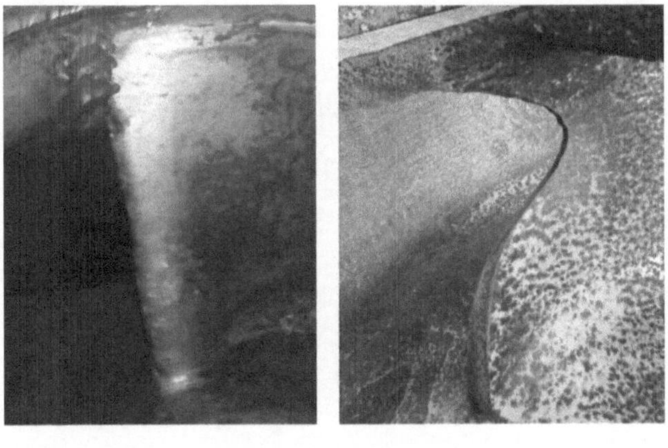

Figure 1. Erosive wear on dredging pump impellers.

impellers made of high chromium cast iron. In Figure 1, the left image shows an impeller that was used to dredge silt for 1 year, and the right image shows an impeller that was used to dredge medium-coarse sand for 8 months. It can be found that sand particles with large diameters damage the impeller heavily, despite the latter having thick blades and made of wear-resistant material. Impeller is the most important part of the pump. For improving the performance or increasing the life of pump, predicting the location and degree of erosive wear in impeller is inevitable (Roco & Addie 1987).

The process of erosion prediction is to find out the way in which the particle impact the wall and then to solve the erosion rate (ER) via experimental or semi-experimental expressions. Simulating the sand–water interaction of the flow and tracing the trajectories of sand are the key steps. Several numerical models are compared in the following erosion simulations of sand-water flow to determine a strategy that can meet the engineering demands of accuracy and cost.

2 COMPARISON OF THE NUMERICAL MODEL

The commercial CFD code ANSYS Fluent is used to investigate the influences of turbulence model, near-wall grid treatment, wall roughness, and considering particle volume fraction on the sand-water flow. In ANSYS, the discrete phase model (DPM) combines a particle motion equation (Eq. 1), which is described via the Lagrangian method, with the Navier–Stokes equation, which is described via the Euler method, and achieves the particle tracing:

$$m_d \frac{du_d}{dt} = F_{drag} + F_g + F_{other} \tag{1}$$

The right-hand side of the equation contains drag force, gravity force, and other forces, such as pressure gradient force, virtual mass force, and lift force.

The flow in pump is too complex to describe the direct mechanism of the flow-field differences, so a pipe with continuous bending is chosen as the object of study. Figure 2 shows the diagram of the pipe of total length 13.4 m, inner diameter 0.66 m, turning radius of bending 0.99 m, and bend angles 60°, 90°, and 120°. Owing to continuous bending, inertia force and wall constraint leads to secondary flow that significantly affects the whole flow field and enhances the inhomogeneity of the flow in the pipe.

The velocity of water on the inlet is set to 10 m/s, and the particle diameter is set to 0.75 mm uniformly. The *Re* number of the flow is approximately 6.6×10^6, and *St* number is approximately 1.25, both are similar to the flow in an actual engineering case, in which medium-coarse sands ($d_{50} = 0.0075$) are dredged by pump DRC257-900. A fully reflected wall and an erosion formula, in which only the influence of the magnitude of impacting velocity

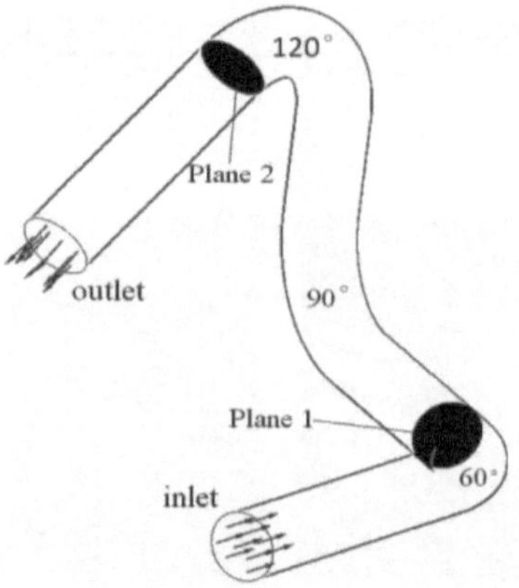

Figure 2. Geometry of the bending pipe.

is considered, are used in this section. It makes the erosion rate not physical, but meaningful to display the count and velocities of impacting particles.

2.1 *Turbulence model and near-wall treatment*

In general, two-equation turbulence models satisfy the precision requirement in CFD calculations and have high "cost–benefit" ratio. Among them, the realizable k–ε model (Shih et al. 1995) and the SST k–ω model (Menter 2012) are recommended in the flow simulations of turbomachinery. The former usually combines with the scalable wall functions, and the latter is combined with the enhanced wall function. By determining Δy, the distance from the first grid to the wall and some meshes with different y+ are created and presented in Table 1. Through these meshes, water flows in the pipe are steady-simulated using the above two models and the corresponding wall functions. Velocity inlet, 0 Pa pressure outlet, and no slipping wall are given as boundary conditions. The velocities on plane 1 and the pressure on plane 2 are shown in Figure 3. The results with mesh 4 are omitted as they are very similar to those with meshes 3 and 5.

In Figure 3a, the numerical results with the same turbulence model look similar, in which the y+ is not significantly greater than 400 (meshes 3 and 5). The results with overlarge y+ (mesh 1 & 2) are not physical: the near-wall flow occurs at the wrong place as the result of the artificial thickness of viscous sublayer, near the wall; the secondary flow is also influenced by the imprecise distribution of pressure gradient. The mesh with y+ no larger than 400 is acceptable. Otherwise, there are obvious differences in Figure 3b: the magnitude of velocity with the SST k–ω model is more uneven than the one with realizable k–ε model. It might be because the overestimation of turbulent viscosity in the k–ε model quickly dissipates the secondary flow caused by the 90° elbow.

Spherical sand particles are individually released with the same velocity of water at the inlet face center and other four positions, which are 0.16 m away from the center. The particles are tracked using a steady-state and one-way coupling method, in which the effect of particles on water is not considered and is used to study the influence of the water flow field. Figure 4 shows the comparisons of the particle trajectories between mesh 3 and mesh 5 with two models. The discrete random walk model is not used to ensure that the contrast is effective.

Table 1. Meshes adopted in simulations.

Number	Δy m	No. of elements	Average y+ of walls
1	0.021	90,517	7455
2	0.008	114,832	3744
3	0.002	292,128	400.4
4	0.0001	372,864	20.4
5	0.00002	497,184	4.2

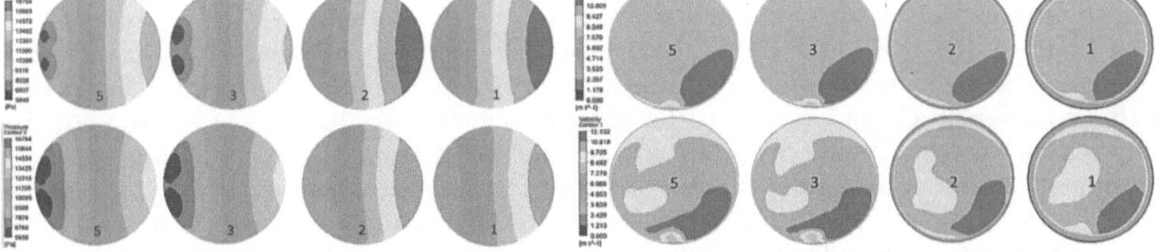

a. Contour of static pressure on plane 1 b. Contour of velocity on plane 2

Figure 3. Numerical results with two turbulence models (top: realizable k–ε model; bottom: SST k–ω model; numerals indicate the number of meshes).

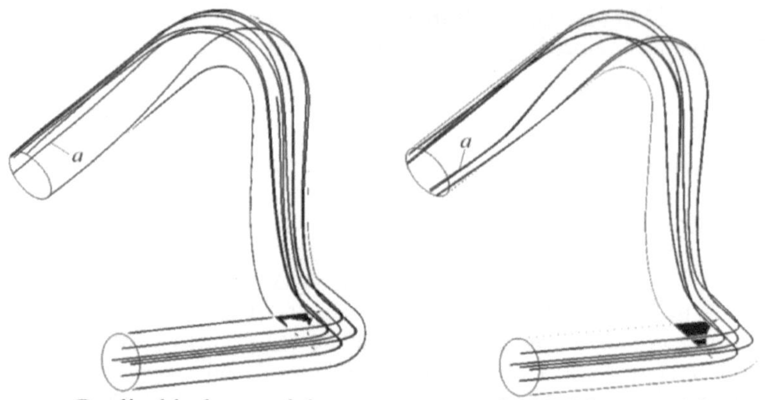

a. Realizable k–ε model   b. SST k–ω model

Figure 4. Particle trajectories in the two turbulence models.

In Figure 4, the particle trajectories in meshes 3 and 5 are similar; however, there are some differences between the two models, especially on a, the trajectory of particle released 0.16 m below the center of inlet. Compared with the realizable k–ε model, the tuning radius of trajectory a in the SST k–ω model is smaller at the third bending and the particle escapes at a lower position. Furthermore, three particles stagnate nearby the boundary layer zones in mesh 5 with the realizable k–ε model to make the track incomplete. It might be because the particle is too large when compared with the near-wall grid or because the scalable boundary treatment does not properly deal the low y+ zones. Mesh 3 and the SST k–ω model will be adopted in the following section, and the difference of the flow velocities can be kept for a distance after the elbow and similar particle trajectories can be obtained with low computing requirement.

2.2 *Wall roughness*

Wall roughness influences the impact angles of particles and then affects the prediction of sand trajectories. Solnordal & Chong (2015) did not observe the "V"-type erosion scar but an elliptic scar at the elbow in their gas-sand erosion experiment. The "V"-type scar is a characteristic numerical result of the elbow erosion simulations (Vieira et al. 2016), and is considered as the results of secondary flow (Peng & Cao 2016). Solnordal showed that the reason might be the influence of wall roughness: a rough wall leads particles to reflect with random angles and then avoid the aggregation of particles.

As shown in Table 2, three types of wall roughness have been defined on the pipe wall to investigate their influences on numerical results. The volume fraction of the uniform-size particles released at inlet is 7% (mass flow rate is 725 kg/s). The shape factor of particles is set to 0.738 (between a cube and a regular tetrahedron) according to the SEM measuring results of the sand sample in Xiamen project. A total of 16,320 trajectories are tracked in steady state during the calculation. Figure 5 shows the *ER* results under these wall conditions.

In Figure 5, the position where the maximum *ER* occurs slightly changes with different wall roughness. The maximum magnitude of the *ER* with rough wall is lower than the one with smooth wall. Considering that the particle velocities in these cases are similar, the only reason for the decrease in the *ER* is the wide distribution of the impacting position. Compared with that of wall 2, the *Rsm* of wall 1 is smaller, which means that the mean value of inclination angles, which obeys Gaussian distribution, is large. The directions of the velocities of the reflected particles are more diverse with a larger distribution of inclination angles, and then the maximum value of the *ER* is lower. The values of the *ER* for walls 2 and 3 are the same, which means that the rough wall in the reflection model affects only the inclination angles. The wall roughness weakens the aggregation effect by

Table 2. Three types of wall roughness.

	Mean roughness (*Ra*)	Standard deviation (*Rq*)	Mean roughness slope (*Rsm*)
Number	m	m	m
1	8E-5	2E-5	1.2E-4
2	8E-5	2E-5	2.4E-4
3	4E-5	1E-5	1.2E-4

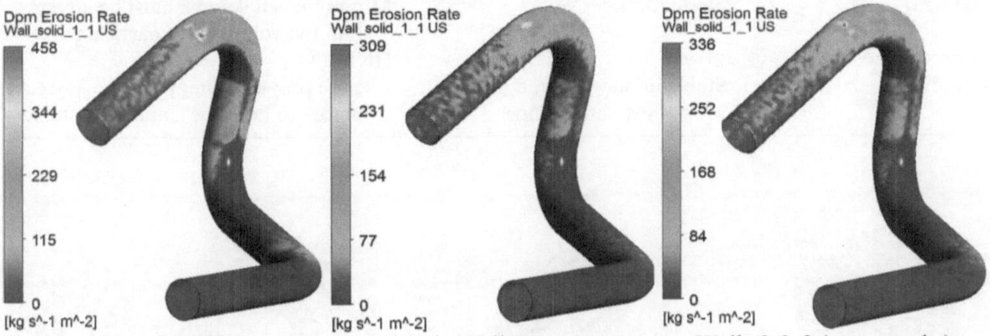

a. Smooth wall b. Wall 1 c. Walls 2 & 3 (same results)

Figure 5. *ER* results with different wall roughness conditions.

increasing the diversity of reflection, and can be used as an effective method in the DPM to fit the practical situation.

2.3 *Particle volume fraction*

In the DPM, the volume of the particles is not concerned in the flow simulations. This treatment does not meet the requirement, so the DPM can only be used in the cases that the particle–fluid flow is dilute (the particle volume fraction is not greater than 10–12%). In the actual flow in dredging pump, the average volume fraction of solid is approximately 20%, and the maximum value in some region is even above 50%. In this case, the influence of solid volume cannot be ignored, and the dense discrete phase model (DDPM) should be used to consider the coupling of the fluid and solid.

Table 3 presents the main differences between the DPM and the DDPM. Although the DDPM considers the volume fraction of particles in the simulation, it has many drawbacks that limit its wide application in rotating machinery. The volume of a particle parcel must be smaller than the minimum grid volume in the DDPM; otherwise, the volume fraction of particles might exceed 1. Therefore, the flow near the wall boundary may not be well predicted if the first grid thickness does not satisfy the requirement of y+. Periodic and interface boundaries are also not recommended, because a false velocity might be caused when a particle crosses them. It is worth mentioning that the DDPM requires more calculation as two-way, transient strategies are needed to keep the simulation stable and accurate.

In order to study the differences between the DDPM and the DPM, the sand-water flow of 7% concentration of sand is simulated. In the DDPM simulation, 500,000 particle parcels, the diameter of which is set to 0.011 m, are uniformly distributed in the pipe initially. Mesh 1 in Table 1 is used to ensure that the minimum cell volume is obviously larger than the parcel volume. The KTGF model is used to simulate the "viscosity effect" of particles caused by collision and interactions with water. The packing limit of the sand is set to 63%, and the transition factor is set to 0.75, to disperse particles where they are too dense. Syamlal-O'Brien granular viscosity model (Syamlal et al. 1993) and Wen–Yu drag law (Wen

Table 3. Main differences between the DPM and the DDPM.

Aspects	DPM	DDPM
Fluid–solid coupling	One-way or two-way	Always two-way
Coupling equations	Momentum (partly)	Mass and momentum (fully)
Collision	Not considered	Based on the KTGF (kinetic theory of granular flow) or DEM (discrete element method)
Drag force	Based on the *Re* number	Based on particle volume fraction and *Re* number
Grid size	No limitation	Minimum cell volume must be larger than the volume of a particle parcel
Theoretical accuracy	Low	High
Stability	Stable in one-way and most two-way simulations	Use the proper strategy and under-relax factors to keep the simulation stable

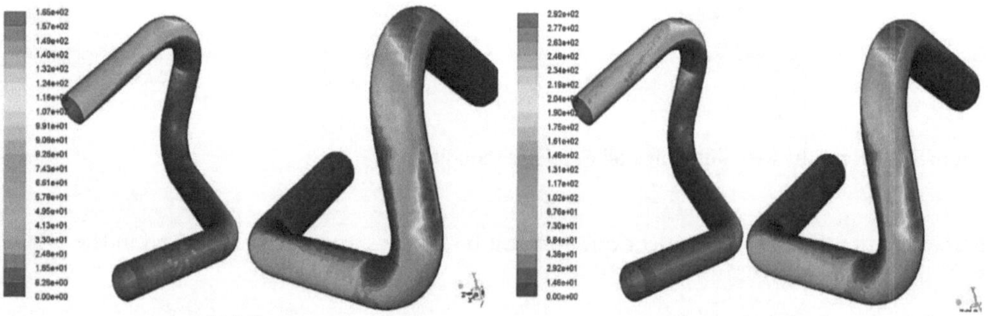

a. At 1.5 s　　　b. At 2.5 s

Figure 6. Numerical results with the DDPM for the erosive wear in the bending pipe.

&Yu 1966) are used in equations. Phase-coupled SIMPLE scheme and QUICK discretization are used to solve them. The under-relaxation factors for pressure, momentum, and the DPM source are set to 0.7, 0.3, and 0.8, respectively. Node based average method is used for the DPM source terms and the DDPM variables. The total simulation time is 2.5 s, and the time step is set to 0.0025s. In every time step, approximately 1950 parcels are injected into the pipe from inlet and there are 570,000 parcels at the most in the pipe. The discrete phase iterates 3 times, and the continuous flow iterates 23 times in one step. The whole calculation takes approximately 30 h, that is hundreds of times larger than the DPM, with an Intel I7-4790 CPU (8 cores). Figure 6 shows the predicted erosive wear on the pipe wall using the DDPM at 1.5 and 2.5 s.

In Figure 6, the locations of the erosive wear do not change obviously during the 1 s interval from 1.5 to 2.5 s. The difference is that a "V"-type scar can be observed on the wall which is just behind the bending of 120°, which might be the result of development of the secondary flow. Compared with the DPM results in Figure 5, the location where the maximum erosive wear appears in Figure 6 is close to the bending, and the maximum value of the *ER*, which is only approximately 127, is only 30–40% of the DPM result. It might be because that some influence of particles on the wall will be suppressed for the collisions with other particles. Particles cannot accumulate limitlessly if the local concentration is high. This "shelter" effect does not exist in the DPM because the volume fraction of discrete phase is not considered.

3 PREDICTION OF THE EROSION IN THE IMPELLER

The flow in dredging pump is more complex than that in bending pipe. Considering the shortcomings of the DDPM with respect to cost and stability, a problem-solving strategy based on the DPM for high-concentration flow is used. It is found that large particles destroy wall more severely than small particles in both practice and experiments. The *St* number of small particles in dredging flow is less than 1, which means that small particles tend to accompany with water and not impact the wall violently. Therefore, the mixture of small particles and water is assumed as a Newtonian flow with homogeneous density and dynamic viscosity (Thomas 1965). Only large particles with volume fraction less than 10% are considered as the discrete phase. With this assumption, a fine mesh with periodic boundary can be used to solve the flow field and then to predict the erosive wear. The flow and erosion in a single channel of the impeller is simulated using this strategy for the dredging pump DRC257-900. The mesh with 154,360 elements is shown in Figure 7. The average y+ of the wall boundary is 357.

In dredging works being conducted in Xiamen, the average concentration of sand slurry is approximately 22%. The diameter distribution of particles is obtained by screening test. The volume fraction of sand with diameter greater than 1 mm in discrete phase is less than 30%. A similar diameter distribution is set for these sand particles in the simulation. Their shape factor is set to 0.738 as above. The mixture of the other sand and water is assumed as homogeneous fluid with density of 1264 kg/m^3 and dynamic viscosity of 1.793×10^{-3} Pa·s. The SST k–ω model and rough wall defined as wall 1 in Table 1 are used in the simulation. In addition to drag force, virtual mass force and pressure gradient force are considered in particle motion equations. Reflection coefficients presented by Grant and Tabakoff (2012) are used to compute velocities of the reflected particles. A polynomial fitting function, which is shown below, is used to consider the influence of the impact angle on the *ER* according to an experiment conducted by Yao (1988). Figure 8 shows the comparison between Equation 2 and the experimental result.

$$f(\theta) = 0.0117 + 1.444\theta + 0.453\theta^2 - 1.776\theta^3 + 0.679\theta^4 \quad (2)$$

The velocity exponent for the erosion rate equation is set to 2.41. One-way and two-way coupling methods are compared with respect to cost and accuracy. The one-way coupling simulation takes approximately 8 min, and the two-way coupling simulation takes approximately 30 min with the same computing environment as described above. Figure 9 shows the *ER* results predicted using one-way and two-way couplings individually and a photograph of the actual erosive wear in a worn impeller, which has been used in Xiamen for 8 months.

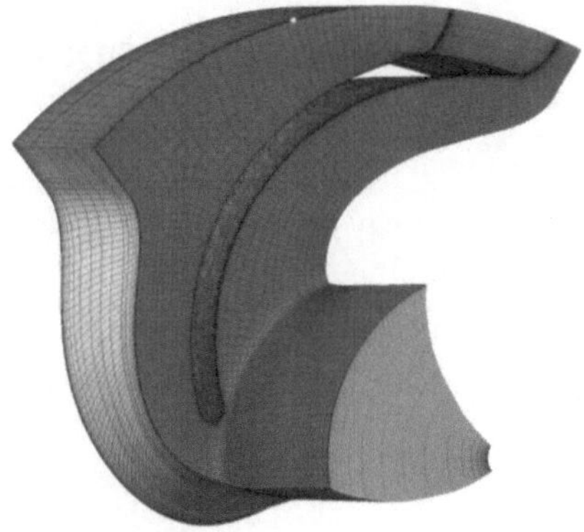

Figure 7. Mesh of a single passage in the impeller of DRC257-900.

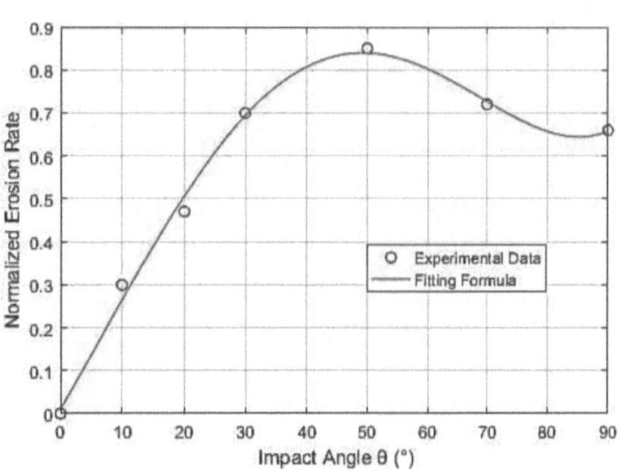

Figure 8. Impact angle function and experimental data of high-chromium cast iron.

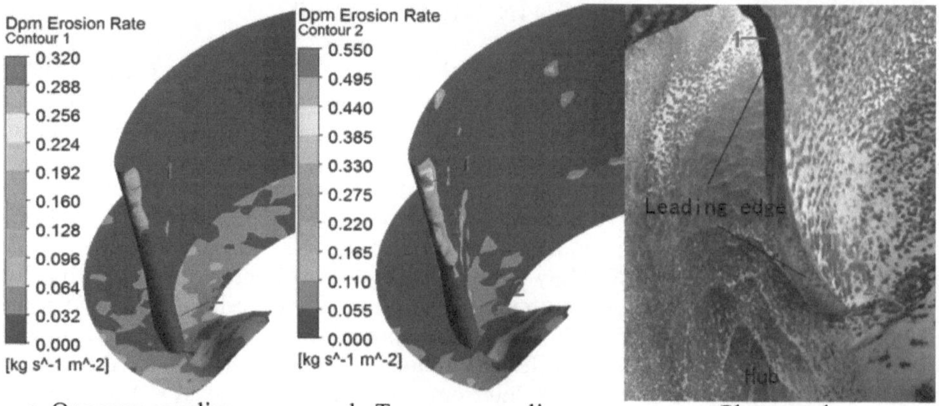

a. One-way coupling b. Two-way coupling c. Photograph

Figure 9. Erosive wear in the numerical results and the actual impeller.

As shown in Figure 9c, severe erosion scars appear in the upper part of the leading edge (location 1) and at the junction zone (location 2) between the leading blade and the hub wall. The main difference is in the distribution of the *ER*. Similarly to the actual situation shown in Figure 9c, the maximum *ER* appears at location 1 in Figure 9b, and the amplitude is larger than the maximum one in Figure 9a. It illustrates that the two-way coupling method in this strategy can predict the *ER* more accurately than one-way coupling. The time spend is much less than that in the DDPM.

4 CONCLUSIONS

In this study, the influences of the turbulence model, wall roughness, and the volume fraction of particles on the prediction of the *ER* in sand-water flow were studied using a bending pipe case. Based on the comparison results presented in this study, a DPM-based solving strategy was used to predict the *ER* of a dredging pump impeller, considering only the erosion caused by the effect of large particles. The following conclusions can be drawn from this study:

1. The SST k–ω model possesses good numerical accuracy and adaptability to different near-wall treatments. The y+ value of 400 is also acceptable if the near-wall flow is not of utmost concern in the slurry flow.
2. Wall roughness disperses the impact angles of particles and can be used to decrease the *ER* value.
3. The numerical simulation of erosion in the bending pipe simulation using the DDPM, which has a high theoretical accuracy, consumes more time, whereas there were no obvious differences in the results for the location of erosion scar between the DDPM and the DPM.
4. The erosive wear numerical result with the problem-solving strategy and the two-way coupling method agrees well with the actual impeller.

The erosion rate formula will be studied on the basis of this experiment and the whole pump erosive wear will be predicted in the future work.

REFERENCES

Ahmad, K., Baker, R.C. & Goulas, A. 1986. Computation and Experimental Results of Wear in a Slurry Pump Impeller. *Nihon Jinzo Gakkai Shi,* 200, 439–445.

Bikbaev, F.A., Maksimenko, M.Z., Berezin, V.L., Krasnov, V.I. & Zhilinskii, I.B. 1972. Wear on branches in pneumatic conveying ducting. *Chemical & Petroleum Engineering,* 8, 465–466.
Dong, X., Zhang, H.L. & Wang, X.Y. 2009. Finite element analysis of wear for centrifugal slurry pump. *Procedia Earth & Planetary Science,* 1, 1532–1538.
Gandhi, B.K., Singh, S.N. & Seshadri, V. 2002. Variation of Wear along the Volute Casing of a Centrifugal Slurry Pump. *Jsme International Journal,* 44, 231–237.
Grant, G. & Tabakoff, W. 2012. Erosion Prediction in Turbomachinery Resulting from Environmental Solid Particles. *Journal of Aircraft,* 12, 471–478.
Li, Y., Zhu, Z., He, Z. & He, W. 2011. Abrasion characteristic analyses of solid-liquid two-phase centrifugal pump. J*ournal of Thermal Science,* 20, 283–287.
Liu, J., Xu, H., Qi, L. & Li, H. Study on Erosive Wear and Novel Wear-Resistant Materials for Centrifugal Slurry Pumps. ASME 2004 Heat Transfer/Fluids Engineering Summer Conference, 2004. 101–104.
Menter, F.R. 2012. Two-equation eddy-viscosity turbulence models for engineering applications. *Aiaa Journal,* 32, 1598–1605.
Pagalthivarthi, K.V., Furlan, J.M. & Visintainer, R.J. Effect of particle size distribution on erosion wear in centrifugal pump casings. ASME 2013 Fluids Engineering Division Summer Meeting, 2013. V01CT20A005-V01CT20A005.
Pagalthivarthi, K.V. & Gupta, P.K. 2009. Prediction of Erosion Wear in Multi-Size Particulate Flow through a Rotating Channel. Fdmp *Fluid Dynamics & Materials Processing,* 5.
Peng, G.J., Luo, Y.Y. & Wang, Z.W. Research on wear properties of centrifugal dredge pump based on liquid-solid two-phase fluid simulations. Materials Science and Engineering Conference Series, 2015. 042048.
Peng, W. & Cao, X. 2016. Numerical prediction of erosion distributions and solid particle trajectories in elbows for gas–solid flow. *Journal of Natural Gas Science & Engineering,* 30, 455–470.
Roco, M.C. & Addie, G.R. 1987. Erosion wear in slurry pumps and pipes. *Powder Technology,* 50, 35–46.
Shih, T.H., Liou, W.W., Shabbir, A., Yang, Z. & Zhu, J. 1995. A new k-[epsilon] eddy viscosity model for high reynolds number turbulent flows. *Computers & Fluids,* 24, 227–238.
Solnordal, C.B., Chong, Y.W. & Boulanger, J. 2015. An experimental and numerical analysis of erosion caused by sand pneumatically conveyed through a standard pipe elbow. *Wear,* s 336–337, 43–57.
Syamlal, M., Rogers, W. & Obrien, T.J. 1993. Multiphase flow with interphase exchanges (MFIX) documentation theory guide. *Nasa Sti/recon Technical Report N,* 94.
Thomas, D.G. 1965. Transport characteristics of suspension: VIII. A note on the viscosity of Newtonian suspensions of uniform spherical particles. *Journal of Colloid Science,* 20, 267–277.
Venkatesh, E.S. 1986. Erosion Damage in Oil and Gas Wells. *Spe Rocky Mountain Regional Meeting.*
Vieira, R.E., Mansouri, A., Mclaury, B.S. & Shirazi, S.A. 2016. Experimental and computational study of erosion in elbows due to sand particles in air flow. *Powder Technology,* 288, 339–353.
Walker, C.I. & Bodkin, G.C. 2000. Empirical wear relationships for centrifugal slurry pumps: Part 1: side-liners. *Wear,* 242, 140–146.
Wen, C.Y. & Yu, Y.H. 1966. Mechanics of Fluidization. *Chem.engng Prog.symp.ser,* 62, 100–111.

Hydraulic Engineering V – Hong, Liu & Xie (Eds)
© 2018 Taylor & Francis Group, London, ISBN 978-0-8153-7471-8

Numerical simulations of flow past a submerged vertical cylinder with porous structure

Binglian Hu & Liquan Xie
College of Civil Engineering, Tongji University, Shanghai, China

Tsung-Chow Su
Department of Ocean and Mechanical Engineering, Florida Atlantic University, Florida, USA

ABSTRACT: The flow past a submerged vertical cylinder with porous structure in open channels was investigated numerically using OpenFOAM. The numerical simulations were conducted to compare the horseshoe vortex and the vortex shedding for cases of cylinders with and without porous bottom. The flow fields were calculated by solving the Three-Dimensional (3-D) Reynolds-Averaged Navier-Stokes (RANS) equations with the SST k-omega turbulence model for incompressible viscous flow by PISO algorithm. The flow mechanism was investigated by the results of velocity fields and velocity contour. The form of porous structure influenced the flow structure near the bed. The horseshoe vortex and the vortex shedding were weakened as the height of the porous section became higher, while the domain of influence of the vortex behind the cylinder was getting narrower, which may lead to a weaker scouring near the cylinder than that without porous bottom.

1 INTRODUCTION

The problem of viscous incompressible flow past a cylinder has for a long time received much attention both theoretically and numerically. Numerical solutions have been obtained for steady viscous flow over a cylinder at Reynolds numbers up to 300 based on Newton's method (Fornberg, 2006). As vorticity starts to recirculate back from the end of the wake region, this region becomes wider and shorter. Tritton (2006) had described the measurements of the drag force on circular cylinders and advanced experimental evidence for a transition in the vortex street in the wake of a cylinder at a Reynolds number around 90. Lim simulated at Reynolds number 20,000 which is large enough that many aspects of the flow depend on Reynolds number effectively (Lim et al, 2009). Both the inflow boundary layer and the flow field around the cube are developed by measurements and LES simulations, the detailed comparison between them is presented. The LES model of flow over the cylinder at Reynolds number 3900 was established by a second-order finite volume method with central-difference schemes by OpenFOAM (Lysenko et al, 2012). The results were found to be in good agreement with the previously obtained experimental data.

In most of the civil engineering applications, the Reynolds number is usually much larger than 400. Therefore, the flow is normally turbulent (Zhao et al, 2009). Zhang & Dalton (1998) established three-dimensional models to investigate the flow past a circular cylinder for a series of Reynolds numbers. The transition of the flow to turbulence they found agrees well with those in experiments (Williamson, 1992). Lei et al (2001) investigated the influence of the size of computational domain in the 3-D simulation of the flow past a circular cylinder. They finally draw a conclusion that only if the size in the spanwise direction is at least four times the cylinder diameter, the accurate 3-D wake flow can be simulated.

When the flow goes over a surface-mounted circular cylinder, the flow is complex and three-dimensional. Its essential features include (Figure 1): (1) a horseshoe vortex and the downflow appear in front of the cylinder; (2) a vortex flow pattern occurs at the sides of the cylinder; and (3) the streamlines contract and the vortex starts to shed at the cylinder downstream (Roulund

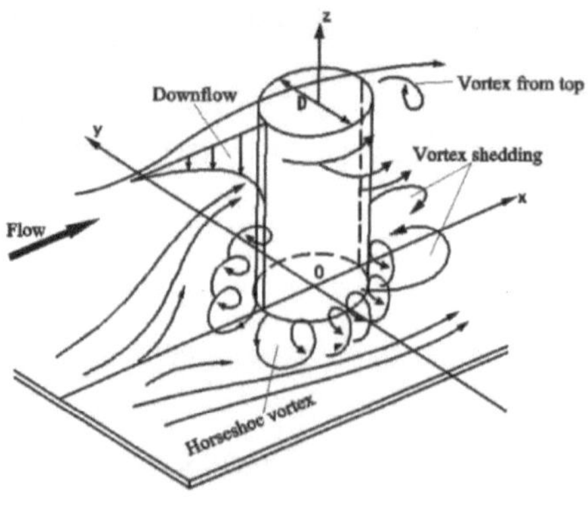

Figure 1. Definition sketch.

et al, 2005). The main flow mechanisms, the horseshoe vortex, the vortex shedding and the streamline contraction are understood (Karniadakis & Triantafyllou, 1992).

The horseshoe vortex influences the local scour in front of the cylinder. To diminish the horseshoe vortex, the action of hydrodynamic must be weakened in order to decrease the strength of scouring. Experiments conducted in clear water conditions showed the use of a slot alone with a width equal to one fourth of the pier diameter can reduce equilibrium sour depth by as much as 20% by Chiew (1992). Wang et al (2011) proposed a ring columns placed upstream of a bridge pier to protect it from local scouring.

The present study focuses on the flow field around the submerged vertical cylinder with porous structure to investigate the degree of potential scouring. To address the problem, we have selected the open source software for computational fluid dynamics, OpenFOAM. The size and the strength of the horseshoe vortex were compared in the simulation.

2 CALCULATION MODEL

An open source tool, OpenFOAM (Open Field Operation and Manipulation), was used to calculate the flow. It's the foremost a C++ library used to solve partial differential equations and ordinary differential equations. The directions are in terms of the Cartesian co-ordinates, x, y and z, which are respectively in the direction of the flow to the cylinder, perpendicular to the flow, and parallel to the cylinder. Turbulent models are mostly developed based on a high Reynolds number situation. The SST model was selected to be the turbulence model because of its large Reynolds number and weak pressure gradient. The algorithm needs to couple pressure and velocity in the incompressible cases, so the PISO (Pressure-Implicit with Splitting of Operators) type algorithm was used in this simulation. The velocity is predicted using the momentum predictor. Then, the pressure and the velocity is corrected until a predefined number of iterations is reached. Afterwards, the transport equations of the turbulence model are solved.

2.1 *Boundary conditions and algorithm*

At all wall boundaries, the fixedFluxPressure boundary condition adjusting the pressure gradient is applied to the pressure field, so that the boundary flux matches the velocity boundary condition. At the inflow boundary, constant profiles for velocities and turbulence quantities are specified. At the outflow boundary, the zero normal gradients of all quantities except for pressure are implemented.

PISO algorithm (Pressure implicit with splitting of operator), an extension of the SIMPLE algorithm to solve the N-S equations has been used in this paper. PISO is a pressure-velocity calculation procedure. In each iteration step, the latest predicted velocity will be adapt in the velocity of adjacent grid, which is better than the SIMPLE algorithm.

2.2 *Simulation cases*

Seven cases with and without porous structure were carried out in this paper. A flow past a complete vertical cylinder with the ratio of height to diameter equals 1 at Re number 5800 was simulated in case A (see Figure 2). Case B and case C have the same Re number but the different form of substructions from case A. Case B and case C have three smaller cylinders as the porous structure at the bottom. The height of the porous structures ranged from 1 cm to 3 cm, and the total height remained 6.7cm (see Table 1).

The diameters of the main upper structure were all 5.7 cm, while the smaller cylinders below were 2cm. The model of case B2 is shown in Figure 3. The block meshes of case B2 and C2 are shown in Figures 4–5.

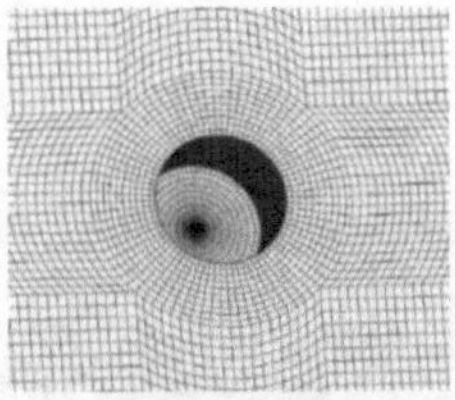

Figure 2. The local mesh of case A.

Table 1. The simulation cases of Case B and Case C.

Cases				Case B	Case C
Porous structures				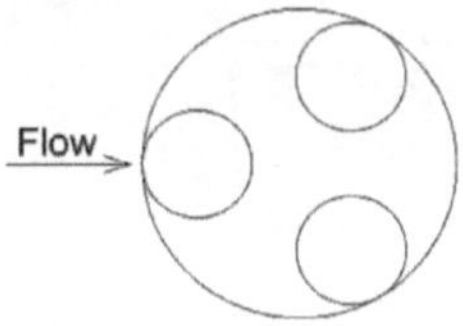	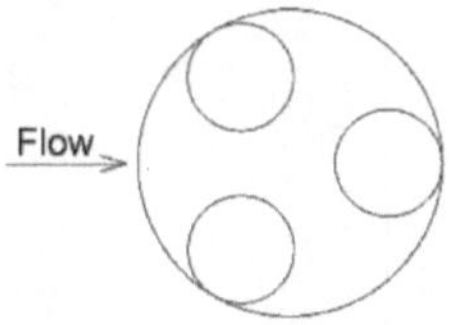
H_{lower}	1 cm	H_{upper}	5.7 cm	B1	C1
	2 cm		4.7 cm	B2	C2
	3 cm		3.7 cm	B3	C3

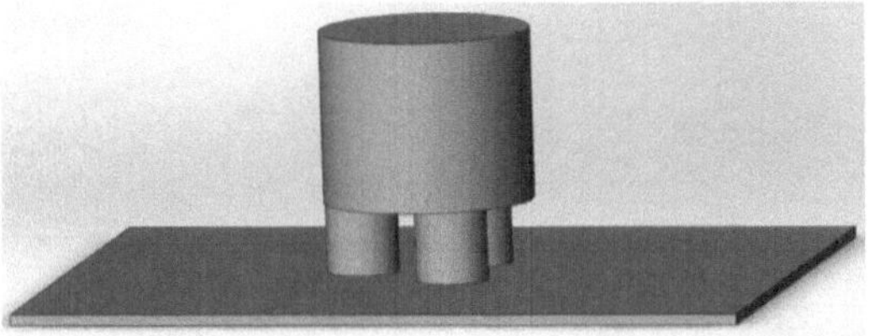

Figure 3. The model of case B2.

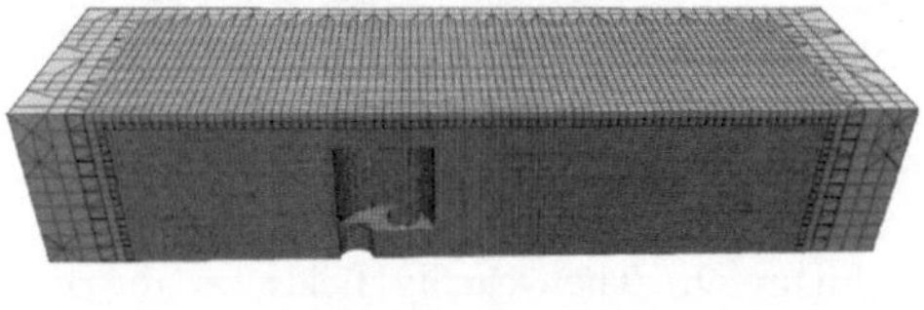

Figure 4. The block meshes of case B2.

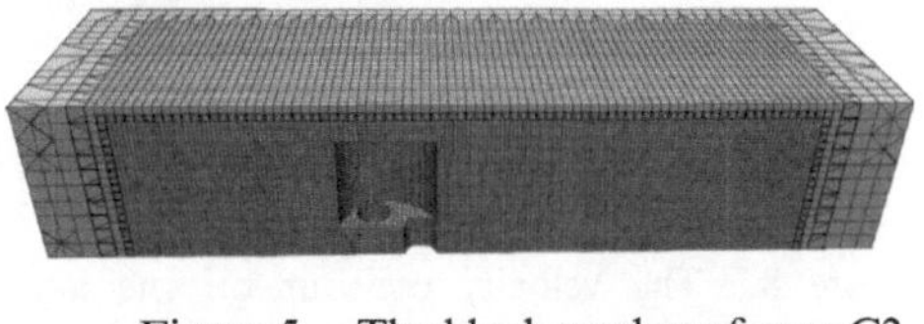

Figure 5. The block meshes of case C2.

3 CALCULATION RESULTS

Figures 6–11 showed the velocity contour and the velocity fields on the x-y plane near bed in case A, B and C. The streams contract at the sides of the cylinder and the vortex shedding

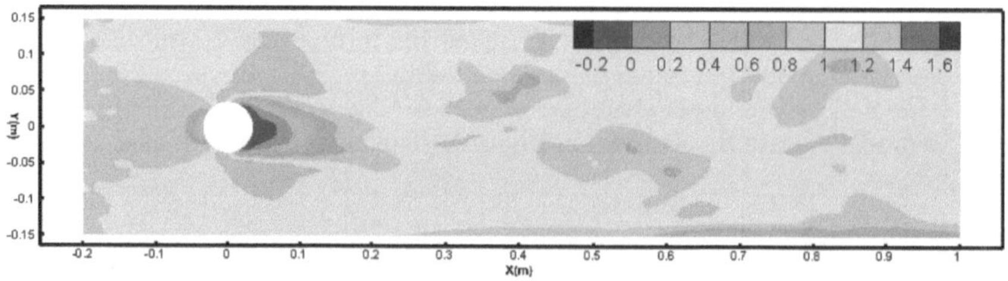

Figure 6. The velocity contour on the x-y plane near bed in case A.

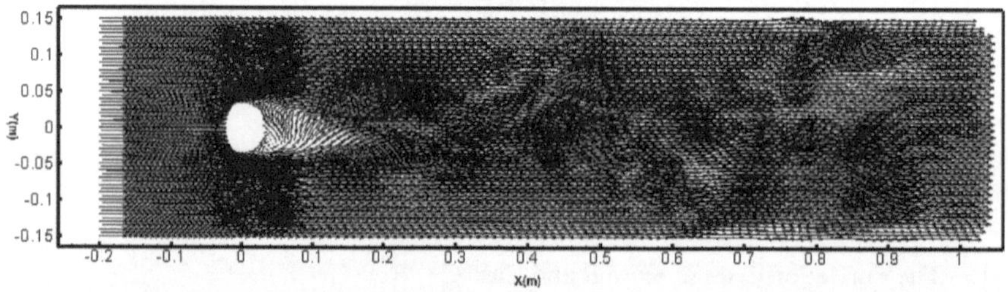

Figure 7. The velocity fields on the x-y plane near bed in case A.

a.
b.
c.

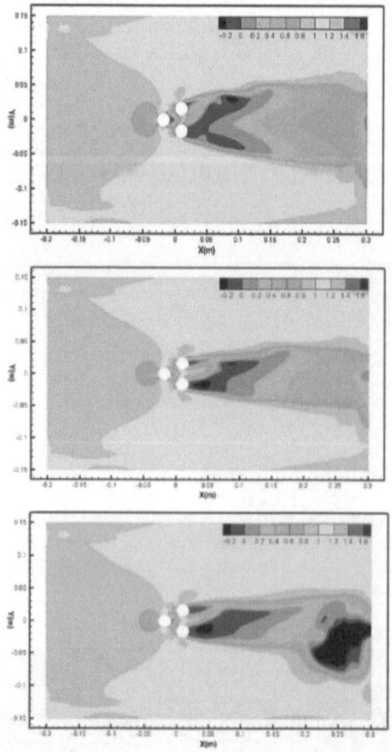

Figure 8. The velocity contour on the x-y plane near bed in case B (a: B1, b: B2, c: B3).

a.
b.
c.

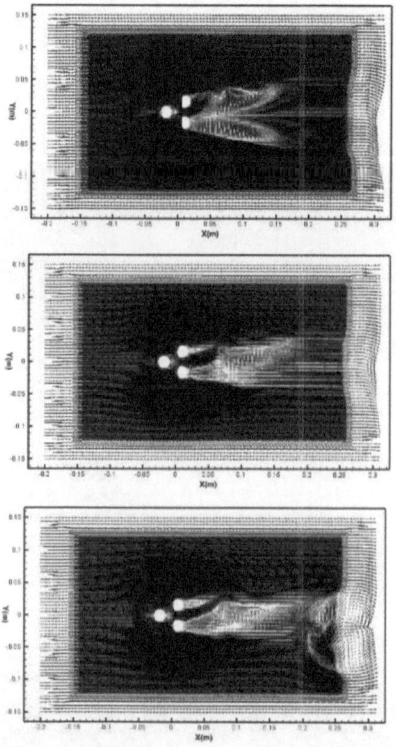

Figure 9. The velocity fields on the x-y plane near bed in case B (a: B1, b: B2, c: B3).

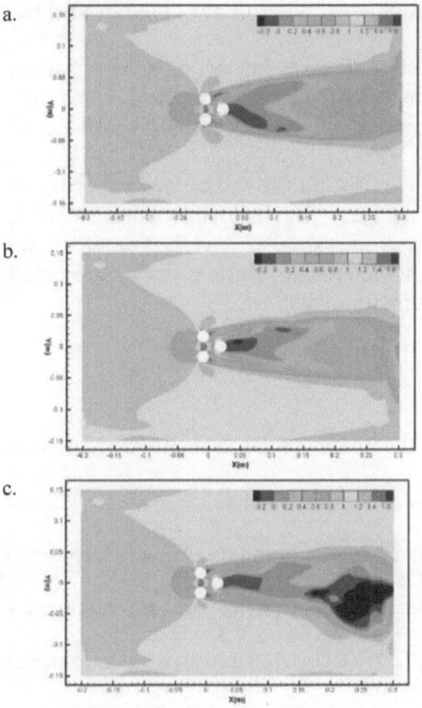

Figure 10. The velocity contour on the x-y plane near bed in case C (a: C1, b: C2, c: C3).

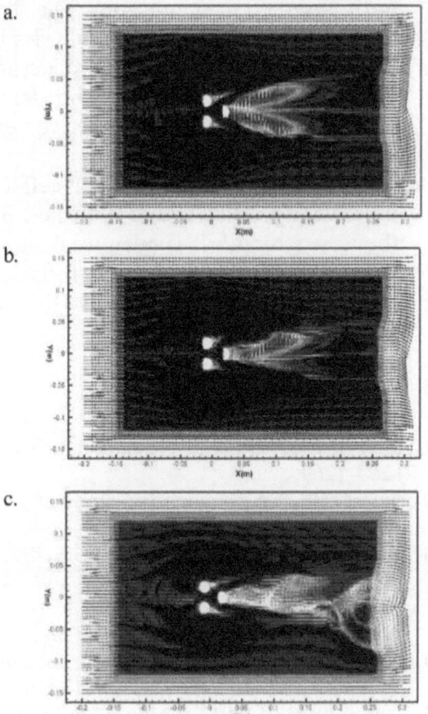

Figure 11. The velocity fields on the x-y plane near bed in case C (a: C1, b: C2, c: C3).

behind the cylinder is visible in the simulation. The horseshoe vortex in front of the cylinder is observed. The results show a reverse flow both in the upstream and downstream areas near the cylinder. The vortex on the top of the cylinder and the cylinder's spanwise vortex are observed clearly. As the porous structure become higher, the scope of the vortexes become narrower and the horseshoe vortex and the vortex shedding were weaker in both case B and C.

Compared with case B, the influenced domains of vortex behind and at sides of the cylinder were narrower in case C, but the velocity in front of the cylinder in case C was usually larger than that in case B. The porous structures of case C is better in the aspect of scouring behind and at the sides of the cylinder, but case B can cause a smaller scouring in front of the cylinder. However, both case B and case C have weaker horseshoe vortex and vortex shedding than a normal submerged cylinder in case A.

4 CONCLUSION

This paper presents the study on the flow mechanism past a submerged vertical cylinder with and without porous structures numerically based on OpenFOAM. The flow mechanism was investigated by the analysis of velocity fields and the velocity contour. The main conclusions are summarized as follows:

1. The flow field past a submerged vertical cylinder with porous structure and the vortex in front of and behind the cylinder are simulated using the Reynolds-averaged Navier-Stoke equations by the SST k-omega turbulence model based on the PISO algorithm.
2. The numerical simulation illustrates the main characteristics of the flow around the submerged cylinder. The horseshoe vortex in front of the cylinder and the vortex shedding are

both observed in simulations. The streamlines contract at the sides of the cylinder and the vortex shedding behind the cylinder is also visible.

3. The influence of the porous structure on the vortex is of great importance. The porous section separated the flow and decreased its intensity. An increase of the porous section height can cause a weaker vortex shedding, which may lead a decreased scour depth around the cylinder.
4. The flow direction has a delicate effect on the flow field around the submerged vertical cylinder with porous structure. In case B, the flow was separated by the former small cylinder, the flow strength had been weakened, which may lead to a smaller vortex. Therefore, we must consider the more dangerous case in case B, in which the structures may be subjected to local scouring by the erosive flow with higher intensity of vortex.

REFERENCES

Chiew, Y.M. 1992. Scour protection at bridge piers. *Journal of Hydraulic Engineering,* ASCE 118(9), 1260–1269.

Fornberg, B. 2006. A numerical study of steady viscous flow past a circular cylinder. *Journal of Fluid Mechanics,* 98(04), 819–855.

Karniadakis, G.E. & Triantafyllou, G.S. 1992. Three-dimensional dynamics and transition to turbulence in the wake of bluff objects. *Journal of Fluid Mechanics,* 238, 1–30.

Lei, C. et al. 2001. Spanwise length effects on three-dimensional modelling of flow over a circular cylinder. *Computer Methods in Applied Mechanics and Engineering,* 190, 2909–2923.

Lim, H.C. et al. 2009. Flow around a cube in a turbulent boundary layer: LES and experiment. *Journal of Wind Engineering and Industrial Aerodynamics,* 97(2), 96–109.

Lysenko, D. et al. 2012. Large-Eddy Simulation of the Flow Over a Circular Cylinder at Reynolds Number 3900 Using the OpenFOAM Toolbox. *Flow, Turbulence and Combustion.,* 89(4), 491–518.

Roulund, A. et al. 2005. Numerical and experimental investigation of flow and scour around a circular pile. *Journal of Fluid Mechanics,* 534, 351–401.

Tritton, D.J. 2006. Experiments on the flow past a circular cylinder at low Reynolds numbers. *Journal of Fluid Mechanics,* 6(04), 547.

Wang, C. et al. 2011. Ring columns as pier scour countermeasures. *International Journal of Sediment Research.* 26(6), 353–363.

Williamson, C.H.K. 1992. The natural and forced formation of spot-like vortex dislocations in the transition of a wake. *Journal of Fluid Mechanics,* 243, 393–441

Zhang, J. & Dalton, C. 1998. A three-dimensional simulation of a steady approach flow past a circular cylinder at low Reynolds number. *International Journal for Numerical Methods in Fluids,* 26(9), 1003–1022.

Zhao, M. et al. 2009. Direct numerical simulation of three-dimensional flow past a yawed circular cylinder of infinite length. *Journal of Fluids and Structrues,* 25(5), 831–847.

Hydraulic Engineering V – Hong, Liu & Xie (Eds)
© 2018 Taylor & Francis Group, London, ISBN 978-0-8153-7471-8

Effects of land forming on hydrodynamics in the Yangtze Estuary

X.P. Dou, M.X. Cao, Q. Xu, X.F. Luo, H.C. Wang & C.T. Lu
Key Laboratory of Port, Waterway and Sedimentation Engineering of MOT, Nanjing Hydraulic Research Institute, Nanjing, China

ABSTRACT: In this study, the recent change in the Yangtze Estuary is described using the results of the reclamation project of Hengsha Shoal. A 2D numerical model of tidal currents for the estuary is established. Some changes in the tidal capacities, the peak velocities of flood, and the ebb currents in the North Channel and the North Passage were found after land-forming projects at Hengsha East Shoal (HES). The diversion ratio of the upper section in the North Channel decreased by 0.49%, while that of the lower section in the North Passage decreased by 0.09%. The high tide level in the North Channel lowered by 5–15 cm in some regions, and the high tide level in the North Passage slightly increased by 6–15 cm. The effects of land forming at HES on the flow field mainly concentrate on the North Channel and the North Passage, and the effect on the North Channel is larger than that on the North Passage and less on other reaches.

1 INTRODUCTION

Among the existing beach resources in the Yangtze Estuary, Hengsha East Shoal (HES) is a region that combines many advantages such as geographical position, land, shoreline, and channel. It lies between the 12.5 m deep-water channel of the North Passage in the south and the North Channel in the north (designed 10 m navigation channel). The land area can be increased to approximately 480 km^2 through hydraulic reclamation projects. If a large-scale dug-in harbor district is constructed in the region, then the shoreline can be increased to longer than 100 km and can directly enter the deep-water region to realize the breakthrough of 20 m deepwater port. The development of HES can solve issues related to land shortage, water depth, and shoreline restrictions in Shanghai Port and other problems faced by the ever-expanding Shanghai City currently. Besides, knowing that sediment from Yangtze River has the tendency to decrease, taking full advantage of dredging silt in deep-water channel of the Yangtze Estuary can not only provide valuable sediment resources for land forming at HES but also reduce pollutions of dumped silt in the sea area.

Some scholars have studied the development of HES. Cheng J.Y. et al. (2013) and Bao Q.F. et al. (2013) pointed out that land forming at HES provided new opportunities for the development of Shanghai Port. Yu Z.Y. et al. (2013), Shao R.S. et al. (2013), and Cheng Z.L. et al. (2013) demonstrated the necessity and possibility of port construction at HES. Ge J.Z. et al. (2013a, 2013b) investigated the influence of port construction at HES on the surrounding water areas. On the basis of these studies, here we analyze the feasibility of land forming at HES in the aspects of natural general conditions and influences of land formed through the reclamation project on the surrounding water areas and waves, tides, and so on, and provide technical support for the utilization of land of HES. Natural conditions of the engineering region are described as follows:

Located in the east of Hengsha Island, HES is a large-scale shoal naturally formed at the Yangtze Estuary. The natural water depth is –2 to 0 m. Since 2003, the reclamation project of HES, both constructed and under construction, has been divided into eight phases, the first phase of which spanned from December 2003 to December 2004 with promoted siltation of

53,000 mu. The second phase was from April 2006 to May 2008, and its promoted siltation was 47,000 mu. The third phase spanned from September 2006 to November 2008 with a reclaimed land of 26,000 mu. The fourth phase was from December 2007 to June 2009 and promoted the siltation of 22,600 mu. The fifth phase was from March 2009 to January 2011 with a newly constructed levee of length 19 km. The sixth phase spanned from 2006 to November 2015 whose reclaimed land was 26,000 mu. The seventh phase was from January 2015 to December 2017 whose reclamation of land was up to 20,200 mu. The eighth phase commenced in November 2016 and was planned to be completed in 2020 with claimed land covering 63,600 mu. At present, HES has developed the region to approximately 23 km in the east of Hengsha Island with the total exploited area of about 110 km^2. The research object of this study is the shoal formed above −5 m elevation of HES with an area of 370 km^2 (shaded area in Figure 1).

1.1 *Hydrodynamic conditions*

The annual mean runoff volume at the Yangtze Estuary is 28,700 m^3/s, and the volume of runoff in the flood season (May to October) constitutes 70% of the annual value. The annual average tidal range (Zhongjun station) is 2.66 m, and the extreme tidal range is 4.62 m. The total tidal influx at the entrance of the estuary ranges from 1.3 billion m^3 (neap tide in dry seasons) to 5.3 billion m^3 (spring tide in flood seasons). The annual average wave height at Niupijiao Station of the lower entrance at the North Passage of the estuary is 0.7 m, and the extreme wave height is 4.11 m. Tides outside the estuary are regular semi-diurnal tides, and tides within the estuary are irregular half-day shallow sea tides.

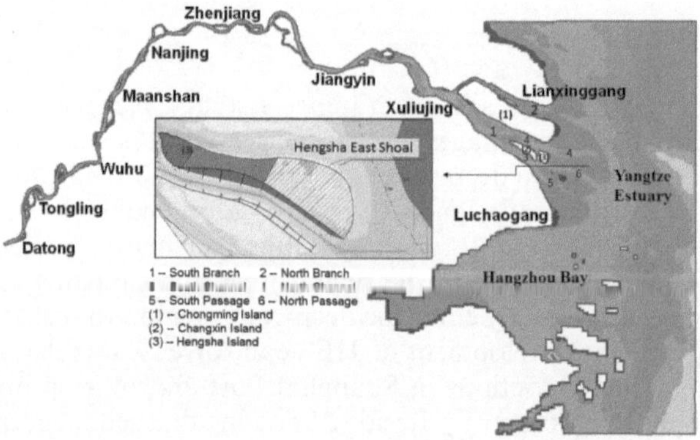

Figure 1. Engineering research area.

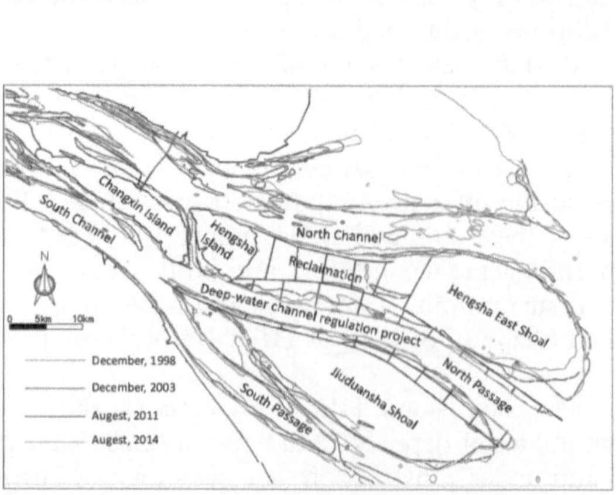

Figure 2. Recent changes in isobath of −5 m in the Yangtze Estuary.

1.2 *Sediment*

The sediment at the Yangtze River estuary is mainly transported from the Yangtze River Basin into the sea by means of suspension. The coarse part deposits at the delta outside the estuary, and the fine particles are carried to the open sea. Since the application of water storage of Three Gorges Project in 2003, the sediment amount transported into the sea has further reduced and the annual average sediment runoff is approximately 100 million t. A large number of shoals distribute at the sand bar area. Under the effect of wind waves and tides, the sediment between flat and channel is exchanged frequently and is the main source of back-silting sediment in the deep-water channel.

1.3 *Evolution of the river reach at the sand bar area*

After implementing the deep-water channel regulation project at the Yangtze Estuary in 1998, the 12.5 m deep channel at the North Passage has been dredged. The form of riverbed at the North Passage is transformed from U shape to V shape, and the sediment deposits between spur dikes. The South Passage is influenced by the deepwater channel project that the river bed scour of its upper main channel extends downstream and the sand bar shortens and deposits higher at its lower section. The sand bar of the North Channel remains in the natural evolution, which changes little in channel morphology, water depth at the top of shoal, and so on. The crest level of each sand bar beach has small annual variation of amplitude with features of their seasonal changes such as sedimentation in flood seasons and scouring in dry seasons. After implementing the deepwater channel project at the Yangtze Estuary, HES has deposited higher with stable sand bodies in recent years. Since 2003, the substantial reduction of sediment flux into the sea has not yet shown its influence on the evolution of river reaches at the sand bar (Figure 2).

2 ESTABLISHMENT OF THE NUMERICAL MODEL

The model included the whole Yangtze Estuary and Hangzhou Bay (Figure 3). The upper boundary extended to Datong of Anhui Province at the tide limit of the Yangtze River, the open sea boundary extended to –50 m isobaths outside the Yangtze Estuary, the north boundary extended to near the Yancheng Port of Jiangsu Province, and the south boundary extended toward Ningbo, Zhejiang Province. The actual measured topography in 2015 was adopted as the terrain within the scope of –20 m isobaths from Jiangyin to outside the estuary. The topography ranging from Jiangyin upward to Datong was measured in 2003, and the rest were added with the newest sea chart. The model adopted the triangular net with 138,281 units and 70,587 nodes. The length of the side of a grid was approximately 150 m on average, and the minimum grid in the engineering region was 38 m.

The model was verified with synchronic hydrologic data at the Yangtze Estuary measured from July 30 to August 30, 2015 (for measured points, see Figure 3). Their tidal level (Figure 4), current velocity, and direction (Figure 5) were compared. The tidal-level proc-

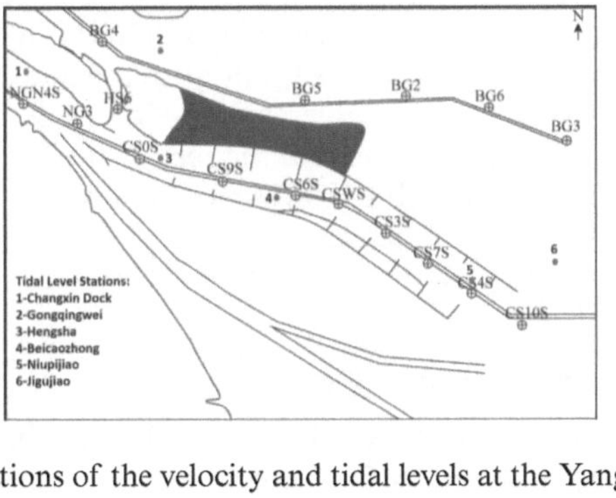

Figure 3. Measurement stations of the velocity and tidal levels at the Yangtze Estuary.

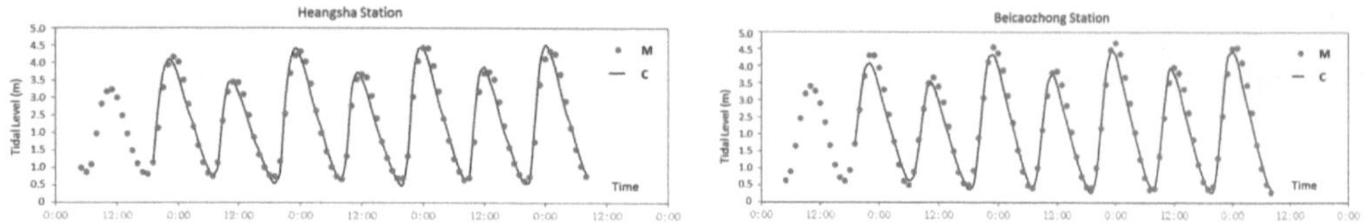

Figure 4. Process validation of tidal levels at the Yangtze Estuary (M—measured values, C—calculated values).

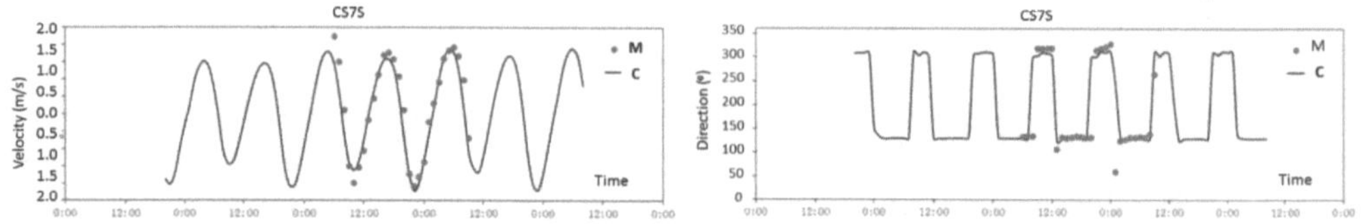

Figure 5. Process validation of the current velocity and direction at the Yangtze Estuary.

esses of the model well matched the measured values. The errors of the high and low tidal levels are controlled within 10 cm, and the phase error was less than 30 min. The peak value of the velocity and its phase correspond to actual measurement. The deviation of average sediment concentration of all the stations is < 20%. This model is used to investigate hydrodynamic changes caused by land forming at HES and sediment deposition after harbor construction.

3 INFLUENCE OF LAND FORMING AT HES ON SURROUNDING AREAS

3.1 *Tidal capacity variation in typical section*

After land–forming projects at HES, the ebb tide capacity and flood tide capacity of the upper and lower sections in the North Channel decreased in varying degrees; both the ebb tide capacity and the flood tide capacity of the upper sections in South Channel increased in varying degrees; the ebb tide capacity of the upper sections in the North Passage slightly decreased, and the flood tide capacity slightly increased; both the ebb tide capacity and the flood tide capacity increased on the lower sections in the North Passage. Calculations of the diversion ratio of each section show that the diversion ratio of the upper section in the North Channel decreased by 0.49%, while that of the lower section in the North Passage decreased by 0.09%.

3.2 *Flow-field changes in surrounding areas*

It can be seen from the difference values of flood and ebb peak velocity changes (Figure 8) before and after the project that the flood and ebb peak velocity in the deep-water channel of the North Passage increased in varying degrees, but both them decreased between spur dike at the lower section of the North Passage; except for the lower section of the North Channel where the flood peak velocity slightly increased, the rest of the North Channel all decreased and the ebb peak velocity decreased with the influence area of the flood peak velocity larger than that of the ebb peak velocity. Influences of the land forming at HES on the flow field mainly concentrate on the North Channel and the North Passage and less on other reaches.

3.3 *Changes in tidal levels in surrounding areas*

It can be seen from the tidal-level changes (Figure 8) that the high tide level in the water area in the north of HES has lowered by 5–10 cm in most regions and by 10–15 cm in some

Figure 6. Sections of the diversion ratio at the Yangtze Estuary.

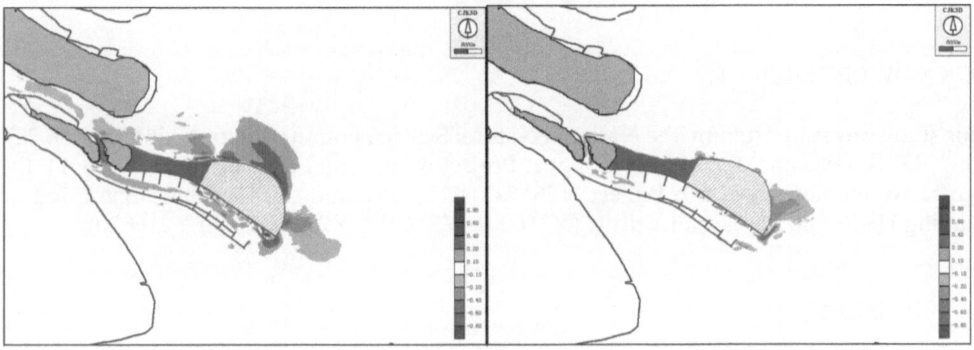

Figure 7. Difference values of flood and ebb peak velocities after land–forming projects at Hengsha (left – flood peak; right – ebb peak).

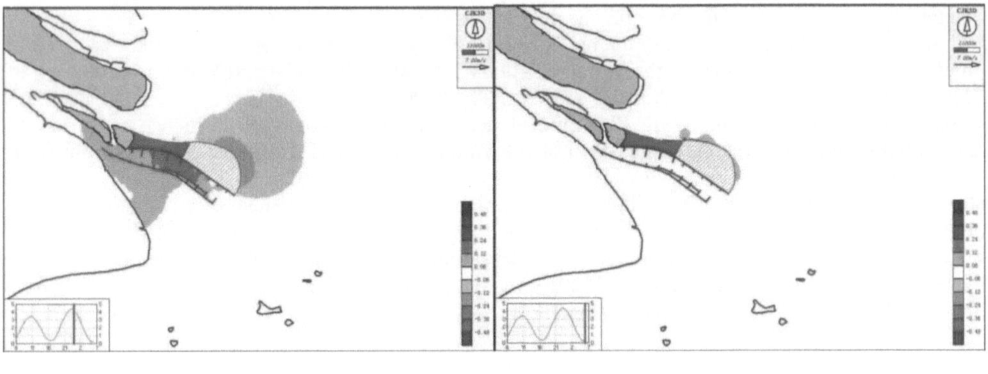

Figure 8. Tide-level changes before and after land-forming projects at Hengsha (left – high tide level; right – low tide level).

regions. However, the high tide level in the North Passage has slightly increased by 6–15 cm. The scope and extent of the influence brought by the land forming at HES on the high tide level is in general larger than that on the low tide level, which is because the shoals above −5 m after land–forming projects at HES blocked floodplain flows at high tide levels. Because the water areas are wide in open seas, flows disperse toward other directions after being blocked by the reclamation projects. Therefore, high and low tide levels have slightly changed, and the average intermediate tide level of all tides have hardly changed.

4 CONCLUSION

1. The −5 m isobaths at Hengsha Shoal are stable. The implementation of channel regulation at the Yangtze Estuary and the siltation circle engineering at HES provided a better boundary for the development and utilization of land resources at HES.
2. After land-forming projects at HES, the diversion ratio of the upper section in the North Channel decreased by 0.49%, while that of the lower section in the North Passage decreased by 0.09%. The high tide level in the water area in the north of HES lowered by 5–10 cm in most regions and by 10–15 cm in some regions. However, the high tide level in the North Passage slightly increased by 6–15 cm.
3. The scope and extent of the influence of land-forming construction at HES is mainly concentrated on the waters of southern and northern ports and passages with the split ratio of less than 1%, with little influence on the deep-water channel at the North Passage.
4. The above results are calculated on the basis of fixed-bed conditions without making topographic adjustment after the engineering project.

ACKNOWLEDGMENTS

This study was supported by the National Natural Science Foundation of China (51479122), the "333" Science and Technology Support Project of Jiangsu Province (BRA2015459), the Central Water Resources Fee Project (126153 – 0210283), and the Central Public Fund of Nanjing Hydraulic Research Institute (Y216001, Y216012, Y216019, and Y216020).

REFERENCES

Bao Qifan, Jiang Xia. 2013, roblems of Shanghai Port's development and the research on space expansion in futures. *Chinese Engineering Science*, 15(6): 35–40.

Cheng Jiyu, Jiang Xuezhong and He Qing. 2013, Prospects for the new marine city and deepwater harbor in the Yangtze Estuary, Shanghai. *Chinese Engineering Science*, 15(6): 11–13.

Cheng Zekun, Shao Rongxun. 2013, Technological possibility to develop the large-scaled port based on Hengsha shoal. *Chinese Engineering Science*, 15(6): 41–47.

Ge Jianzhong, Guo Wenyun, Ding Pingxing. 2013a, Hydrodynamic influence of proposed excavated-in harbor in the Hengsha Shoal of the Yangtze Estuary I: Numerical Model and Validations. *Journal of East China Normal University (Natural Science)*, 04: 79–90.

Ge Jianzhong, Guo Wenyun, Ding Pingxing, et al. 2013b, Hydrodynamic influence of proposed excavated-in harbor in the Hengsha Shoal of the Yangtze Estuary II: Influence on the current field. *Journal of East China Normal University (Natural Science)*, 04: 91–105.

Shao Rongsun, Shi Xiongbiao, Yu Ling. 2013, Reliminary analysis for the necessity of building a new harbor in Shanghai. *Journal of East China Normal University (Natural Science)*, (4): 10–16.

Yu Zhiying, Zhang Zhilin, Jin Liu, et al. 2013, Morphological stability analysis of the planned excavated-in harbor and its channel in the Hengsha Shoal of the Yangtze Estuary. *Journal of East China Normal University (Natural Science)*, (4): 55–71.

Hydraulic Engineering V – Hong, Liu & Xie (Eds)
© 2018 Taylor & Francis Group, London, ISBN 978-0-8153-7471-8

Response of salinity intrusion to winds in the Yangtze Estuary

L. Ding, X.P. Dou, X.Y. Gao & J. Jiao
Key Laboratory of Port, Waterway and Sedimentation Engineering of MOT, Nanjing Hydraulic Research Institute, Nanjing, China

J. Hu
Nanjing Water Planning and Designing Institute Co. Ltd., Nanjing, China

ABSTRACT: Salinity intrusion restricts the exploration and utilization of estuarine freshwater, which is influenced by an important factor, namely the wind. In this study, a numerical model capable of simulating salinity transportation in the Yangtze Estuary is established. Northerly winds prevail in this area in winter, and different wind directions have different impacts on the North Branch and the North Channel. The northeast wind intensifies salinity intrusion in the North Branch, followed by the north wind; however, the northwest wind has a weakening effect on salinity intrusion in the area. The northerly wind aggravates salinity intrusion in the North Channel. The increase in salinity intrusion caused by strong northerly winds has a serious adverse effect on the safety of the Qingcaosha Reservoir. This study provides a basis for the safety of water sources in the Yangtze Estuary.

1 INTRODUCTION

Yangtze River has the largest estuary in China (length ~182 km), starting from Xuliujing. It is fan-shaped and in the pattern of three-order bifurcations with four outlets into the sea (Figure 1). Chongming Island divides Yangtze River into two parts, namely the North Branch (NB) and the South Branch (SB). Changxing Island and Hengsha Island subdivide the SB into the North Channel (NC) and the South Channel (SC). Jiuduan Shoal divides the SC into the North Passage (NP) and South Passage (SP) (Figure 1). Yangtze River flows into the sea directly through the NB, NC, NP, and SP. Tides outside the estuary are regular semi-diurnal tides, and those in the estuary are irregular half-day shallow sea tides. Yangtze River has abundant water; however, because of the uneven distribution in time, the runoff in the dry season is much smaller than that during the flood season. The annual mean runoff in the Yangtze Estuary is 28,700 m^3/s, and that in the dry season (October to April) occupies only about 30% of the whole year. When the runoff is relatively small, saltwater will intrude into the estuary. Salinity intrusion in the dry season is the main threat to the exploitation and utilization of freshwater (Ding 2015).

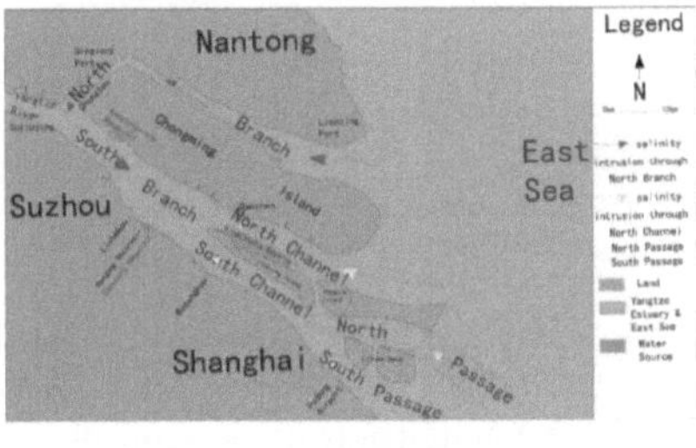

Figure 1. Map of the Yangtze Estuary.

Shanghai is the most densely populated and economically advanced city in China, which lies in the Yangtze River Delta and mainly takes water from the estuary. At present, the water supply scale of the three major water sources, from upstream to downstream in the Yangtze Estuary: Dongfengxisha water source, Chenhang water source, and Qingcaosha water source (Figure 1), has reached approximately 80% of the water supply in Shanghai. The quality of water sources should meet the "Standard of Surface Water Environmental Quality" (GB 3838–2002), which indicates that the chloride concentration of water should not be higher than 250 mg/L (i.e., salinity 0.45‰).The longest salinity intrusion endangering water sources in the Yangtze Estuary occurred from 3 to 25 February 2014, affecting the drinking water of 2 million people. Studies conducted worldwide on this issue indicate runoff and tide as the main factors that influenced salinity intrusion (Bowden 1966, Uncles 1980). However, in addition to small runoff (about 10,000 m^3/s), continued strong northerly winds (with speed exceeding 10 m/s for a long time) should have also contributed to the severe salinity intrusion in this period. The salinity intrusion in the NB is the most severe among the four outlets (Mao 1993, Zhao 2015), and the saltwater spilling over from the NB is one of the important sources endangering the water sources of Shanghai, especially Dongfengxisha Reservoir and Chenhang Reservoir, which take water from the SB. The Qingcaosha Reservoir takes water from the NC; besides saltwater spilling over from the NB, salinity intrusion directly from sea through the NC was also an important influence factor.

For this reason, research on the response of salinity intrusion to winds in the Yangtze Estuary is important. In this study, a numerical model is established with Delft3D to simulate salinity intrusion in the Yangtze Estuary, and the study is focused on the SB and NC, which serve as source for water sources. Furthermore, the mechanism of the wind affecting salinity intrusion and the degree of influence are analyzed in order to provide a reference for the intake of water sources in the Yangtze Estuary and guarantee the rational utilization of freshwater.

2 NUMERICAL MODEL

2.1 *Model establishment*

Numerical simulation is one of the main methods to simulate and analyze salinity intrusion. The type of mixing of most area in the Yangtze Estuary in the dry season is partial mixing, and the NB is well mixed (Ding 2015). This means that the difference in a vertical line is not obvious. Therefore, a two-dimensional horizontal numerical model can simulate salinity intrusion well in the Yangtze Estuary. The flow module in Delft3D is used to establish the 2D numerical model of the tidal current and salinity in the Yangtze Estuary to carry out simulation research of the salinity transport process (Ding 2017).

The model calculation range and grid are shown in Figure 2. The model includes the Yangtze Estuary, Hangzhou Bay, and adjacent sea. The farthest north boundary is at 34.67°N; the farthest south boundary is at 29.33°N; and the farthest east boundary is at 124.24°E. The tidal open boundary is obtained through the harmonic analysis of the main tidal constituents by applying water-level boundary conditions. The upstream boundary is the runoff control

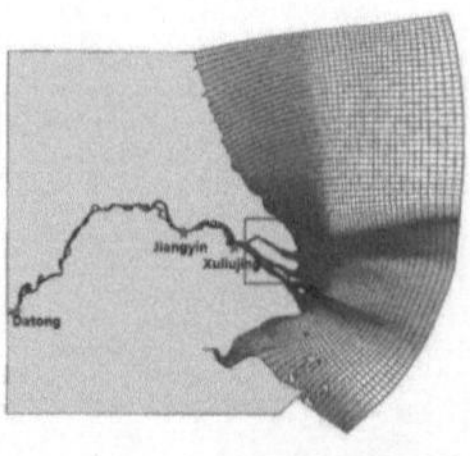

Figure 2. Model grid.

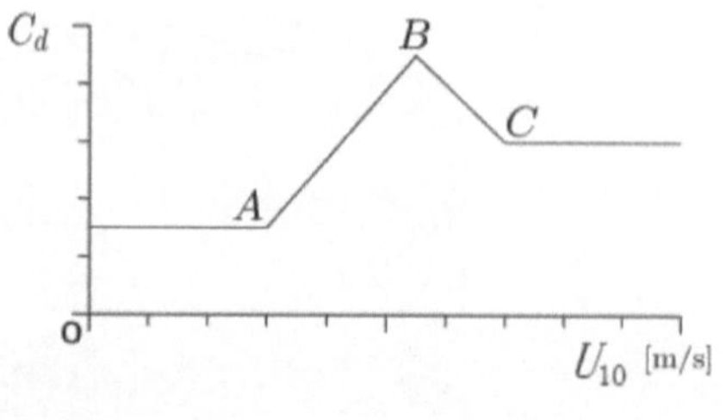

Figure 3. Relationship between the drag coefficient and the wind speed.

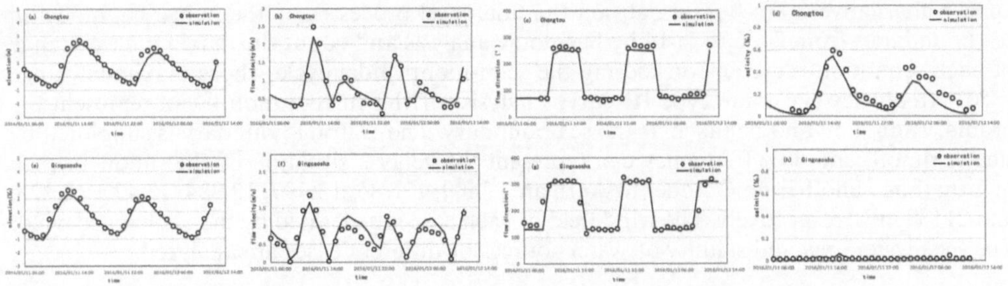

Figure 4. Validation results for (a)–(d) elevation, flow velocity, flow direction, and salinity near Chongtou, respectively; (e)–(f) elevation, flow velocity, flow direction, and salinity of near Qingcaosha, respectively.

condition and is taken in Datong. Datong Hydrological Station is longer than 600 km from the upper reaches of the Yangtze Estuary. It is the nearest hydrological station in the lower reaches of Yangtze River from the estuary, and is not influenced by the tidal dynamics. The model grids are body-fitted orthogonal grids, with 1431 × 163 elements. The grid size in the open sea is large, reaching up to 2 km × 2 km, and the grids are locally densified in the estuary, with the minimum size of 70 m × 60 m. The terrain from Jiangyin down to estuary is the measured terrain of 2011, and the local terrain is updated according to reclamation projects. The terrain from Jiangyin to Datong is the measured terrain of 2006. The time step is set at 15 s. The roughness of the model is provided by the Chezy coefficient, and the value varies from 80 to 200 $m^{1/2}/s$ according to the terrain. The average salinity diffusion coefficient is 250 m^2/s. The effect of the wind is expressed by wind stress τ_w on the water surface in this model:

$$\overrightarrow{\tau_w} = \rho_a C_d \overrightarrow{U_{10}} \left|\overrightarrow{U_{10}}\right| \cdot \overrightarrow{\tau_w} = \rho_a C_d \overrightarrow{U_{10}} \left|\overrightarrow{U_{10}}\right| \quad (1)$$

where ρ_a is the air density, C_d is the drag coefficient of winds, and U_{10} is the wind speed 10 m above the free surface. C_d is influenced by U_{10}, as shown in Figure 3. The wind field can be specified pursuant to the measured wind speed. According the wind field above the Yangtze Estuary and adjacent sea, in this model, when the wind speed is 0 m/s, C_d is 0.0012; when the wind speed is 4 m/s, C_d is 0.0021; and when the wind speed is 0–4 m/s, they are related to primary function. During model calculation, the field after running for 2 months is taken as the initial field, and then the cases are simulated.

2.2 *Model validation*

The model used the results of the large-scale full-tide hydrologic survey in the Yangtze Estuary conducted in January 2016 to validate the numerical model by verifying the water level, velocity, flow direction, and salinity at each measuring point. Because of limited space, only the validation results of the spring tide at the measuring point near the entrance of the NB (Chongtou) and near the Qingcaosha Reservoir were given (Figure 4), and the model can well simulate the hydrodynamic force and saltwater transport in the Yangtze Estuary.

3 CALCULATION AND ANALYSIS

The operation of the Three Gorges Dam, which was built in the middle reach of Yangtze River, has changed runoff remarkably since 2006 (Cao 2006). In this study, salinity intrusion characteristic under hydrologic condition due to the operation of the Three Gorges Project were investigated. In the Yangtze Estuary, spring tide, moderate tide, and neap tide appear alternatively. We obtain data on the tidal level processes of the numerical model in sea boundaries from 2006 to 2014 by harmonic analysis and count the maximum tide range of each semi-month cycle. Considering the semi-month tidal cycle whose maximum range is 50% frequency (Pearson Type III curve analysis) of the current month can represent the ordinary situation and be taken as the sea boundary. The Datong boundary is simplified by steady runoff, and 50% frequency can represent the general situation of the runoff in each month; thus, runoff from October to April are 23409, 17,135, 13,797, 12,384, 13,422, 18,225, and 21137 m^3/s, respectively. The wind field of each case is generated by the unchanged wind. The freshwater area and salinity of water sources of different cases are analyzed.

3.1 *Freshwater area*

The runoff of Yangtze River is the smallest in January, and April is the last month of a continuous dry season, whose runoff is relative larger. In this study, we take these two months as examples to study the influence of the northerly wind on the freshwater area of the Yangtze Estuary. Here, freshwater means the water with salinity no larger than 0.45‰. Four typical events in a semi-month cycle are discussed, namely spring tidal flood slack, spring tidal ebb slack, neap tidal flood slack, and neap tidal ebb slack. Freshwater areas in different cases are listed in Table 1; in theses simulations, the wind speed is 15 m/s.

When the Yangtze Estuary is not influenced by the wind, the area of freshwater in the spring tide is smaller than that in the neap tide, and that in the flood is smaller than that in the ebb slack. As the runoff in April is marked larger than that in January, the freshwater area is also larger in April. Comparing the same moment affected by different winds with the same speed (15 m/s), both the north wind and northeast wind reduce the freshwater area in January and April. The influence of the northwest wind is more complex, because it enlarges the freshwater area in January, whereas reduces it in April. Therefore, northerly winds aggravate salinity intrusion as a whole; however, similar to the northwest wind in April, different combinations of winds and runoff may have different results.

3.2 *Salinity duration of water source intake*

In order to carry out an in-depth analysis of the influence of the north wind on salinity intrusion in the Yangtze Estuary, salinity duration of water source intakes will be discussed hereinafter.

Figure 5 shows the semi-month salinity duration of Dongfengxisha Reservoir intake (DRI), Chenhang Reservoir intake (CRI), and Qingcaosha Reservoir intake (QRI) of each month in the dry season without the wind. The split ratio of the North Branch is even less than 5%; thus, the salinity in the NB is the highest in north south horizontal direction. The Yangtze Estuary is in the Northern Hemisphere and thus the tide in the northern part rises earlier and southern part ebbs earlier under the effect of the Coriolis force. The water in the

Table 1. Freshwater areas under the influence of northerly winds.

		Freshwater area (km^2)			
Month	Wind direction	Spring tidal flood slack	Spring tidal ebb slack	Neap tidal flood slack	Neap tidal ebb slack
January	No wind	37	122	263	372
	North	0	91	32	146
	Northeast	0	80	8	90
	Northwest	465	512	688	761
April	No wind	952	1059	1196	1201
	North	221	453	635	690
	Northeast	26	98	32	120
	Northwest	783	863	965	991

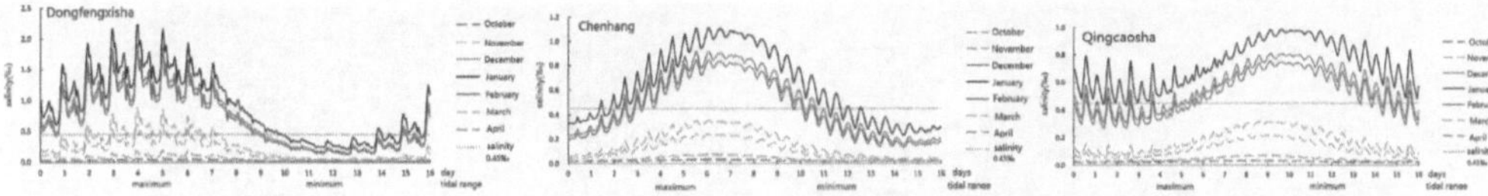

Figure 5. Semi-month salinity duration of water source intake in the Yangtze Estuary (left: Dongfengxisha Reservoir intake; middle: Chenhang Reservoir intake; right: Qingcaosha Reservoir intake).

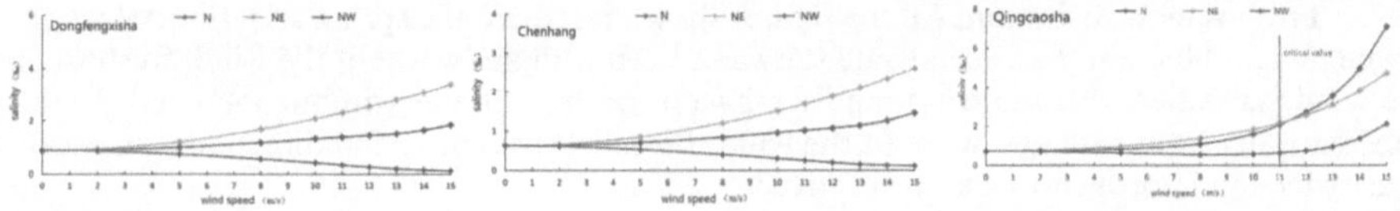

Figure 6. Half-moon average salinity of water source intake in the Yangtze Estuary (left: Dongfengxisha Reservoir intake; middle: Chenhang Reservoir intake; right: Qingcaosha Reservoir intake; N: north wind; NE: northeast wind; and NW: northwest wind).

NB spills over into the SB; when the salinity intrusion is severe, the high-concentration saltwater will flow into the SB and affect the downstream (Ding 2016). As the high-salinity water flows through different water sources in different times, DRI suffers first, then CRI, after about 2–3 days, and finally QRI, after another 3–4 days.

Taking January for example, we simulate the salinity under the influence of the northern wind. Half-moon average salinities of DRI, CRI, and QRI are shown in Figure 6.

For DRI and CRI, when the wind direction is north or northeast, the wind will aggravate salinity intrusion and salinity is positively correlated with the wind speed; the influence of the northeast wind is more obvious, but the northwest wind decreases the salinity.

For QRI, the relationship between salinity and wind is more complicated. The same as that of DRI and CRI, salinity has positive correlation with the north wind and the northeast wind; however, salinity is higher under the influence of the northeast wind within a critical wind speed, whereas the north wind has a significant effect when the wind speed exceeds this critical value. In the case of the northwest wind, with the increase in the wind speed, salinity first decreases and then increases rapidly. The lowest salinity has a corresponding wind speed, which is the same as the critical value mentioned above. In the result of this numerical model, the critical value equals 11 m/s. Therefore, when the wind speed exceeds 11 m/s, regardless of the direction, the northerly wind results in the intensification of salinity intrusion.

3.3 *Mechanism of the wind influencing salinity intrusion*

According to the numerical simulation results in Sections 3.1 and 3.2, both freshwater distribution and salinity duration in different observation points in the Yangtze Estuary are complicated. This phenomenon is due to the complex river regime. There are four outlets to the sea in the Yangtze Estuary. Moreover, the wind influences each of these outlets, especially the NB and NC differently. As shown in Figure 7, under the influence of the north wind and the northeast wind, freshwater areas of both the NB and NC decrease in degree. However, under the influence of the northwest wind, freshwater area in the NC is less than that without the wind; however, in the NB, more freshwater is brought by the northwest wind. The influence of the wind on the NC and NB is different, and the relationship between salinity duration and wind speed shows different characteristics.

Determining the effect of the wind on salinity intrusion includes both the local drag effect and wind-driven circulation (Li 2012). In the former, the wind affects the flow in the estuary by surface shear stress, and the flow velocity is changed by the wind, thereby changing salinity transportation. The wind-driven circulation is caused by the wind over the sea; more high-salinity sea water will be brought into the estuary by the north wind in the Northern Hemisphere. Therefore, the local drag effect of the north wind and northeast wind will

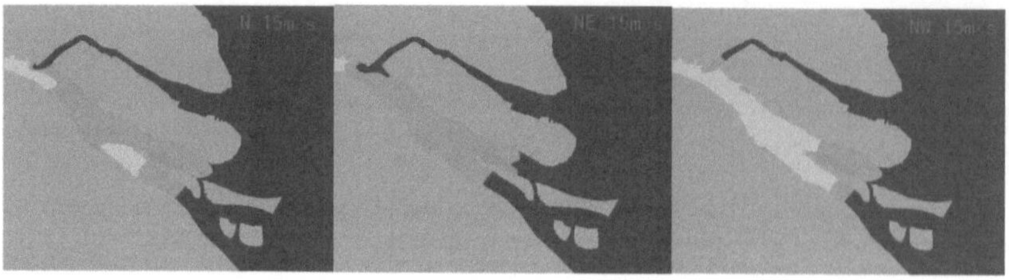

Figure 7. Freshwater distribution. All the figures shown represent the spring tidal flood slack in April. Dark blue to light blue water areas indicate saltwater both with and without the wind, freshwater without the wind, saltwater with the wind, and freshwater both with and without the wind. Light green indicates saltwater both with and without the wind. From left to right, wind directions are north wind, northeast wind, and northwest wind, respectively.

aggravate the salinity intrusion of the Yangtze Estuary, and that of the northwest wind weakens the salinity intrusion. The wind-driven circulation of the northeast, north, and northwest winds will increase the salinity in the estuary.

The influence of the local drag effect is more significant than the wind-driven circulation in the NB. The north wind and the northeast wind aggravate the saltwater spilling over from the NB, and the degree of the northeast wind becomes greater, while the northwest wind releases the spilling. On the contrary, the northwest wind will reduce the salinity of water flowing into the SB from the NB. In the NC, the wind-driven circulation is the main manifestation of the influence of the wind. The northerly winds produce the circulation, increasing the salinity intrusion in the NC. Therefore, in the Dongfengxisha Reservoir and the Chenhang Reservoir, which mainly suffered threat from saltwater from the NB, the salinity of intakes in different cases, from large to small, influenced by the northeast wind, north wind, no wind, and northwest wind is considered. In the Qingcaosha Reservoir, which is also affected by saltwater directly from the sea, the northwest wind makes salinity of water from the NB lower but that of the water from the sea larger. When the wind speed is smaller than a certain value (here, 11 m/s), salinity spilling from the NB becomes important, the northwest wind decreases the salinity of QRI decrease, and the northeast wind increases the salinity more than the north wind at the same speed. When the wind speed is higher than 11 m/s, the wind-driven circulation makes the salinity intrusion of the NC extremely severe and the influence becomes more obvious. Northerly winds in all directions rapidly increase salinity with increasing the wind speed, thereby having the highest impact.

Major of scholars in China have proposed to build a tidal gate in the NB to isolate the water sources from high-salinity water. On the basis of the results of this study, tidal will put an end to saltwater spilling over from the NB and have clear benefits, especially for the Dongfengxisha Reservoir and the Chenhang Reservoir. However, the tidal gate cannot eliminate the influence of salinity intrusion. When persistent strong winds blow, the Qingcaosha Reservoir will still be badly exposed. However, the Qingcaosha Reservoir is much larger than the other two reservoirs. It can supply water for 68 days without intake, and in the daily operation, the stored water can provide supply for approximately 30 days (Zhu 2013). As a result, if the wind field is accurately predicted, salinity intrusion warning can be precisely provided; therefore, providing the reservoir enough time to store more water is also an effective way to guarantee the water supply.

4 CONCLUSION

The Yangtze Estuary has a complex river regime that makes the wind–flow–salinity dynamics more complicated. The saltwater in the South Branch and the North Channel mainly comes from the North Branch and directly from the sea. Northerly winds from different directions have different effects on saltwater from these two sources. The salinity intrusion of the North Branch is aggravated by north and northeast winds but weakened by the northwest wind due to the wind drag effect. In contrast, the wind-driven circulation caused by the northerly wind makes the salinity intrusion of the North Channel severe. A prolonged period of strong northerly winds will have a serious adverse effect on the Qingcaosha Reservoir, thereby requiring immediate attention.

ACKNOWLEDGMENTS

This study was supported by the Central Water Resources Fee Project (126153-0210283) and the Central Public Fund of Nanjing Hydraulic Research Institute (Y216012, Y216019, Y216020, and Y217010).

REFERENCES

Bowden K.F. 1966, The circulation, salinity and river discharge in the Mersey Estuary. *Geophysical Journal of the Royal Astrophysical Society*. 10, 383–400.

Cao Y., Chen J.Y., Zhang E.F., et al., 2006, Influence of Three Gorge reservoir filled with water on freshwater resource in the Yangtze River estuary. *Advances in Water Science*, 04:554–558.

Ding L., Dou X.P., Gao X.Y., et al., 2015, The research of saltwater intrusion in Yangtze Estuary. China Ocean Engineering Academy. *The 17th China Ocean and Coast Engineering Symposium (volume two)*. China Ocean Engineering Academy.

Ding L., Dou X.P., Gao X.Y., et al., 2016. Analysis of saltwater intrusion into Yangtze estuary during dry seasons of 2013 and 2014. *Hydro-science and engineering*, 4, 47–53.

Ding L., Gao X.Y., Dou X.P., et al., 2017. Research of the influence of salinity intrusion on Qingcaosha water source. *37th IAHR World Congress*. International Association for Hydro-Environment Engineering and Research. (in press).

Li L., Zhu J.R., Wu H., 2012, Impact of wind stress on saltwater intrusion in the Yangtze Estuary. *Science China Earth Sciences*, 55:1178–1192.

Mao Z.C., Shen H.T., & Yao Y.D., 1993, Analysis of sources of saltwater intrusion along south bank of south branch of Yangtze River Estuary. *Marine Science Bulletin*, 03:17–25.

Uncles R.J. & Radford P.J. 1980, Seasonal and spring-neap tidal dependence of axial dispersion coefficients in the Severn—a wide, vertically mixed estuary. *Journal of Fluid Mechanics*, 98(4):703–726.

Zhao J., He Q., Wang X., et al., 2015, Field observations on the characteristics of current and sediment of the south and north branches in the Yangtze Estuary. *Resources and Environment in the Yangtze Basin*, 01:21–29.

Zhu J., Gu Y., Wu H., 2013. Development of determination of the period not suitable for taking domestic water supply to the Qingcaosha Reservoir near Changjiang River Estuary. *Oceanologia Sinica*, 44(5): 1138–1145.

Hydraulic Engineering V – Hong, Liu & Xie (Eds)
 ISBN 978-0-8153-7471-8

Reliability analysis of a slope-type seawall armor structure

Qingjun Liu
Nanjing Hydraulic Research Institute, Nanjing, Jiangsu, China
Hohai University, Nanjing, Jiangsu, China

Qihua Zuo, Dengting Wang & Tianting Sun
Nanjing Hydraulic Research Institute, Nanjing, Jiangsu, China

ABSTRACT: This study analyzes the safety of a seawall slope armor based on the reliability theory. By taking a typical slope-type seawall in Guangdong Province as an example to establish a performance function for the slope armor panel, we first calculate the probability of panel failure under the existing conditions and then conduct a sensitivity analysis of factors such as wave height, period, water depth, and plate thickness. The results of this study can provide a technical reference for the design, construction management, and operation and maintenance of the armor structure for similar seawalls in the future.

1 INTRODUCTION

As an important structure preventing marine disasters, especially storm surge disasters, the environment, which a seawall, is affected by numerous influential factors of safety, such as wave, wind, tidal level, and tidal current. These influential factors are characterized by random changing. Therefore, it is more appropriate to use the reliability method on the basis of probability theory to conduct seawall safety analyses.

W.T. Bakker and J.K. Vrijling (1980) introduced reliability theoretical design into seawall engineering design and laid a good foundation for seawall reliability analysis. Later, K.O. Ronold (1990), considering the influences of the environment and terrain, used the second-moment method to perform reliability calculations to seawall wave overtopping under the action of storm surge at a high tidal level and conducted a sensitivity analysis of influential factors. On the basis of Hudson's formula and van der Meer's formula, Can Elmar Balas et al. (2002) established an armor stone performance function for some seawalls in Turkey and used different methods to calculate the reliability of the armor stone. C. Mai Van et al. (2007, 2009) conducted a research on the seawall failure mechanism by selecting some seawalls in Vietnam to calculate the failure probabilities in six failure modes, including overflow, overtopping, armor block and proposed an optimum seawall design. Zhang Xiangdong (2012) and Tian Wenhua (2012) studied seawall overtopping reliability.

The above-mentioned studies were mainly focused on seawall overtopping and armor block reliability analysis, ignoring slope armor panel. In this study, we take some seawalls in Guangdong as examples to establish a performance function for the armor panel, calculate the reliability of the armor structure, and conduct the sensitivity analysis of the wave height, period, water depth, and panel thickness.

2 RELIABILITY THEORY AND SIMULATION METHOD

The working state of the structure can be described by using the following performance function:

$$Z = g(x_1, x_2, \cdots x_n) \tag{1}$$

where x_1, x_2, and x_n are the random variables influencing the structure functions.

The working state of the structure can be divided into the following three classes according to different structure performance functions:

1. When $Z > 0$, the structure is in a reliable state.
2. When $Z = 0$, the structure is in a limit state.
3. When $Z < 0$, the structure is in a failure state.

The structure failure probability (P_f) can be expressed as:

$$P_f = P(Z < 0) = \iint \cdots \int_{z<0} f_X(x_1, x_2, \cdots, x_n)\, dx_1\, dx_2 \cdots dx_n \tag{2}$$

where $f_X(x_1, x_2, \ldots, x_n)$ is the probability density function of the structure performance function.

As it is difficult to obtain the probability density function and to realize a direct calculation method for the structure failure probability using formula (2), the Monte Carlo method can be used in engineering applications.

The basic idea of the Monte Carlo method is taking into account random variable samples according to random variable distribution type to produce several simple subsamples (assuming that the number of subsamples is N). Then, we substitute each group of simple subsamples into the structure performance function to determine whether the structure fails or not and to count the failure times (n_f times).

It is known according to the law of large numbers that the structure failure probability is the unbiased estimation of failure probability, that is

$$\hat{P}_f = \frac{n_f}{N} \tag{3}$$

The sampling variance is

$$\sigma^2 = \frac{1}{N}\hat{P}_f\left(1 - \hat{P}_f\right) \tag{4}$$

To ensure the sampling error, let the confidence be $1-\alpha$, the relative error ε is expressed as

$$\varepsilon = \frac{\left|\hat{P}_f - P_f\right|}{P_f} \approx \frac{\left|\hat{P}_f - P_f\right|}{\hat{P}_f} \le z_{\alpha/2} \cdot \sqrt{\frac{1}{N}\hat{P}_f\left(1 - \hat{P}_f\right)} \tag{5}$$

It is known from formula (5) that the simulation accuracy of the Monte Carlo method is dependent on sampling times and variance σ^2 and not on variable dimensions. To reduce the simulation error, the simulation times N can be increased.

3 RELIABILITY ANALYSIS OF THE ARMOR STRUCTURE

3.1 *Basic information on work*

The study is conducted at the downstream estuary in Jianjiang River, Guangdong Province, being a slope-type seawall with an overall length of 4.95 km. The tide prevention standard is based on a 50-year return period. At the base, cement mortar masonry is used. For the front slope armor, 450 mm thick concrete slabs are used, whose slope is 1:2. Wave wall is provided on the seawall top. The inclination of the rear slope is 1:2. The sectional form is shown in Figure 1.

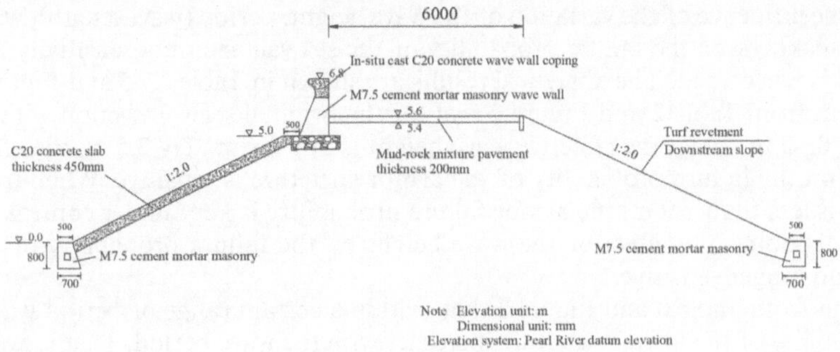

Figure 1. Seawall section.

3.2 *Establishment of performance functions*

According to the relationship between the actual armor thickness and theoretical safe thickness, the following armor performance function is established:

$$Z = W_R - W_S \tag{6}$$

where W_R is the actual armor panel thickness and W_S is the theoretical safe thickness of the armor panel.

For the theoretical safe thickness of seawall armor panel, the method recommended by the *Seawall Engineering Design Code* of China (2014) is used together with the actual situation of Chinese seawalls. For the concrete or reinforced concrete board support with a visible joint, the panel thickness meeting the stability requirement is low when the slope rate (m) is 2–5:

$$W_S = 0.07\eta H^3 \sqrt{\frac{L}{B}} \frac{\rho_\omega}{\rho_c - \rho_\omega} \frac{\sqrt{m^2+1}}{m} \tag{7}$$

Therefore, formula (6) can be further expressed as:

$$Z = W_R - 0.07\eta H^3 \sqrt{\frac{L}{B}} \frac{\rho_\omega}{\rho_c - \rho_\omega} \frac{\sqrt{m^2+1}}{m} \tag{8}$$

$$L = \frac{gT^2}{2\pi} tanh \frac{2\pi d}{L} \tag{9}$$

where W_R is the front slope armor panel thickness, which is 0.45 m; η is a coefficient, whose value is 1.0; H is the design wave height (m), which is $H_{1\%}$ and can be obtained by conversion through the effective wave height *Hs*; ρ_c is the board density, which is 2.2 t/m³; ρ_ω is the water density, which is 1.0 t/m³; L is the wave length (m), which is calculated by formula (9); B is the armor panel length along the slope direction (perpendicular to water edge), which is 5.6 m; m is the slope rate, which is 2.0; T is period (s); and d is water depth (m).

3.3 *Reliability calculation and sensitivity analysis of influential factors*

We consider water depth before seawall (*d*), effective wave height (*Hs*), and average period (*T*) as random variables. The long-term distribution type and statistical parameters are shown in Table 1. In the calculation in this study, the dependence among various random variables is not considered temporary.

We use the Monte Carlo method to develop MATLAB calculating program in order to calculate the failure probability of this seawall armor. The result shows that the slope armor failure probability of this seawall section is 12.4%.

To study the influence of the variation of the wave height, period (wave length), water depth, and armor thickness on the failure probability of the seawall armor, a sensitivity analysis of these factors is performed. The analytical results are shown in Tables 2–5 and Figures 2–5.

It is evident from Table 2 and Figure 2 that for the original seawall section with an armor thickness of 0.45 m, when the effective wave height is approximately 2.5 m, the effect of the wave height on the failure probability of the armor structure is sensitive. When the effective wave height is less than 1.9 m, the armor failure probability is kept under control within 5% largely. At this point, the effect of the wave height on the failure probability of the armor structure is no longer sensitive.

It is evident from Table 3 and Figure 3 that, within a certain range of period variation, the failure probability of the armor structure increases with the wave period. That is, waves with a long period are more unfavorable to the armor structure. Besides, the influence of the period variation on the failure probability of the armor structure is not as sensitive as the wave height.

It is evident from Table 4 and Figure 4 that, within a certain range of water depth variation, the failure probability of the armor structure increases slightly with water depth. Besides, in relation to wave height and period, the effect of the water depth on the failure probability is not sensitive.

It is evident from Table 5 and Figure 5 that when the armor thickness is 0.25–0.40 m, the effect of the variation of the armor thickness on the failure probability of the armor structure is very sensitive. A slight increase or decrease in the armor thickness can result in a significant variation of the failure probability. The armor panel thickness of the seawall used is 0.45 m, and this sensitive area is avoided.

Table 1. Statistical characteristic parameters of random variables.

Random variable	d (m)	H_s (m)	T (s)
Average value (μ)	4.26	2.1	9.4
Standard variance (σ)	0.68	$C_v = 0.2$, $C_s = 0.5$	$C_v = 0.3$, $C_s = 0.6$
Distribution type	Extreme I distribution	P-III distribution	P-III distribution

Table 2. Influence of the wave height (H_s) on the failure probability.

Wave height (m)	1.0	1.2	1.5	1.8	2.0	2.2	2.5	2.8	3.0
P_f(%)	0.001	0.01	0.3	3.1	8.8	17.3	35.3	54.9	66.9

Table 3. Influence of the wave period (T) on the failure probability.

Wave period (s)	4.0	5.0	6.0	7.0	8.0	9.0	10.0	11.0	12.0	13.0
P_f(%)	0.4	1.3	3.1	5.3	7.8	11.3	14.0	18.2	21.5	26.0

Table 4. Influence of the water depth (d) on the failure probability.

Water depth (m)	3.0	3.2	3.4	3.6	3.8	4.0	4.2	4.4	4.6	4.8	5.0
P_f(%)	8.14	8.8	9.35	10.1	11.1	11.3	11.7	13.0	13.7	13.9	15.4

Table 5. Influence of the armor thickness on the failure probability.

Armor thickness (m)	0.15	0.20	0.25	0.30	0.35	0.40	0.50	0.55	0.60	0.65	0.70
P_f(%)	99.9	98.9	91.5	73.6	48.5	27.6	5.1	1.8	0.6	0.2	0.002

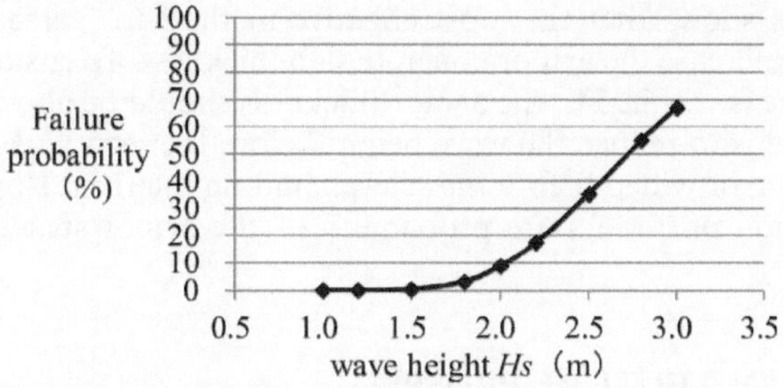

Figure 2. Variation of the failure probability with wave height (H_s).

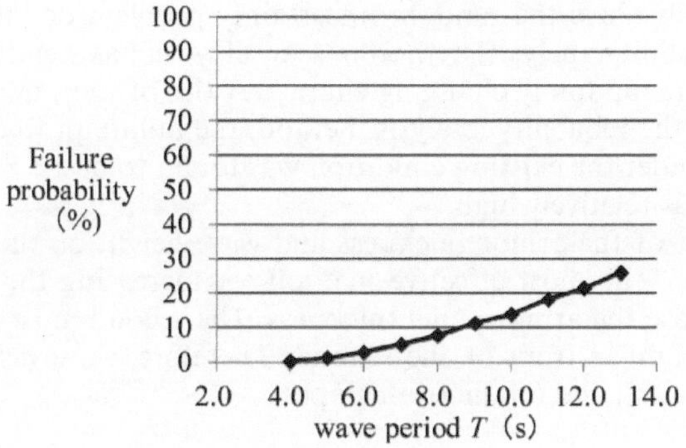

Figure 3. Variation of the failure probability with wave period.

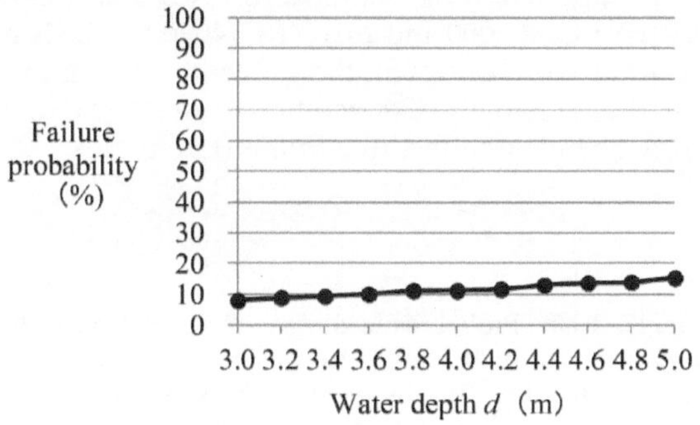

Figure 4. Variation of the failure probability with water depth.

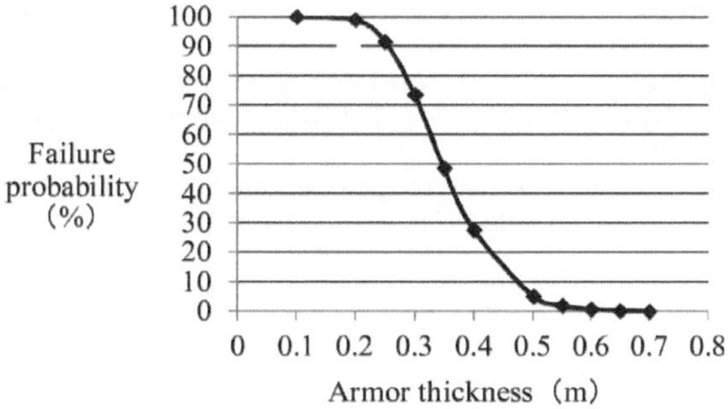

Figure 5. Variation of the failure probability with armor thickness.

The above analysis shows that the most effective method to increase the safety of the armor structure is to increase the armor concrete slab thickness. To ensure that the reliability of the armor structure is within 5%, the armor thickness should be above 0.5 m. The second most effective method is to reduce the wave height before the seawall. For this reason, wave dissipation piles can be provided before the seawall and on the front slope. The effect of the water depth and period on the failure probability of the armor structure is relatively not sensitive.

4 CONCLUSION AND FUTURE PROSPECTS

The environment of the seawall is affected by numerous influential factors of safety. However, the reliability analysis method based on the probability theory and the mathematical statistics theory can satisfactorily solve the random uncertainty problem of influential factors. The application of the reliability analysis method in seawall work has significant advantages.

After considering the randomly changing characteristics of the wave height, water depth, and period and using the reliability analysis method, the failure probability of a slope-type seawall armor panel under the existing condition was found to be 12.4%. The failure risk of the armor structure was relatively high.

Moreover, the effect of the armor thickness and wave height on the safety of the armor structure was sensitive. The most effective method for increasing the safety of the armor structure was to increase the armor panel thickness. The second most effective method was to reduce the wave height in front of the seawall. Therefore, wave dissipation piles can be provided before the seawall and on the front slope.

ACKNOWLEDGMENTS

This study was financially supported by the National Key Research and Development Program of China (Grant Nos. 2016YFC1402000 and 2016YFC1402002), the National Natural Science Foundation of China (Grant No. 51579156), the Fifteenth Session Program between China and Bulgaria Scientific and Technological Cooperation Committee (No. 15–13), and the Major Project of Nanjing Hydraulic Research Institute Funds (Grant Nos. Y216024 and Y216017).

REFERENCES

Bakker, W.T. & Vrijling, J.K. 1980. Probabilistic design of sea defences. *Proceedings International Conference on Coastal Engineering* 1980:2040–2059.

Can Elmar Balas. & Aysen Ergin. 2002. Reliability-Based Risk Assessment in Coastal Projects: Case Study in Turkey. *J. Waterway, Port, Coastal, Ocean Eng* 128(2):52–61.

Knut O. Ronold. 1990. Reliability Analysis of a Coastal Dike. *Coastal Engineering* 14(1990):43–54.

Mai Van C., et al. 2007. Failure mechanisms of sea dikes inventory and sensitivity analysis. *Proceedings of the Coastal Structures* 2007:550–561.

Mai Van C., et al. 2009. Probabilistic Design and Reliability Analysis of Coastal Structures-a Vietnam Case. *Asian and Pacific Coasts* 2009:201–209.

Seawall Engineering Design Code (GB51015T-2014). 2014. Beijing: China Planning Press. (Chinese).

Tian Wenhua. 2012. Slope Type Seawall Overtopping Risk Analysis. Degree of Master. China Marine University. Qingdao. (Chinese).

Zhang Xiangdong et al. 2012. Seawall Overtopping Reliability Analysis under Permissible Overtopping Conditions. *Water Transportation Engineering* 468(7) 15–18. (Chinese).

Hydraulic Engineering V – Hong, Liu & Xie (Eds)
© 2018 Taylor & Francis Group, London, ISBN 978-0-8153-7471-8

Wave run-up influenced by the protective facings of the sloping breakwater

T.T. Sun, D.T. Wang & Q.J. Liu
Nanjing Hydraulic Research Institute, Nanjing, China

V. Penchev
Black Sea—Danube Association of Research and Development (BDCA), Varna, Bulgaria

ABSTRACT: The calculation methods used for the wave run-up stipulated in Chinese codes have some defects, especially under long-wave conditions. The formulas with relatively limited applicability also do not take long-wave conditions into account. Therefore, there is a large deviation between the calculated value and the actual result, which may result in some potential safety problems. In this study, we investigate the influences of different types of protective facings of the sloping breakwater on the wave run-up under irregular wave conditions by performing a series of sectional physical model tests. The main contribution of this study is to propose the correlation coefficient of protective facings and provide a theory reference for the design of bank revetment on the river-side slope, which is on open-water areas.

1 INTRODUCTION

The sloping breakwater is a traditional and widely adopted structure in the port and coastal engineering, which is an important engineering facility to protect coastal areas from wave strike. Owing to the easy and violent break of waves on the slope surface, the required natural stones are very large when the water is deep. As a result, approximately 100 types of concrete artificial blocks with different shapes and high stability have been developed, such as dolosse, accropode, concrete tetrapod, and hollow square, which are broadly leveraged. With the rapid development of marine economy and human demand for marine energy, several coastal and marine projects are in or about to function. The engineering construction is developing to the deep sea, and the construction and wave conditions are becoming increasingly worse. Therefore, determination of the armor block forms plays an important role in the research and design of the sloping breakwater as well as the safety standards of engineering.

Wave run-up has a significant effect on coastal erosion, scouring, siltation, sediment transport, and seawall safety. The influence factors of wave run-up on the sloping breakwater is complex, mainly including the incident wave elements, such as wave height, wave length, wave steepness, and wave period; direction of incident wave; wind speed and direction, platform properties of complex slope; slope roughness and seepage properties; and the slope of the breakwater. Because the sloping breakwater has many types of cross-sectional forms, such as single slope and compound section with platform, research results with different application conditions are obtained. Literature review shows that systematic studies on the wave run-up influenced by the protective facings of the sloping breakwater are relatively few.

1.1 *Calculation method and numerical simulation of the wave run-up*

In May 2002, the Technical Advisory Committee on Water Defense summarized several results of recent experimental and theoretical analyses of wave run-up and analyzed the effect of the slope roughness, the slope of the breakwater, platform characteristics, the wave incident angle, and the vertical wall on it. Lu (2007) compared the calculation method of the wave run-up proposed in the *Code for design of levee project* to that of Van der Meer, which is widely recognized in Western countries, and proposed suggestions for modification and use of the latter Qi (2012) used the source function wave method to study the effect of roughness on the wave run-up with barrier board, which is widely used in seawall projects as an example.

The formulas derived by Putian, Hedi, Guanting, and the recommended formula proposed in the *Code for design of levee project* are widely used in the wave run-up calculation. The mean wave height and mean wave period are suggested to be calculated by the formula proposed by Putian. For hills and plain areas, the wave height and mean wave length can be calculated by the formula derived by Hedi.

1.2 *Physical model study of the wave run-up*

The physical model has been developed for more than 100 years since the proposal of the model of estuary model of Mensey by Reynolds in 1885. The physical model is intuitive, and the flow pattern is realistic. It can provide validation data for numerical wave simulation.

Saville (1955) plotted the relationship curves of wave run-up and wave steepness, water depth, and architectural forms through sectional physical model tests. De Waal (1992) estimated the empirical formula of maximum wave run-up through a series of scaled model tests. However, it can be only applied to some special section forms and has some limitations as well. Chen (1992, 2005, and 2010) studied the effects of different protective facings and the height and width of complex slope platform on the wave run-up under irregular waves, determining the probability distribution of wave run-up and putting forward the relevant calculation method.

In view of the complexity of the wave flow phenomenon on the sloping breakwater, the theoretical calculation for a preliminary design is generally adopted and then verified by physical model tests in engineering projects. Wave run-up has been a research focus of scholars worldwide. In conclusion, it has important engineering significance and academic value to study the wave run-up influenced by different protective facings of the sloping breakwater.

2 TEST SETUP

In the experiment, we use a normal model and simulate it in accordance with the Froude number similarity. According to the factors of design water level, wave elements, test section, and equipment condition, the model geometric scale is taken as 1:15. This experiment, performed on the basis of *Wave Model Test Regulation*, adopted the JONSWAP spectrum.

Sectional physical model tests are conducted in a long-wave flume of Nanjing Hydraulic Research Institute. The flume is 175 m long, 1.2 m wide, 1.8 m deep, and it can produce waves, currents, and winds simultaneously. The working section of the wave flume is divided into two parts of size 0.6 and 0.6 m along the width direction. One part is used to place the model section and carry out the model test, and the other part is used to diffuse reflected waves produced by the wave maker. One end of the flume is provided with wave elimination gentle slope, and the other end is equipped with computer-controlled irregular wave generators to produce the required waves.

To reduce errors, experiments are conducted at least three times for the same wave element when watching the wave run-up on the sloping breakwater with different protective facings. If gaps between experiment results are not significant, an average could be taken as the final experiment result. Otherwise, the number of experiments should be increased until the three groups of experiments produce relatively close result.

The test section is shown in Figure 1. It is a single-slope breakwater with a slope of $m = 3$. The bottom elevation is 9.0 m, the crest elevation is 14.8 m, and there is a breakwater wall with roof height of 15.1 m.

The experimental wave parameters are calculated according to the *Code of Hydrology for Harbour and Waterway* and the *Code for design of levee project*, as given in Table 1.

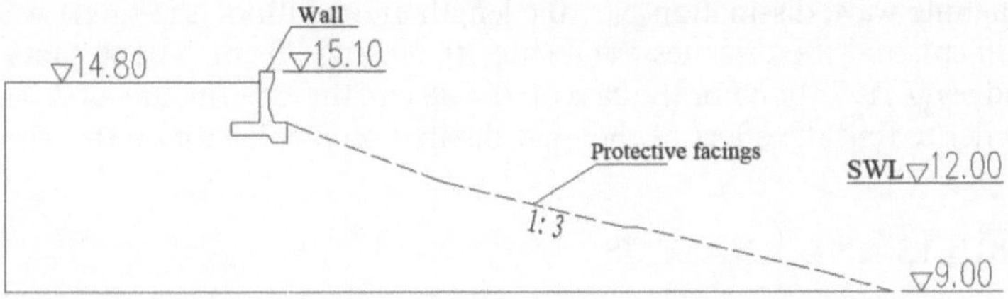

Figure 1. Cross section of the breakwater model.

Table 1. Experimental wave parameters (1.5 and 2.0 times the average annual maximum wind speed).

Mean depth (m)	Wind distance (km)	Wind speed/ ($m\ s^{-1}$)	$H_{1\%}$ (m)	$H_{5\%}$ (m)	$H_{13\%}$ (m)	H_m (m)	T_m (s)	L (m)
3.0	11.4	19.7	1.14	0.96	0.82	0.54	3.26	14.32
3.0	11.4	26.2	1.36	1.15	0.99	0.66	3.61	16.56

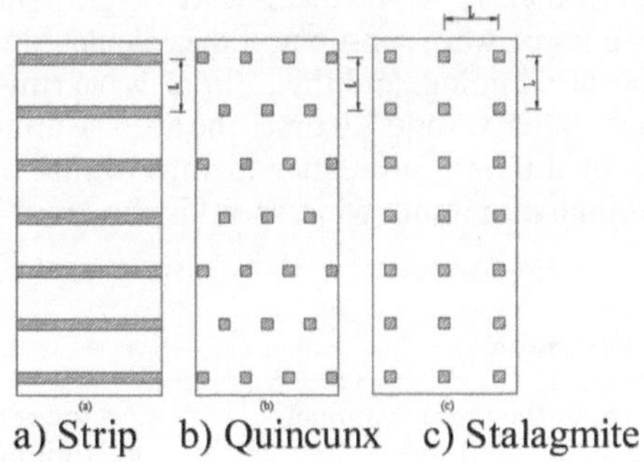

Figure 2. Different wave dissipation piers (top view).

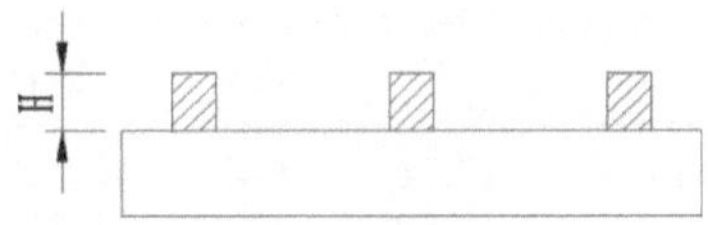

Figure 3. Stalagmite wave dissipation pier (front view).

3 EXPERIMENTAL GROUP ARRANGEMENT

First, the wave run-up values of concrete slab under 1.5 times the average annual maximum wind speed (19.7 m/s) and 2.0 times the average annual maximum wind speed (26.2 m/s) are measured as reference values. Then, the wave run-up values of different protective facings under 1.5 and 2.0 times the average annual maximum wind speed are measured and their range of decrease was calculated compared with concrete slab to evaluate the wave-absorbing effect of each protective facing.

The different protective facings of the sloping breakwater are (1) barrier board, (2) hollow square, (3) accropode, (4) strip wave dissipation pier (Figure 2 a), (5) quincunx wave dissipation pier (Figure 2 b), and (6) stalagmite wave dissipation pier (Figures 2 c and 3).

For the strip wave dissipation pier, the width of a single wave dissipation pier is 30 cm, the thickness is 20 cm, and the distances between the adjacent two pier centers L are 1 and 2 m.

For the quincunx wave dissipation pier, the thickness of the single-wave dissipation pier is 20 cm and the length and width are both 30 cm. The distance between two adjacent pier edges is 50 cm in the horizontal direction, and the distances between two adjacent rows of pier centers (L) are 1 and 2 m.

For the stalagmite wave dissipation pier, the length and width of the single-wave dissipation pier are both 30 cm, and the thicknesses (*H*) are 30, 60, and 90 cm. The distance between two adjacent pier edges (*L*) is 75 cm. For the case of $H=30$ cm, the experiments of $L=30$ and 120 *cm* are added in order to find the effect of the wave dissipation pier layout on the wave run-up.

4 TEST RESULTS AND ANALYSIS

4.1 *Wave run-up*

Wave run-up of the forward wave on a single slope under the direct action of the wind can be determined according to the formula given in the *Code for design of levee project*:

$$R_P = \frac{K_\Delta K_V K_P}{\sqrt{1+m^2}}\sqrt{H_m L} \tag{1}$$

where R_P is the wave run-up, K_Δ is the roughness and permeability coefficient of slopes, K_V is the empirical coefficient, K_P is the cumulative frequency conversion coefficient of wave run-up, m is the slope of the breakwater, H_m is the mean wave height, and L is the wave length.

Table 2 presents the test values of wave run-up and the calculated values of common protective facings, the test elevation of climbing, and the ratio of wave run-up of different protective facings to that of concrete slab under 1.5 and 2.0 times the average annual maximum wind speed.

The ratio of wave run-up of different protective facings to that of concrete slab under 1.5 and 2.0 times the average annual maximum wind speed is shown in Figure 4 (a) and Figure 4 (b), respectively.

Table 2. Test results of the wave run-up.

Forms of protective facings		1.5 times the average annual maximum wind speed: Wave run-up (m) Test value	Calculated value	Elevation (m)	Ratio of concrete slab	2.0 times the average annual maximum wind speed: Wave run-up (m) Test value	Calculated value	Elevation (m)	Ratio of concrete slab
Concrete slab		2.26	2.19	12.71	1.00	2.73	2.68	13.18	1.00
Barrier board		1.35	–	11.80	0.60	1.54	–	11.99	0.56
Hollow square		1.11	1.27	11.56	0.49	1.40	1.41	11.85	0.51
Accropode		0.97	1.08	11.42	0.43	1.31	1.20	11.76	0.48
Strip wave dissipation pier	$L=1$ m	1.07	–	11.52	0.47	1.23	–	11.68	0.45
	$L=2$ m	1.33	–	11.78	0.59	1.52	–	11.97	0.56
Quincunx wave dissipation pier	$L=1$ m	1.11	–	11.56	0.49	1.49	–	11.94	0.55
	$L=2$ m	1.54	–	11.99	0.68	2.08	–	12.53	0.76
Stalagmite wave dissipation pier	$H=30$ cm $L=75$ cm	2.16	–	12.61	0.95	2.47	–	12.92	0.90
	$H=60$ cm $L=75$ cm	1.66	–	12.11	0.73	2.11	–	12.56	0.77
	$H=90$ cm $L=75$ cm	1.99	–	12.44	0.88	2.18	–	12.63	0.80
	$H=30$ cm $L=30$ cm	1.76	–	12.21	0.78	2.25	–	12.70	0.82
	$H=30$ cm $L=120$ cm	2.25	–	12.70	1.00	2.55	–	13.00	0.93

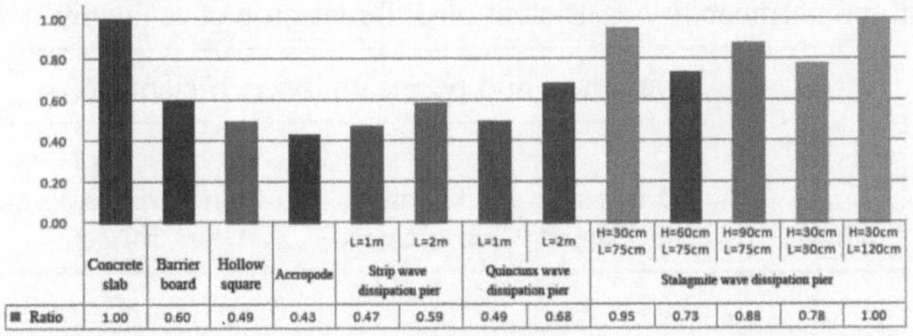

a) 1.5 times the average annual maximum wind speed

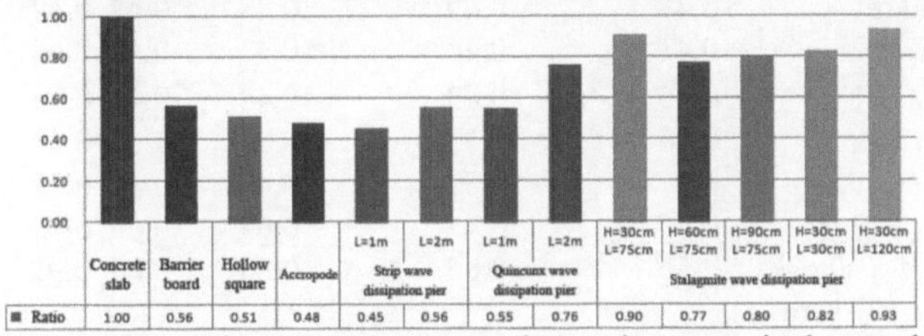

b) 2.0 times the average annual maximum wind speed

Figure 4. Ratio of the wave run-up of different protective facings to that of concrete slab under the average annual maximum wind speed.

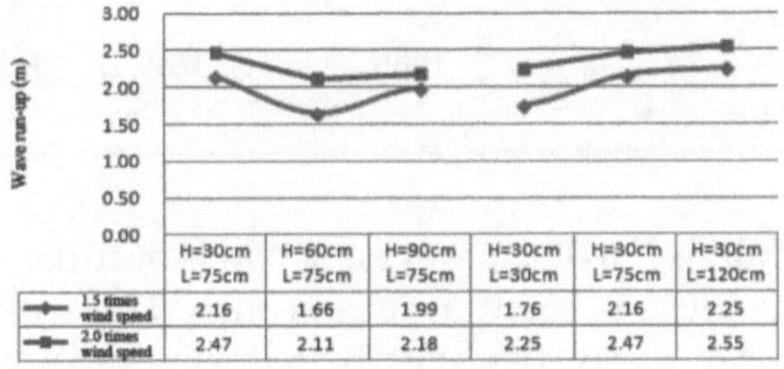

Figure 5. Wave run-up of the stalagmite wave dissipation pier with different densities and thicknesses of the wave dissipation pier.

It can be seen from Figure 4 that the accropodes reduce the wave run-up amplitude significantly in different forms of protective facings and that the wave elimination effect is the best. The second best methods of different protective facings are the strip wave dissipation pier and the hollow square. The wave elimination effect of the quincunx wave dissipation pier and the barrier board is slightly lower than that of the strip wave dissipation pier and hollow square. The stalagmite wave dissipation pier reduces the wave run-up amplitude of the smallest and the wave elimination effect is the worst.

For the special wave dissipation pier (strip, quincunx, and stalagmite), with the increase in the wave dissipation pier density, that is, with the decrease of L, the decrease in the amplitude of the wave run-up is greater and the effect of wave elimination is more obvious.

However, for the stalagmite wave dissipation pier, with the increase in the wave dissipation pier thickness, that is, with the increase of H, the decrease amplitude of wave run-up is non-monotonic, and the specific reasons will be explained in subsequent analysis.

Wave run-up of the stalagmite wave dissipation pier with different densities and thicknesses of the wave dissipation pier is shown in Figure 5.

It is evident from Figure 5 that, for the stalagmite wave dissipation pier, with the increase in the wave dissipation pier thickness, the wave elimination effect is not monotonically increasing. Beyond certain thickness, the wave elimination effect declines and the wave run-up increases significantly.

This may be due to the fact that the wave elimination effect is related to not only the layout density of the wave dissipation pier, that is, the wave elimination effects of both sides of the pier, but also the thickness of the wave dissipation pier, which directly determines the decrease amplitude of wave run-up by the top of the pier.

Table 3. Roughness and permeability coefficient of different protective facings.

Forms of protective facings		Roughness and permeability coefficient (K_Δ)				
		1.5 times wind speed	2.0 times wind speed	Mean value	Approximate range	Standard recommended value
Concrete slab		0.95	0.95	0.95	0.95	0.95
Barrier board		0.57	0.54	0.55	0.54–0.57	–
Hollow square		0.47	0.49	0.48	0.47–0.49	0.55
Accropode		0.41	0.46	0.43	0.41–0.46	0.47
Strip wave dissipation pier	$L = 1$ m	0.45	0.43	0.44	0.43–0.45	–
	$L = 2$ m	0.56	0.53	0.54	0.53–0.56	–
Quincunx wave dissipation pier	$L = 1$ m	0.47	0.52	0.49	0.47–0.52	–
	$L = 2$ m	0.65	0.72	0.69	0.65–0.72	–
Stalagmite wave dissipation pier	$H = 30$ cm $L = 75$ cm	0.91	0.86	0.88	0.86–0.91	–
	$H = 60$ cm $L = 75$ cm	0.70	0.73	0.72	0.70–0.73	–
	$H = 90$ cm $L = 75$ cm	0.84	0.76	0.80	0.76–0.84	–
	H = 30 cm $L = 30$ cm	0.74	0.78	0.76	0.74–0.78	–
	$H = 30$ cm $L = 120$ cm	0.95	0.89	0.92	0.89–0.95	–

When the mean wave height is fixed, the wave can climb over the top of the pier only when the wave dissipation pier thickness is below a certain value. This will increase the bottom friction resistance, causing a wave energy loss, thereby reducing the wave run-up.

When the wave dissipation pier is thicker than the set value, the wave will not climb over the top of the pier. Only the wave elimination effects of both sides of the pier will contribute to the reduction of wave run-up; thus, the wave run-up will increase.

4.2 *Roughness and permeability coefficient*

Formula (1) shows that only the roughness and permeability coefficient will be influenced by different protective facings on the condition that mean wave height (H_m), wave length (L), empirical coefficient (K_V), cumulative frequency conversion coefficient of wave run-up (K_P), and the slope of the breakwater (m) are fixed.

Therefore, assuming that the roughness and permeability coefficient of concrete slab is 0.95 according to the *Code of Hydrology for Harbour and Waterway* and the *Code for design of levee project*, the roughness and permeability coefficient of different protective facings can be relatively determined according to the test data, and are given in Table 3.

Considering the discreteness of the experimental data, it is suggested to take the mean value of the roughness and permeability coefficient of different protective facings under 1.5 and 2.0 times the average annual maximum wind speed as the analysis object.

5 CONCLUSION

It is evident from Table 3 that for the common protective facings, such as concrete slab, barrier board, hollow square, and accropode, their roughness and permeability coefficient were close to the corresponding coefficients given by the *Code of Hydrology for Harbour and Waterway* and the *Code for design of levee project* after comparison.

The test value of the roughness and permeability coefficient of the hollow square was 0.48, and the standard recommended value given by the *Code of Hydrology for Harbour* was

0.55, which was 12.7% smaller than the standard value. The test value of the roughness and permeability coefficient of accropode was 0.43, and the standard recommended value given by the *Code of Hydrology for Harbour* was 0.47, which was 8.5% smaller than the standard value. Considering the partial safety of the code, the proposed value was relatively large, so the experimental value is acceptable.

For the special wave dissipation pier (strip, quincunx, and stalagmite), the roughness and permeability coefficient were not given by the relevant codes, and therefore can be referred to practical engineering, as given in Table 3.

In this study, we investigated the effects of different types of protective facings of the sloping breakwater on the wave run-up under irregular wave conditions by performing a series of sectional physical model tests. The relationship between the roughness and permeability coefficient and the wave run-up was discussed, to complement the roughness and permeability coefficient range of different protective facings and to improve the calculation formulas of the wave run-up in relevant codes.

ACKNOWLEDGMENTS

This study was financially supported by the National Natural Science Foundation of China (Grant No. 51579156), the Fifteenth Session Program between China and Bulgaria Scientific and Technological Cooperation Committee (No. 15–13), and the Major Project of Nanjing Hydraulic Research Institute Funds (Grant Nos. Y216024 and Y216017).

REFERENCES

Chen G.P., Wang Z., Yuan W.X., et al. 2013. Calculation of wave run-up under the irregular wave action. *Port and Waterway Engineering*: 23–30. Beijing, China.

Chen G.P., Yu G.M., Zhang J.C., 1992. The effect of berm width and elevation on irregular wave run-up. *The Ocean Engineering*: 59–67. Nanjing, China.

Chen G.P., Zhou Y.R., Ju L.H., 2005. Influence of Protective Styles of Seawall on Wave Runup and Overtopping. Port and Waterway Engineering: 28–30. Beijing, China.

Lu Y.J., He Y.S., Liu Y., 2007. Analysis on wave run-up calculation and discussion on ovtertoppong design criteria for sea dike. *Water Resources and Hydropower Engineering*: 30–34. Beijing, China.

Ministry of Transport of the People's Republic of China (MOT), 2001. Wave Model Test Regulation (JTJ234-2001T), *China Communication Press*, Beijing, China.

Ministry of Transport of the People's Republic of China (MOT), 2015. Code of Hydrology for Sea Harbour (JTS 145-2015), *China Communication Press*, Beijing, China.

Ministry of Water Resources of the People's Republic of China (MOW), 2013. Code for design of levee project(GB50286-2013, *China Planning Press*, Beijing, China.

Qi G.J., Yuan X.M., 2012. Numerical simulation of wave run-up on fence panels of sloping dikes. *Advances in Science and Technology of Water Resources*: 16–19. Nanjing, China.

Saville T., 1955. Laboratory data on wave run-up and overtopping. *Lake Okeechobee levee sections*, U.S. Army, Corps of Engineers, Beach Erosion Board, Washington D.C., America.

Technical Advisory Committee on Water Defense, 2002. Wave run-up and wave overtopping at dikes. *Technical Advisory Committee on Water Defense,* Delft, Netherlands.

Waal J.P., Meer J.W., 1992. Wave run-up and overtopping on coastal structures. *Proc. Coastal Engineering Conference,* Venice, Italy.

Hydraulic Engineering V – Hong, Liu & Xie (Eds)

Comparative experiment of the influence of PVD and EVD on vacuum preloading effects

Zhaohua Sun
School of Architecture Engineering, Nantong University, Nantong, Jiangsu, China

Mingjun Gao
Key Laboratory of Ministry of Education for Geomechanics and Embankment Engineering, Hohai University, Nanjing, Jiangsu, China

ABSTRACT: The design philosophy of Electric Vertical Drains (EVDs) is employed as electrodes with drainage abilities. The feasibility of using EVDs as vertical drains needs to be further researched. This paper summarizes the laboratory investigations of the vacuum preloading methods in kaolin clay. These tests consisted of the installation of different vertical drains in the soil. The performance of EVDs was compared to Prefabricated Vertical Drains (PVDs). The vacuum preloading tests indicate that the cumulative water discharge of EVDs and PVDs is similar. The filter membrane and drainage path of EVDs has an obvious improvement on the drainage effect. The energy consumption of vacuum preloadidng by means of EVDs is acceptable.

1 INTRODUCTION

For several decades prefabricated vertical drains (PVDs) have been used for consolidation of soft clays (Saowapakpiboon 2010). The PVD core and filter provide a drainage path to accelerate the consolidation of soft clays. The properties of PVDs, including the tensile strength, the discharge capacity and the consolidation effect, have been discussed in many tests (Liu 2009; Saowapakpiboon 2011). With the advent of conducting polymers, electric vertical drains (EVDs) have been used as electrodes and drainage paths. The appearance of EVDs is similar to PVDs. Electro-osmosis is potentially more effective in reducing the water content than conventional PVDs with or without surcharge.

Electro-osmosis consolidation is one of the methods of improving soft clay deposits (Lee 2007; Micic 2001). The original technique, which requires applying an electric field to the soil, was developed by Reuss in 1809 (Reuss 1809). Clay particles carry a net negative charge on their surface, balanced by a net positive charge in the pore water. When electrodes are installed in soft clays the ions in the pore water are attracted to the oppositely charged electrodes and drag the surrounding free water molecules with them (Shang 1997). The net water movement is towards the cathode, and if drainage is provided, consolidation occurs in the soil. Extensive research with the electro-osmosis method has been conducted on soft clayey soil in many countries (Casagrande 1949; Bjerrum 1967; Lo 1991; Wu 2014; Chien 2010).

Recently, some trials have been conducted to consolidate soft clays utilizing the electro-osmosis method using EVDs (Chew 2004; Kaniraj 2011; Karunaratne 2011; Sun 2015). The drainage performance of EVDs has important influence on the treatment effect. However, this characteristic was not quite obvious in these tests. This paper summarizes the laboratory investigations of EVDs used in vacuum preloading tests in kaolin clay compared to PVDs.

2 DRAINAGE PERFORMANCES OF EVDS COMPARED WITH PVDS

2.1 *Test scheme*

Vertical drains provide considerably shortened horizontal drainage paths for pore water flow in soft ground improvement, thereby accelerating the soil consolidation. PVDs have a series of merits, such as low cost, easy quality control and small disturbance to the foundation. Thus PVDs have been used successfully in soft soil foundation treatment. EVDs consist of a fluted core, two strands of copper wire and a filter membrane. EVDs can also be used as drainage paths. The basic performance indexes of PVDs are shown in Table 1 according to the industry standard (JT/T 521–2004). The indexes of EVDs were obtained from Huahong Yingfei Electro-osmosis Technologies Co., Ltd. It was clear that there are some differences in the two types of vertical drains such as the material, processing technology and cross sectional dimension. The drainage performances of PVDs and EVDs were compared in vacuum preloading tests under the same conditions.

The experiment device is mainly composed of an rectangular cell and a vacuum drainage system as shown in Fig. 1. The cell is made of Plexiglas with the dimensions of 230 mm long, 160 mm wide and 170 mm high. The vacuum drainage system is composed of vertical drains, a sand cushion, a sealing membrane, a steam-water separator, a steam-gasoline separator and a vacuum pump connected by a pneumatic hose. One end of the pneumatic hose is connected to the vacuum pump, and the other is embedded in the sand cushion. The surface of the steam-water separator has the scale division line. A vacuum meter, which can be used to monitor the vacuum degree of the vacuum drainage system, was connected to the steam-water separator. The ultimate pressure provided by the vacuum pump (FY-C-2 N) is 0.1 MPa. The vacuum pump uses 250 watts of power, and the pumping rate is 7.2 m3/L.

The soil used in this study was obtained from the Nanjing Chuncheng Kaolin Clay, Co., Ltd. Kaolin clay, which is creamy white, is a soft clay with low permeability. The physical

Table 1. Basic performance indexes of PVDs and EVDs.

Vertical drain	Core-making materials	Cross sectional dimension Width (mm)	Cross sectional dimension Thickness (mm)	Tensile strength at 10% axial strain at dry state (kN/10 cm)
PVD	Polyethylene, polypropylene	100	4	>1.2
EVD	Polyethylene, carbon black, graphite	96	4	>1.2

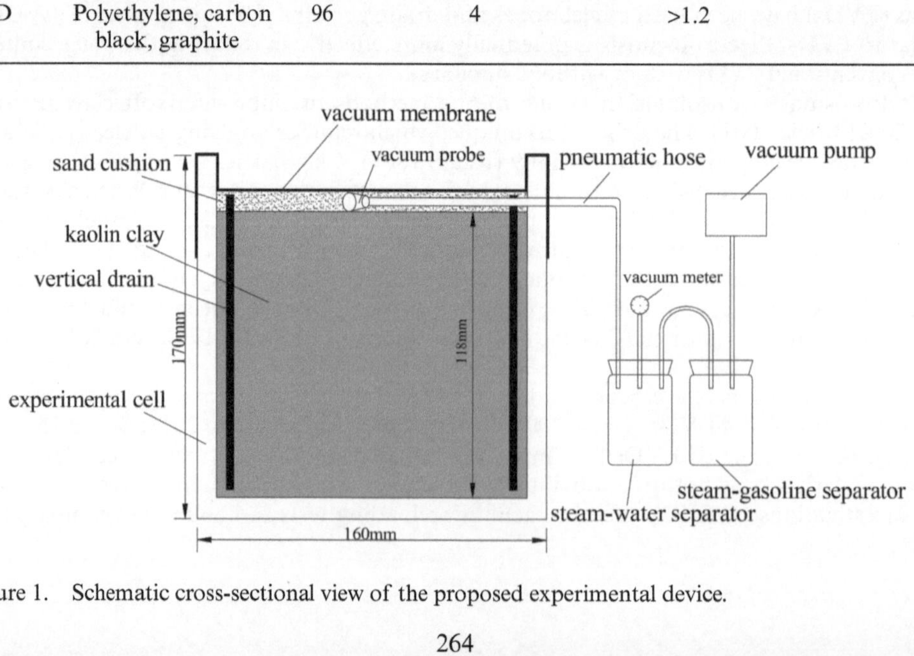

Figure 1. Schematic cross-sectional view of the proposed experimental device.

and mechanical properties and the chemical composition content of the soil are show in Table 2. The grain size distribution was homogeneous, which was less than 15 microns in diameter. More than 95% of the particles were in the range of 0~10 microns. The objective moisture content of the prepared remolded sample was 70%. After standing for 24 hours, the actual moisture content of the soil was measured. The prepared soil sample was placed and compacted in the experimental cell in layers. The quality of the soil sample for each test was 7166 g. The height of the soil was 118 mm. A 20 mm thick layer of sand cushion was added on top of the soil.

Two groups of vacuum preloading tests were conducted in the laboratory. The only difference of the two groups of tests is the vertical drains. Two EVDs or PVDs were inserted into the soil in each test. To prevent the kaolin clay from squeezing into the core of the vertical drains, the bottom needed to be sealed before the vertical drains were inserted into the soil. The vacuum pressure design value was 85 kPa. The cumulative water discharge of each test was collected. The tests were stopped when water was no longer discharged. The vacuum drainage system was then removed. Then, the moisture content of the surface, middle and underlying layers of the soil was tested.

2.2 *Test results*

2.2.1 *Vacuum degree*

Vacuum degree can reflect the sealing condition of the vacuum drainage system and the operation condition of the vacuum pump. After the vacuum pump was opened, the vacuum degree under the membrane reached 85 kPa within 5 minutes and stabilized at this value in these tests.

2.2.2 *Water discharge*

The cumulative water discharge of EVDs and PVDs in the two groups of tests is shown in Fig. 2. The water discharge curve of the EVDs was coincided with the curve of the PVDs for the first three hours. The water discharge capacity of the EVDs was then slightly below that of the PVDs. The vacuum preloading drainage rate decreased with the increasing of vacuum time. The cumulative water discharge of the PVDs was 1021 mL, and there was no longer water discharged after 13 hours. The cumulative water discharge of the EVDs was 998 mL

Table 2. Physical and mechanical properties of the kaolin clay.

Specific gravity	Fineness (mesh)	Liquid limit (%)	Plastic limit (%)	Chemical composition (%)					
				SiO_2	Al_2O_3	Fe_2O_3	K_2O, Na_2O	CaO, MgO	others
2.72	200	43.7	22.2	52	29	1.8	3	1.13	13.07

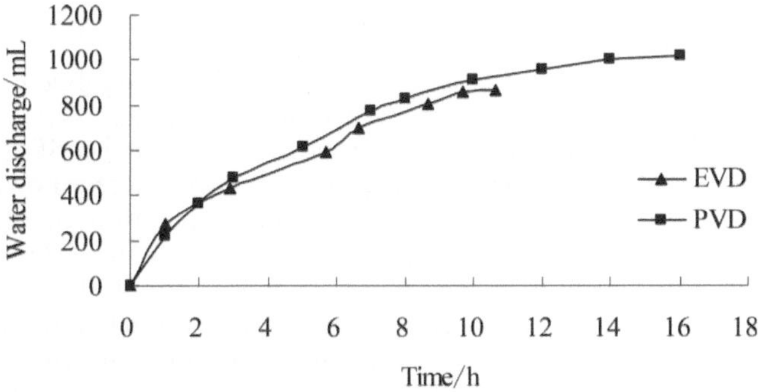

Figure 2. Water discharge curve of the EVDs and PVDs in the vacuum preloading tests.

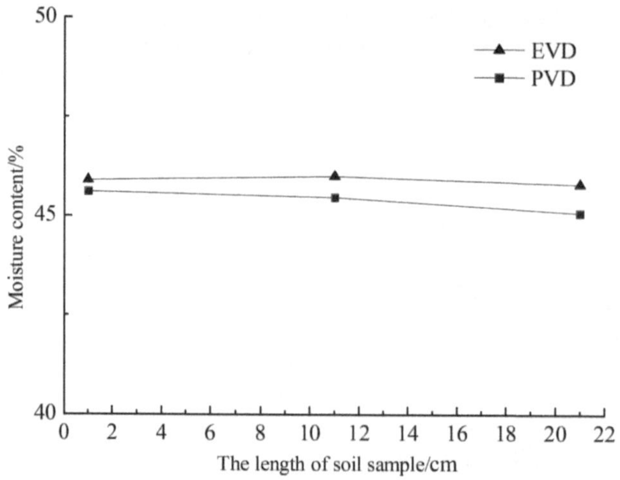

Figure 3. Moisture content distribution.

within 16 hours. Thus, it can be observed that the cumulative water discharge of the EVDs and PVDs was similar. The drainage effect of the EVDs can be guaranteed if there is enough time. This is mainly because of the differences in the basic performance index of the two types of vertical drains, which is consistent with the data reflected in Table 1. The discharge capacity of EVDs was less than that of the PVDs.

2.2.3 *Moisture content distribution*

The actual moisture content of the sample was 69%. The moisture content of the surface, middle and underlying layers of the soil after treatment was tested, as shown in Fig. 3. The average value in the EVD and PVD tests was 45.9% and 45.4%, respectively. The moisture content distribution was uniform in these tests. There was little difference between the two groups of tests.

2.2.4 *Power consumption*

How to meet the need of the engineering strengthening effect with comparatively low energy consumption and costs is of considerable interest. This paper concentrates on the vacuum preloading contrast test of EVD with PVD. The total energy consumption and the energy consumption per unit of the discharged water of the vacuum preloading test can be calculated by (1) and (2), respectively:

$$W_V = Pt \tag{1}$$

$$\varpi_V = \frac{W_V}{Q_V} \tag{2}$$

where P = power of the vacuum pump, and here, P = 250 W; t = treatment time of vacuum preloading; and Q_v = total discharged water of vacuum preloading.

The total energy consumption and the energy consumption per unit of the discharged water of the vacuum preloading test calculated by (1) and (2) were shown in Fig. 4. The total energy consumption and the energy consumption per unit of discharged water of the EVDs test were slightly higher than that of the PVDs test. This is because the energy consumption of the vacuum preloading is proportional to the vacuum time. The vacuum time of the EVDs test was approximately 3 hours longer than that of the PVDs test. The result show that the total energy consumption and the energy consumption per unit of discharged water of EVDs in vacuum preloading test was acceptable in the laboratory.

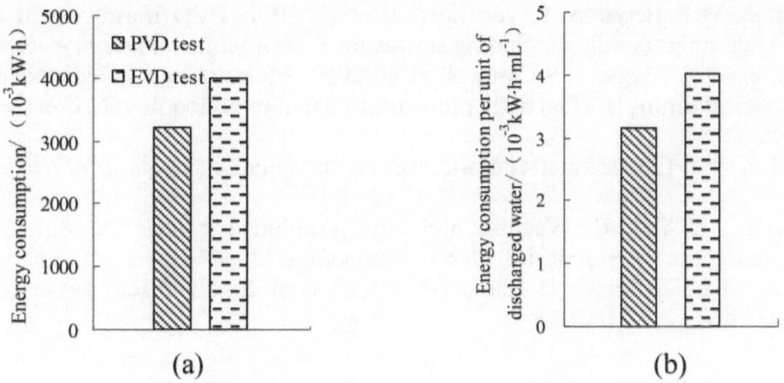

Figure 4. Energy consumption histogram of the vacuum preloading test; (a) energy consumption and (b) energy consumption per unit of discharged water.

3 CONCLUSIONS

The cumulative discharged water of the EVDs and PVDs was close in the vacuum preloading contrast tests. The discharge capacity of EVDs was less than that of PVDs. The drainage effect of EVDs can be guaranteed if there is enough time. The moisture content distribution was uniform and close in the two groups of tests. The total energy consumption and the energy consumption per unit of discharged water of EVDs in vacuum preloading test was acceptable in the laboratory. EVDs can be used as a vertical drain for the soil.

ACKNOWLEDGEMENTS

This work is currently supported by Natural Science Foundation for the University of Jiangsu Province (No. 16 KJB560010) and the Science and Technology Project of the Ministry of Housing and Urban-Rural Development of China (No. 2016-K4–047).

REFERENCES

Bjerrum, L., J. Moum, O. Eide. 1967. Application of electro-osmosis to a foundation problem in a Norwegian quick clay. *Geotechnique* 17: 214–235.

Casagrande, I.L. 1949. Electro-Osmosis in Soils. *Geotechnique* 1: 159–177.

Chew, S.H., G.P. Karunaratne, V.M. Kuma, et al. 2004. A field trial for soft clay consolidation using electric vertical drains. *Geotextiles and Geomembranes* 22: 17–35.

Chien, S.C., C.Y. Ou, Y.C. Lee. 2010. A novel electroosmotic chemical treatment technique for soil improvement. *Applied clay science* 50: 481–492.

Kaniraj, S.R., H.L. Huong, J.H.S. Yee. 2011. Electro-osmotic consolidation studies on peat and clayey silt using electric vertical drain. *Geotechnical and Geological Engineering* 29: 277–295.

Karunaratne, G.P. 2011. Prefabricated and electrical vertical drains for consolidation of soft clay. *Geotextiles and Geomembranes* 29: 391–401.

Lee, M.H. 2007. Electrically induced settling and consolidation behaviour of soft soil. *KSCE Journal of Civil Engineering* 11: 185–191.

Liu, H.L., J. Chu. 2009. A new type of prefabricated vertical drain with improved properties. *Geotextiles and Geomembranes* 27: 152–155.

Lo, K.Y., K.S. Ho, I.I. Inculet. 1991. Field test of electroosmotic strengthening of soft sensitive clay. *Canadian Geotechnical Journal* 28: 74–83.

Micic, S., J.Q. Shang, K.Y. Lo. 2001. "Electrokinetic strengthening of marine clay adjacent to offshore foundations." Paper presented at the annual meeting for the Eleventh International Offshore and Polar Engineering Conference. Stavanger, Norway, June 17–22.

Reuss, F.F. 1809. Charge-induced flow. Proceedings of the Imperial Society of Naturalists of Moscow. Moscow, Russia.

Saowapakpiboon, J., D.T. Bergado, P. Voottipruex, et al. 2011. PVD improvement combined with surcharge and vacuum preloading including simulations. *Geotextiles and Geomembranes* 29: 74–82.
Saowapakpiboon, J., D.T. Bergado, S. Youwai, et al. 2010. Measured and predicted performance of prefabricated vertical drains (PVDs) with and without vacuum preloading. *Geotextiles and Geomembranes* 28: 1–11.
Shang, J.Q., K.Y. Lo. 1997. Electrokinetic dewatering of a phosphate clay. *Journal of Hazardous Materials* 55: 117–133.
Sun, Z.H., M.J. Gao, X.J. Yu. 2015. Vacuum preloading combined with electro-osmotic dewatering of dredger fill using electric vertical drains. *Drying Technology* 33: 847–853.
Wu, H., L.M. Hu. 2014. Microfabric change of electro-osmotic stabilized bentonite. *Applied Clay Science* 101: 503–509.

Hydraulic Engineering V – Hong, Liu & Xie (Eds)
© 2018 Taylor & Francis Group, London, ISBN 978-0-8153-7471-8

Study on the wind-induced vibration of steel bridge members in a mountainous city

Xiaohui Tang & Chengqi Wang
School of Civil Engineering, Chongqing Jiaotong University, Chongqing, P.R. China

Zhichao Li
The Eighth Engineering Company, Construction Group of Yunnan Highway Development and Investment Co. Ltd., Yunnan, P.R. China

Yi Chen
School of Civil Engineering, Chongqing Jiaotong University, Chongqing, P.R. China

ABSTRACT: In this study, a mass-spring-damper system of two-dimensional steel bridge members is simulated by a numerical wind tunnel model to investigate the typical performance of the members using the fourth-order Runge–Kutta method for discrete analysis, and to study the mechanism of wind-induced vibration. In addition, vortex-induced vibration velocity is obtained and factors such as wind attack angle and damping ratio are discussed. Using a real bridge as an example, several conclusions are drawn. First, the frequency of vortex shedding is very close to the natural frequency at a wind speed of 45 m/s and a wind attack angle of 0°. At this wind speed, the displacement reaches maximum and the section vortex-excited resonance occurs, which leads to the gradual appearance of a periodic alternating vortex with a relatively uniform size and space. Second, the maximum displacement and maximum lift will change as the wind attack angle changes at 45 m/s and the maximum displacement occurs at the wind attack angle of 0°. Finally, the maximum displacement decreases with increasing damping ratio, and the damping ratio can effectively limit the displacement response amplitude to a certain degree.

1 INTRODUCTION

Long-span steel bridge members are sensitive to wind speeds. The amplitude of the structure increases with the occurrence of vortex-excited vibration, which significantly affects construction process. Wind speed, wind attack angle, and damping ratio are important parameters for the study of bridge wind resistance. At present, three methods are used for bridge wind engineering: field measurement, wind tunnel test, and theoretical analysis (Chen, 2005). With the rapid development of computer technology, numerical simulation has gradually become an inevitable analytical method, because of its high performance–cost ratio and flow-field visualization using theoretical analysis. Vortex-induced vibration of a typical bridge member is studied from three aspects, including wind speed, wind attack angle, and damping ratio, to investigate the effects of various factors on vibration response through an established numerical wind tunnel model. The mechanism of vortex-induced vibration is analyzed from the vortex shedding flow-field diagram. Finally, reliable conclusions are drawn in favor of engineering applications.

2 NUMERICAL WIND TUNNEL SIMULATION OF THE VORTEX-INDUCED VIBRATION OF TRUSS

2.1 *Numerical wind tunnel model*

On the basis of the fluid theory, we take a typical member of Chaotianmen Grand Bridge across Yangtze River as the object of our study. Numerical simulation of vortex-induced vibration of a steel truss member is done, and the position of the steel truss member is shown in Figure 1. The size of the rectangular section of the outer contour is 1.24 m × 1.6 m. Modal analysis of the member was carried out using ANSYS software to obtain the cross-wind frequency (4.155 Hz) and downwind frequency (5.05 Hz).

The square cylinder is located at the centroid of the origin of the flow field. The distance between the lower boundary and upper boundary of the flow field and the square column is 15 D. The distance from the origin to the left border is 15 D and to the right border is 45 D. The area of encryption is 6D × 6D (where D is the length of the windward side of the square cylinder), and structured mesh is used globally. The grid number is about 76613, and the numerical wind tunnel model is shown in Figure 2.

The entrance uses velocity inlet boundary conditions (Wang, 2004), and the exit uses free outflow boundary conditions. In addition, the rectangular section adopts the wall boundary condition. Symmetrical boundary conditions are adopted for up-down direction boundary layers. On the basis of pressure, steady equation solver, and SST k-ω turbulence model (Yu, 1995), the pressure–velocity coupling mode adopts the SIMPLEC algorithm (Zhang, 2005), and both speed difference and convection are discretized by a second-order upwind scheme. In the calculation process, the time step is 0.001 s, residual control is 1×10^{-5}, turbulence intensity is 5%, and turbulence viscosity ratio is 10%. The Y+ of the surface of the flow around is in the range of 30–300; therefore, the flow is addressed in the turbulent zone (Wang, 2004). Velocity along the normal direction of the wall follows logarithmic distribution. The numerical simulation of turbulent flow with full development can be performed under high Reynolds number.

2.2 *Fluid–solid coupling method*

In this study, the model is simplified as a mass-spring-damper system, and the vibration equation is derived as follows:

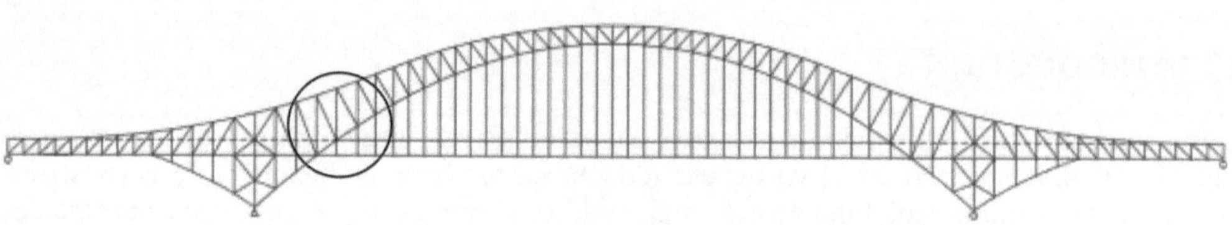

Figure 1. Elevation diagram of the Chongqing Chaotianmen bridge.

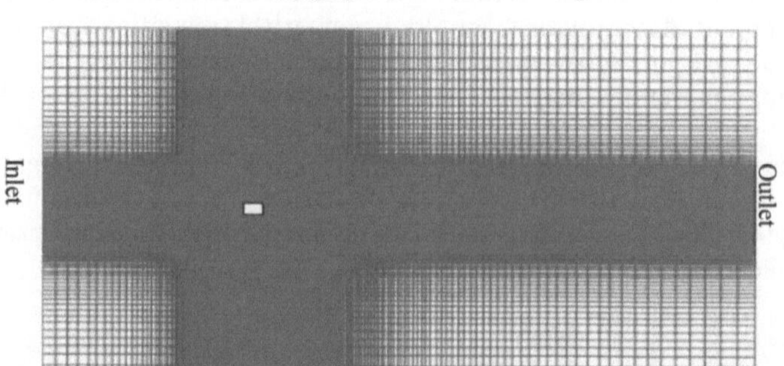

Figure 2. Numerical wind tunnel model.

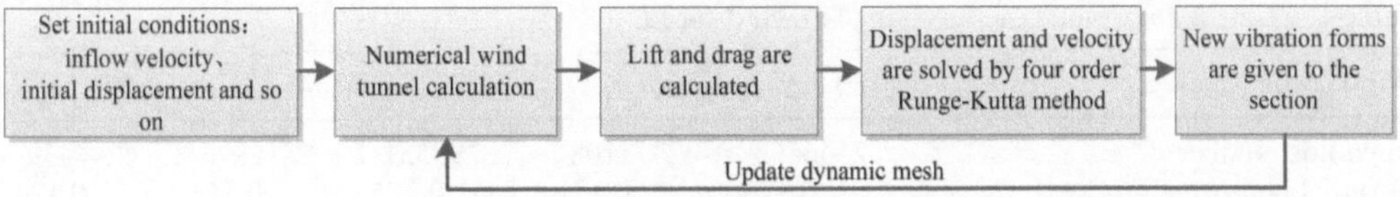

Figure 3. Computational flow diagram of fluid–solid coupling.

$$m\ddot{y}_{(t)} + c\dot{y}_{(t)} + ky_{(t)} = F_{(t)} \tag{1}$$

where m is the structural quality; c is the structural damping; k is the elastic coefficient; F_t is the fluid force; $y_{(t)}$ is the displacement; $\dot{y}_{(t)}$ is the velocity; and $\ddot{y}_{(t)}$ is the acceleration of the structure.

As shown in eq. (2), the implicit method is adopted to decompose the formula:

$$\mathrm{f}(v,y) = f(t)/m - 2\zeta\omega_o\dot{y}(t) - \omega_o{}^2y(t) \tag{2}$$

The vibration velocity and displacement of each time step are obtained using the fourth-order Runge–Kutta method for discrete analysis (1). With the displacement and velocity as the basic parameters of the next calculation, the spring smoothing method is used to update the dynamic grid continuously during the whole period. During calculation, an initial value is required. On the basis of the above method, a self-coded program UDF is embedded in the FLUENT. First, the stationary square cylinder is calculated under unsteady condition when the calculation starts. Second, the calculation result of the above step is used as the vibration mode of the square cylinder to carry out coupling calculation with the fluid. In this way, coupling vibration under different wind speeds is explored, and the calculation process is shown in Figure 3.

3 VORTEX-INDUCED VIBRATION VELOCITY SEARCH AND FLOW-FIELD CHARACTERISTIC ANALYSIS

3.1 *Vortex-induced vibration velocity search*

According to the results of the wind tunnel test (Wang, 2011), wind speed in the simulation of vortex-induced vibration of steel truss varies in the range of 30–50 m/s. The results of the calculation process shown in Figure 3 are presented in Table 1.

The frequency of vortex shedding, lift coefficient, and drag coefficient change with wind speed (and different Reynolds number). As shown in Table 1, the change of Strouhal number is moderate with increasing wind speed.

At 30–40 m/s, the frequency of vortex shedding increased with increasing wind speed, whereas the trend is weak at 40–45 m/s. The frequency of vortex shedding is very close to the natural frequency at the wind speed of 45 m/s. Compared with 40 and 50 m/s, the displacement is significantly increased and the section vortex-excited resonance occurs at this wind speed. The displacement diagram and the spectrogram of the section at 45 m/s are shown in Figures 4 and 5, respectively.

3.2 *Flow-field characteristic analysis*

The vortex cloud chart can reflect the variation of the vortex within the lock-in region under different wind speeds from the analysis of the visual flow-field characteristics. As shown in Figure 6(a)–(f), certain vortices are stripped at the wind speed of 30 m/s, but did not form periodic alternating vortex. At wind speed of up to 35 m/s, periodic vortex street formed gradually. The regularity of the vortex street becomes obvious at 40 and 43 m/s. When section vortex-excited resonance occurs, a vortex street with an obvious periodic alternating vortex

Table 1. Calculated results of the change in wind speed.

Wind speed (m/s)	30	35	40	45	50
Reynolds number	2546663	2971108	3395551	3819996	4244439
Strouhal number	0.123	0.121	0.125	0.115	0.124
Frequency of vortex shedding (Hz)	2.98	3.4	4.02	4.18	5.0
Natural frequency (Hz)	4.155	4.155	4.155	4.155	4.155
Maximum displacement (mm)	9.42	11.00	31.70	65.54	22.133
Maximum lift (N)	1870.80	2290.34	3335.28	3645.39	4522.39

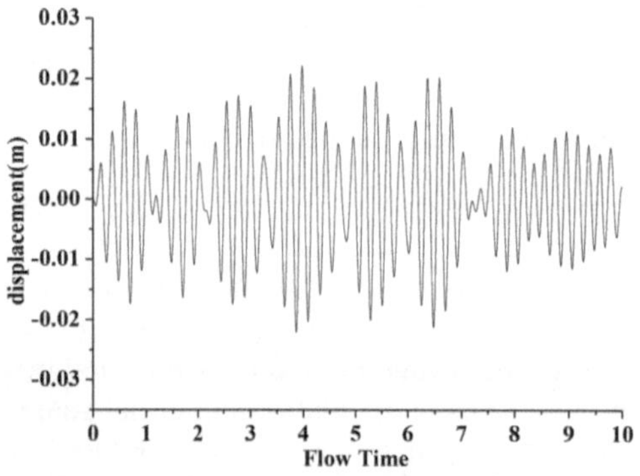

Figure 4. Displacement–time histories of the section at 45 m/s.

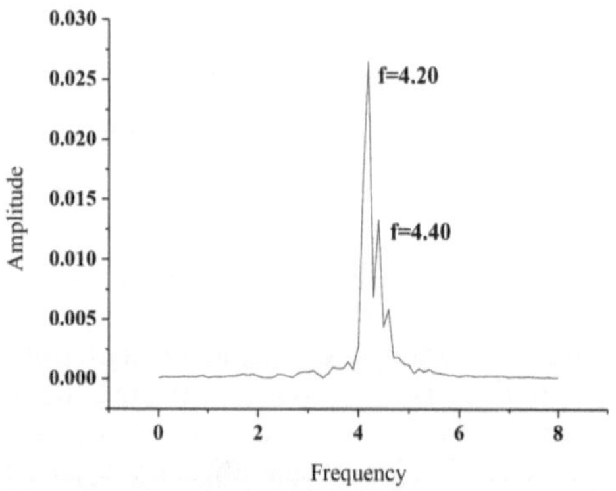

Figure 5. Spectrogram of the section at 45 m/s.

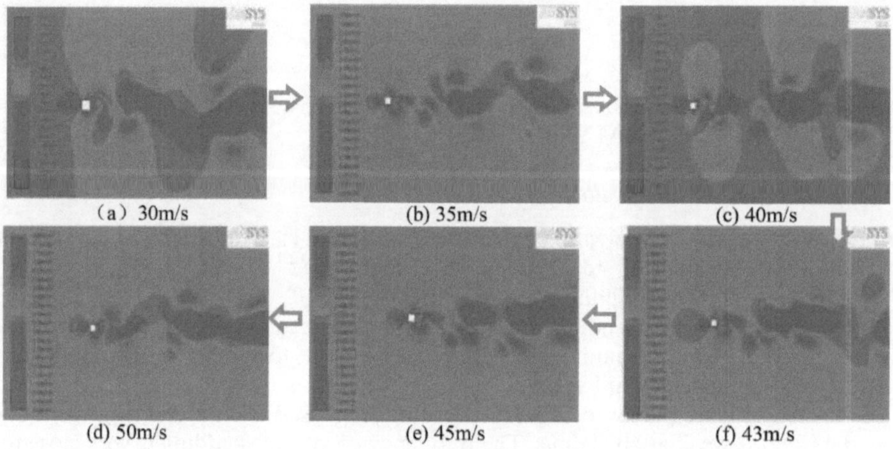

Figure 6. Diagram of vortex shedding at all levels of wind speed.

appears at 45 m/s with a relatively uniform size and space. The periodic vortex street began to disappear at 50 m/s.

4 EFFECT OF THE WIND ATTACK ANGLE AND THE DAMPING RATIO ON VORTEX VIBRATION

4.1 *Analysis of the wind attack angle*

A large number of wind tunnel tests have shown that Strouhal number, which determines the frequency of vortex shedding, changes with the wind attack angle. In this study, we consider

Table 2. Calculated results of the change in wind speed.

Wind attack angle	0°	5°	10°	20°
Reynolds number	3395551	3395551	3395551	3395551
Strouhal number	0.125	0.125	0.125	0.125
Frequency of vortex shedding (Hz)	4.18	4.10	4.32	4.50
Natural frequency (Hz)	4.155	4.155	4.155	4.155
Maximum displacement (mm)	65.54	21.2	31.31	29.8
Maximum lift (N)	3645.39	2152.54	2888.13	2948

Table 3. Calculated results of different damping ratios.

Damping ratios	0.1%		0.5%		0.75%	
Wind speed	30	45	30	45	30	45
Reynolds number	2122694	2547233	2122694	2547233	2122694	2547233
Strouhal number	0.124	0.123	0.121	0.125	124	0.123
Frequency of vortex shedding (Hz)	3.0	4.45	3.4	4.02	3.0	4.45
Natural frequency (Hz)	4.155	4.155	4.155	4.155	4.155	4.155
Maximum displacement (mm)	10.84	68.47	9.42	65.54	8.05	41.608
Maximum lift (N)	1870.81	3756.40	1870.8	3645.39	1870.81	3467.65

the change law of wind attack angle under 0°, 5°, 10°, and 20°. The wind speed was 45 m/s, the turbulence intensity was 5%, and the damping ratio was 0.5%. The results are shown in Table 2.

4.2 *Analysis of the damping ratio*

Compared with the analysis of three types of damping ratio at wind speeds of 30 and 45 m/s, the calculation results of each parameter are presented in Table 3. As shown in the table, maximum displacement showed a decreasing trend with increasing damping ratio. At 30 m/s, the decreasing maximum displacement reached 25.7% when the damping ratio increased from 0.1% to 0.75%. At 45 m/s, the decreasing extent reached 4.3% with the increase of damping ratio from 0.1% to 0.5%. When the damping ratio increased from 0.1% to 0.75%, the decreasing extent was even greater, that is, 39.2%. Therefore, increasing structural damping ratio, to some extent, can effectively limit the displacement amplitude of resonance. Because of the significant effect on the maximum displacement of the structure, it becomes important to consider the control of the damping ratio when designing a bridge.

5 CONCLUSION

This study investigated the vortex-induced vibration of typical steel bridge members in a mountainous city. First, the model was analyzed using a self-coded program UDF in ANSYS-FLUENT. Then, fluid–solid coupling solution and flow-field characteristics were analyzed to obtain the vortex-induced vibration velocity. Moreover, the effects of the wind attack angle and the damping ratio on the vortex-induced vibration were studied. The conclusions drawn from this study are summarized as follows:

The frequency of vortex shedding amplitude increased significantly with increasing wind speed and became very close to the natural frequency. When the section vortex-excited resonance occurred with the increase in wind speed, the vortex street appeared gradually, which became obvious and periodic at the vortex-induced vibration velocity. It is important to avoid vortex-excited resonance when accelerating the structure's destruction during designing.

With the change in the wind attack angle, maximum displacement occurred at 0° and at the same wind speed. An increase in the damping ratio of the structure can effectively limit the displacement amplitude of resonance and affect the maximum displacement of the structure. The control of the damping ratio must be taken into account when designing a bridge.

ACKNOWLEDGMENTS

The authors gratefully acknowledge the support of the Science and Technology Research Project of Chongqing Municipal Education Commission (No. KJ130409).

REFERENCES

Chen Zhengqing. & Xiang Haifan. 2005. Bridge Wind Engineering. Beijing: China communications press.

Wang Chengqi. 2011. Investigation on Wind-Induced Vibration of Long-span Steel Trussed Arch Bridges. Doctoral dissertation of chongqing university.

Wang Fujun. 2004. Computational fluid dynamics analysis: CFD software principles and applications. Beijing: Tsinghua university press.

Yu Ling. 1995, Experimental Study of Wind-Indueed Vibration of Booms of Three Large Arches on Jiujiang's Yangtze River Bridge. Journal of Yangtze River Scientific Research Institute | J Yangtze River Sci Res Inst, 12(3):53–60.

Zhang Lizhu. 2005. Simple Algorithm and Application. Postdoctoral dissertation of Shanghai jiaotong university.

Hydraulic Engineering V – Hong, Liu & Xie (Eds)
© 2018 Taylor & Francis Group, London, ISBN 978-0-8153-7471-8

Research on the social impact assessment index system of dam failure risk

Weiwei Sun, Zhifei Long, Xiaohang Wang & Lei Li
Dam Safety and Management Department, Nanjing Hydraulic Research Institute, Nanjing, China
Dam Safety Management Center, Ministry of Water Resources, Nanjing, China

Mingqing Gan
Xiangshui Reservoir of Pan-Nan Power Plant, Pan City, Guizhou Province, Panzhou, China

ABSTRACT: In the present era of network information communication, the network public opinion, which has received a wide range of attention, is being generated rapidly when significant events occur. As a result, it has put a great deal of pressure on all departments, institutions, and individuals in the society, which cannot be ignored. Previous studies on dam failure risk and social impact simply considered the measurable impacts of important facilities, cultural relics, and so on, ignoring their effect on "people". The main subject of the project is the impact of dam failure risk on the dam engineering community that consists of government, water administrative department, reservoir management organization, as well as the affected downstream public and media. In this paper, research on the social impact's formation mechanism of dam failure risk is clearly demonstrated, and its diffusion hierarchy of dam failure risk is elaborated. The nine roles classified under dam break risk, namely direct encounters, responsible people, handlers, supervisors, related encounters, disseminators, sympathizers, followers, and avoiders, are analyzed. The social impact's evaluation indicators of dam failure risk are also studied, namely personnel loss, publicity, news, and duration. Taking Dahe Reservoir as an example, practical application shows that the assessment results are highly reliable, objective, and reasonable, and have classification accuracy. This assessment method provides a platform for the comparative analysis of the social impact of different dam failure events based on grade evaluation and easy operability. The data obtained is more objective and easily collected, which proves the validity of the assessment results.

Keywords: dam failure risk; social impact; assessment index system

1 INTRODUCTION

In general, dam failure events cause catastrophic incidents, huge losses of lives and economy, as well as environmental impact, which will in turn harm the society. These disastrous consequences have effects on all aspects of our society in several different ways, and the reaction results in social impact. The main focus of the project is the impact of dam failure risk on the dam engineering community that consists of government, water administrative department, reservoir management organization, as well as the affected downstream public and media. When the dam breaks, the consequences directly act on each member of the community and further extend to the outside, resulting in an extensive social impact (Cheng Duqiu, 2014). In turn, because of the huge impact caused by dam failure, the dam engineering community of government, water administrative department and reservoir management organization is bound to make more efforts in dam monitoring and maintenance, making effective decisions timely and actively, in order to mitigate the dam failure risk. This indicates that the social impact can give some feedbacks to the members of the engineering community through the government, reducing the dam failure risk. In China, research on the social impact of dam failure risk is illegal, and therefore the relevant research is scarce. However, currently, with the development of society and the rapid dissemination of information, people are paying more attention to their own living environment. Dam failure events in the past have affected the whole society and therefore it is urgent to carry out research on the evaluation indicators and grading standards of the social impact of dam failure risk. The results of such research can provide the government with the decision-making basis to reduce the risk of dam failure and its social impact.

2 SOCIAL IMPACT'S FORMATION MECHANISM OF DAM FAILURE RISK

2.1 *Temporal changes: Social impact of dam failure risk in different social development stages*

Before the reform and opening up, China had some dam failure events, but the society strictly controlled the information at that time, controlling the impact to a certain degree. However, with the development of society and the rapid dissemination of information, people pay more and more attention to their living environment. With the continuous development of Internet technology, the social media has been fully into the social life, including micro-blogs, WeChat, forums, and QQ. These media have the following characteristics: participation, publicity, communication, socialization and connectivity, enabling everyone to create and disseminate the information (Huang Junhua, 2014). At present, dam failure events can affect the whole society instead of just several parts as in the past.

2.2 *Composition of the social impact and its effect on different members of the engineering community*

The dam engineering community mainly includes the government, water administrative department, reservoir management organization, as well as the affected downstream public and media. We must clarify the relationships between roles, interests, and effects of different members in the community (Zhang Xiuhua, 2008). In modern society, either the government or the public, together with the media, has its own independent interests of economy and policy, especially regarding dam failure risk. During the dissemination of the social impact, media is the main channel of information and the direct link between the government and the public.

"Government" is a broad concept, which means a collection of public sectors, referring the government as the main part. It is an organized system of exercising state power, which includes not only the national administrative organs, but also the organs of the country's legislation and judiciary. From the government's practice in management of dam failure risk and public crisis, it is far from enough to cope with these risks only by the national administrative organs, and the whole national system needs to cooperate with each other (Hu Suping, 2010). As part of the country, the government has dual attributes from the date of birth. On the one hand, it inherits social public management to carry out social "public affairs" from the original community's public administration, linking with public power. On the other hand, a specialized management system emerges, which can realize the fundamental interests of the economically dominant class. In short, the government also has attributes of both the public authority and a ruling tool. In the process of dam failure risk management, to maintain its own legitimacy and derive economic interests from losses, the government will start from its own interests and then choose a scheme to maximize the net income by means of balancing the costs and benefits, on the basis of the internal and external environment conditions. Therefore, at the different levels of economic and social stages, the response measures taken by the government are different considering the influence of value orientation.

Habermas believes that "public" is a more inclusive concept, and "its members are people of personal identities, and they can enjoy those objects that can be discussed through the market as readers, listeners and audience". The public being discussed in this article is broader in the extension than Habermas's definition. It mainly includes all the participants of the dissemination of dam failure risk, and they are receivers of the information as well as the disseminators. The public directly receives the threat of dam failure risk and therefore they demand relevant information. On the one hand, they actively seek information through various approaches, based on their decisions and action. On the other hand, they will testify and evaluate and disseminate all types of information through their own experience and communication. Finally, a strong pressure of public opinion is built to lay supervision and criticism on government policies, measures, and media reports.

Media is the main channel of information in the spread of social impact, and it is the most direct link between the government and the public (Qi Hongxia, 2014). Before the reform and opening up, the public has limited access to media, most of which are newspapers, radio, and television under the government's control. The state monopolized all the social resources at that time and defined the media as the governmental voice. As a result, media is only a derivative of the power without the appeal for independent interests or subjectivity (Wan Xuanao, 2015). Therefore, in the spread of social impact, we can regard it as part of the government; however, it cannot work independently or actively. After the reform and opening up, the whole society gradually differentiated, and the control system of the media has undergone a great change as well. The industrial property becomes independent in economy, which laid the foundation for the media to play an independent role in social affairs. At the same time, under the impetus of media technology, the impact and status of the media in the society increase continuously, and the consciousness of profession and subject begin to recover. All of these provide the media with the space to work alone without the intervention of the government (Zhang Lijuan, 2015). The media discussed here mainly refers to the newspaper, radio, television, and other traditional media in China. At the same time, with the vigorous development of the modern communication technology, the new media based on network technology rapidly rises. The foreign media also accelerates the pace of competition in China. At present, the public has more access to different media and the media can affect more people simultaneously.

2.3 *Direct impact of dam failure events on the engineering community can extend to the whole society*

Once the dam breaks, it causes direct impact on the related engineering community. Then, the government will make active decisions and the media and the public are influenced further. Serious incidents instantly turn houses into ruins, so are many lives. Such incidents also bring enormous psychological trauma to the survivors and the public, turning into their gloomy memory. These will result in mental illness of people (Sheng Yibing, 2015). These incidents can lead to more losses and may seriously harm the social security if much attention has not been paid to. Furthermore, these psychological torture and burden can easily make people despair. Post-disaster reconstruction and recovery will not be easy if these negative emotions are not eliminated or alleviated (Zhang Xiping, 2015). Therefore, social psychology needs to be evaluated to recognize and judge people's mental health and the potential threats of social security after dam failure.

2.4 *Social impact on reducing the risk of dam failure*

As dam failure can produce such a huge social impact, it will guide the government, water administrative department, and the reservoir management organization to put more efforts in monitoring and maintaining dams, in order to make effective decisions timely and actively to mitigate the risk. That is, the social impact will in turn affect the decision makers to make efforts to reduce the risk of dam failure.

3 STUDY ON THE SOCIAL IMPACT ASSESSMENT INDEX OF DAM FAILURE RISK

3.1 *Social impact's diffusion hierarchy of dam failure risk*

In addition to the group that can directly perceive the dam failure risk, the perception of others on failure is obtained through various channels of communication, thereby spreading social impact. From the viewpoints of different organizations, there is a hierarchy in the effects caused by dam failure. The victims and the responsible persons are the first individuals or organizations to perceive the dam failure social impact. Due to the publicity and danger of dam failure, the social impact will spread further and further. As the provider of public goods, the government needs to deal with the problem of dam failure to ensure the stability of the social order and to avoid the social impact from further diffusion. The public and other social organizations will also respond accordingly after obtaining relevant information in different ways. If such impact affects so much that the government cannot handle it in time, the social impact may also spread to other international countries or regions (Wang Renzhong, 2006. Wang Mingchun, 2010). Thus, the number of individuals affected by dam failure gradually increases, and the range within effect is more extensive, with the propagation of information. The levels of the diffusion of social impact of dam failure risk are shown in Figure 1.

3.2 *Role allocation during dam failure risk*

The social impact differs significantly between different groups when a dam breaks. The roles played by different groups during dam failure can be divided into three types: (1) The groups in close relationship with encounters. They have psychological impact and energy loss due to the casualties of their loved ones. Obviously, it is different from other social public, although they do not have any direct loss. (2) The dodgers or people who are indirectly affected. These events are public, so the social impact will lead to changes in the behavior of similar organizations or individuals, such as increasing the input of funds, personnel, technology, time, and others to avoid a similar dam failure incident. (3) The public concern and sympathy. When the incident occurs, the general public often gets information from the disseminators. They pay attention to the incidence and treatments and then give their own feedback. As a result, they have losses of time and donations. For the public, the whole event poses a great threat to the social value system as well as to the public. Much attention from the public can affect the decision makers to give more importance to the problem and review the implied deep-level contradictions, thus avoiding the threat of their own value system. This is also a protection of their interests for the public.

The social impact of dam failure risk generally spreads from inside to outside or from point to surface. The roles can be divided into nine types on the basis of different impacts and reactions of different organizations and individuals, which are presented in Table 1. The impact that different roles cause and suffer from forms a network of social impact, which is shown in Figure 2. The impact flow refers to the flow of impact between dam failure

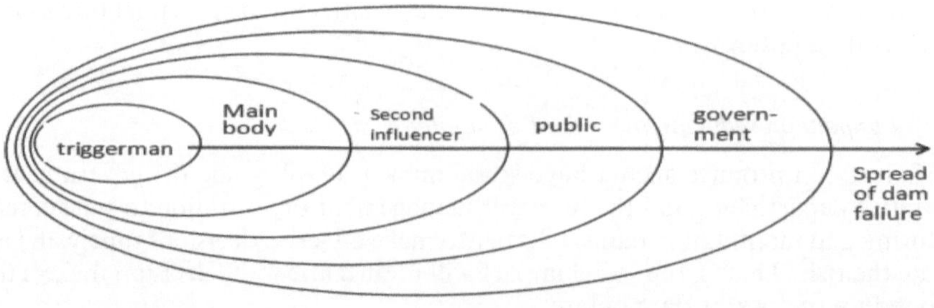

Figure 1. Social impact's diffusion hierarchy of dam failure risk.

Table 1. Role allocation during dam failure risk.

Roles	Definition of roles	Impact of dam failure on roles
Direct encounters	A person or an organization or something that causes a dam failure, which is often subjected to significant damage in dam failure	Casualties, loss of health, loss of time, loss of property, loss of reputation, etc.
Responsible people	An organization or an individual who needs to be responsible for the occurrence of the dam failure	Legal penalties, administrative penalties, economic penalties; the impact of future development of the responsible persons; reputation loss, etc.
Handlers	An organization or a person who needs to reduce the impact after the incident, such as dam management team, ambulance personnel, logistics support staff, and insurance claim agencies	Funds and equipment invested in dealing with the problem; the loss of delaying other incidents; other losses, etc.
Supervisors	Supervise the handlers to deal with the problems in a timely and effective manner and influence other organizations or personnel to avoid such things through their own power	Supervision and management of the time and property and other inputs, along with the inputs to guide and supervise the avoiders
Related encounters	An individual or an organization that has no direct contact with the failure, but it is affected obviously	Loss of time and property caused by relatives' injury or death, penalties, etc.; additional loss caused by the interruption of normal business
Disseminators	The media or other institutions that transmit information about dam failure	Investment and other opportunity costs for obtaining dam information
Sympathizers	The public or an organization that shows sympathy on the encounters in a specific way by understanding the relevant information	Donations out of sympathy, taking care of the encounters
Followers	The public or organization that pays attention to the occurrence and handling of the dam failure through the information disseminated by the disseminators	The time loss and psychological impact caused by concerning about the dam failure
Avoiders	The organization or a person who takes precautionary measures to prevent themselves from damaging	Inputs of funds, personnel, technology, time, and other things to prevent dam failure

and roles. After dam failure, different groups are influenced differently, and thus, they will take different actions. During dam failure, the government, the public, and the media should resume their responsibilities and play their roles. The government should serve as a guide, the public should establish a solid mass foundation and strong support, and the media should become the watcher and the spokesperson of public interest as well as the mouthpiece of the government.

3.3 *Social impact assessment index of dam failure risk*

The public can obtain a wide range of information via the Internet, as well as release information or communicate with others. As a new platform, most news websites have their own "agenda setting". Almost all online news providing sites deliver "Today's hot spots", "important news" to attract the audience, and this is called the agenda setting. The traditional media, such as newspapers, television, and radio, often set up their own websites, like CCTV, people's daily, and China Youth Daily. The content of the carrier is also consistent with the

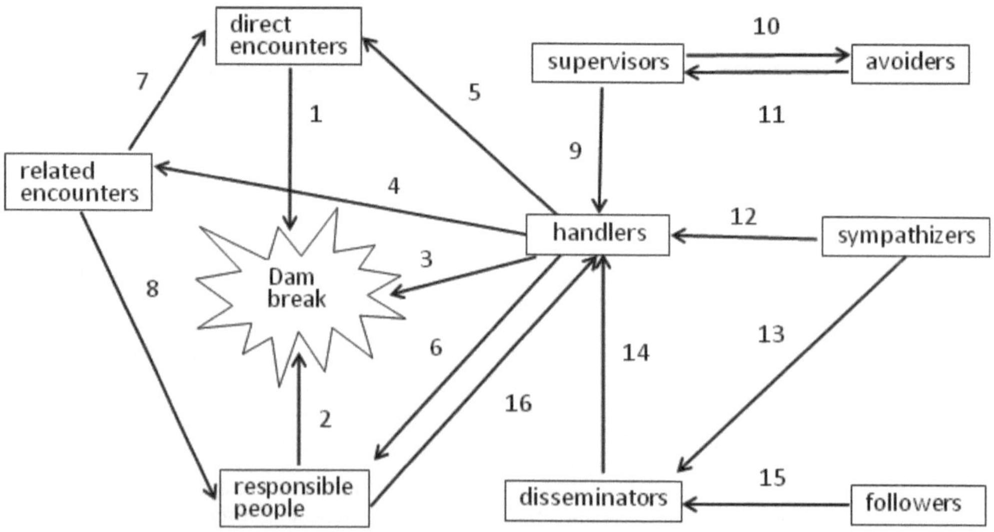

Figure 2. Social impact's network model of dam failure risk.

information on the website. That is, the information on the network includes not only most of the information resources on traditional media, but also their own information resources, referring to news portals and online communities. Many news communities are important places for Internet users to express their thoughts.

Network information storage is massive and easy to search, making it possible to measure the social impact of dam failure events as comprehensive as possible. Internet records and stores the information in the form of web links. Visitors can obtain information through browsing the Web, and therefore, the number of websites recording these events will directly influence visitors' knowledge. In many cases, visitors browse the Internet without purpose. When they visit a website, they often come into contact with all types of information. If there are too many Web pages concerning on certain information, people's impression is bound to deepen and the number of affected visitors will increase (Li Qiang, 2010. Du Bingtao, 2012). Therefore, the number of Web pages recording a certain event on the network can be used as a measure of the social impact of that particular event. The social impact of dam failure events can be divided into three categories as mentioned above. However, from the viewpoint of time, dam failure events last longer, and the public bears the impact for a longer time. The length of the duration is another factor in measuring the social impact of dam failure risk.

According to the above analysis, the social impact assessment index of dam break risk can be evaluated in four aspects, namely personnel loss, publicity, news, and duration.

1. Personnel loss

Dam failure events often produce great harm, including property loss, environmental loss, and casualties. Property loss and environment loss, which are compensated by money, can be eliminated in a short term, whereas psychological loss caused by casualties cannot be compensated by any means. The more the casualties, the more is the emotional loss. Personnel losses are mainly concerned with direct encounters, responsible persons, indirect encounters, processors, communicators, sympathizers, and regulators.

2. Publicity

The most prominent feature of dam failure events is public, which specifically refers to the whole effect on the public after failure. It also involves other public members receiving the threat of dam failure, in addition to the direct victims. The publicity of these events can be measured from two aspects, namely relevance and selectivity. Relevance refers to the range of other social public receiving indirect influence after failure, specifically it is often too difficult to exclude the association of such events and the other social public. The size of the relevance can be described as the scope of certain specific people who are indirectly affected. There are many relevant individuals in these events. The publicity decreases if individuals can get out of the effect at a low cost or it increases if the public cannot take effective measures. The non-selectivity indicates whether the public can avoid the influence, and it can be expressed by the coefficient a: *non-selectable coefficient a = 1 – the number of the indirectly affected people who can avoid the dam failure at a low cost/the total number of the indirectly affected people.* If it is difficult to avoid the impact, the value of a is 1; if it is easy, the value is 0. As a result, the values of the relevance and non-selectivity are bigger, and the dam failure events show more publicity. The roles involved in publicity mainly are avoiders, communicators, handlers, and regulators.

3. News

The propagation of dam failure events needs the news media or the network. The greater the number of news related to a dam failure event, the more likely they are to attract public attention. The roles involved in news mainly are followers, communicators, and regulators.

4. Duration

The duration of dam incident refers to the time from the happening to the end of the incident. The longer the duration of the dam failure, the more the information sources will be. These data will continue to strengthen public awareness and deepen the impression of the dam failure incident, making social impact increasingly greater. The duration T can be estimated as the time elapsed from the beginning of the incident to the end of all treatments. In general, almost all of the aforementioned nine roles are involved in this part.

3.4 *Social impact's classification of dam break risk*

At present, Chinese scholars have developed with a method of grading and classifying a catastrophic event, which is in accordance with the severity of the catastrophic event. In this study, the same method is adopted to grade the social impact assessment indicators, such as publicity, personnel loss, news, and duration. The three values of grades (0, 1, and 2) correspond to three levels (low, medium, and high). Finally, the scores of these four indicators are added by weight to get the comprehensive score of the social impact of dam break risk. The final scores can be divided into five grades (1, 2, 3, 4, and 5), corresponding to five types of social impact of dam break risk (small, general, large, much larger, and huge). On the basis of the above analysis, Tables 2 and 3 present the definitions of all scores and ranks.

According to the above system, based on Table 2, scoring the assessment index, such as publicity and personnel loss, news, and duration, is the foremost task. Then, the scores of

Table 2. Social impact's index score interpretation of dam failure risk.

Index	Level	Score	Interpretation
Publicity	High	2	Occurs in an area where it is difficult to exclude and it is less selective
	Medium	1	Occurs in an area where it is difficult to exclude or it is less selective
	Low	0	Occurs in an area where it is easy to exclude and it has greater selectivity
Personnel loss	High	2	Huge casualties, the number of deaths is more than 30
	Medium	1	The number of deaths is between 10 and 30
	Low	0	The number of deaths is below 10
News	High	2	Similar events rarely occur, 3 times a year or less
	Medium	1	Similar events occur 4–12 times per year on average
	Low	0	Similar events occur more than 12 times per year on average
Duration	High	2	Lasts more than 30 days
	Medium	1	Lasts 10–30 days
	Low	0	Lasts less than 10 days

Table 3. Social impact's grade score of dam failure risk.

Level	Scores	Interpretation
5	8	Social impact is huge, and it declines the values of the whole society
4	6–7	Social impact is great, and most of the social members' psychology is seriously affected
3	4–5	Social impact is large, and it has an impact on the public
2	2–3	Social impact is general, and it only affects the specified organizations or areas
1	0–1	Social impact is small, and it only affects the specified organizations

each index need to be added. Finally, the ranks are divided by the total scores of the social impact of dam break risk.

4 APPLICATION EXAMPLES

On 28 July 2010, Dahe Reservoir experienced a dam break in Changshan Town, Huadian City, Jilin province. About 4 million cubic meters of flood occurred, washing away more than 5 downstream villages (Dahe, Diaoyutai, Kaohe, Xiaoyoufang, and Taipingzhuang). The downstream villagers could not evacuate in time, resulting in collapse of many houses and 38 deaths. The loss of life and property as well as the social impact is huge. One focus of the network is that the affected people blamed the government of Changshan Town. After the tragedy, three cadres were relieved of the post on 31 July 2010 by Huadian government for their poor performance in this tragedy, namely Pan Xueju (Changshan Town Party Secretary), Su Kuo (Deputy Mayor of the town), and Qian Yumin (Changshan Town water station stationmaster). Huadian Municipal Party Committee Organization Department issued a document on 31 July 2010 that Pan was not suitable for the command of the flood relief because of his physical condition and Su Kuo and Qian Yumin were dismissed due to a serious dereliction of duty in the flood relief. In this study, we investigate the social impact assessment index and grading standard of the dam break in Dahe Reservoir.

4.1 *Role allocation during the Dahe Reservoir dam break incident*

In this study, we illustrate the role allocation during the Dahe Reservoir dam break incident, which is presented in Table 4.

4.2 *Social impact assessment of the Dahe Reservoir dam break incident*

After the Dahe Reservoir dam break, all websites started tracking the incident actively. Xinhua, People's Network, Tencent, and other media all involved in a real-time coverage of the events in a short time, up to a sheet of more than 40 pages. These online data can be roughly divided into two categories. The first focuses on the narrative and policy provisions, mainly about the literal description of the incident or the regulations and instructions caused by the dam failure. Therefore, these data do not contain the clear personal viewpoint of the author. The second consists of commentary articles, emphasizing on the discussion and thoughts from the events. These articles reveal the authors' strong viewpoints, showing further impact on authors and readers. More social impact of the dam failure is reflected along with the noticers. On the basis of the interpretation of the evaluating index of dam failure risk in Table 2 and the grade of social impact in Table 3, together with the expert experience, this study grades the dam failure event in Dahe Reservoir in terms of publicity, personnel losses, news, and duration. The final scores will determine the specific level of Dahe's social impact, and the results are shown in Tables 5 and 6. Through the analysis of Dahe's social impact, the level is finally evaluated as 5, indicating that the dam failure in Dahe Reservoir has a huge impact on society.

Table 4. Role allocation during the dam failure state of Dahe Reservoir.

Roles	Performances in Dahe Reservoir dam failure
Direct encounters	38 people were killed; 5 villages were destroyed
Responsible people	Su Kuo (Deputy Mayor of Changshan Town) and Qian Yumin (Changshan town water station stationmaster) were dismissed due to a serious dereliction of duty in the flood relief
Handlers	Contingency rescue and investigation teams consisting of government leaders, medical ambulance personnel, insurance agencies, etc.
Supervisors	Instructions from the national leaders, and the guidance from the relevant State Council departments
Related encounters	Family members and friends of the casualties in this incident
Disseminators	Government departments that transmit information about the dam failure in Dahe Reservoir to the news media
Sympathizers	Donators in this incident
Followers	General social public who concern about the dam safety
Avoiders	Other reservoirs and dams that carry out the inspection

Table 5. Social impact's index assignment of Dahe Reservoir dam failure.

Index	Level	Score	Interpretation
Publicity	High	2	The incident occurs in an area where it is difficult to exclude and it is less selective
Personnel loss	High	2	There are huge casualties, and the number of deaths is more than 38
News	High	2	Similar events rarely occur
Duration	High	2	The social impact of the dam failure in Dahe Reservoir lasts a long time

Table 6. Social impact's grade score of Dahe Reservoir dam failure risk.

Level	Scores	Interpretation
5	8	The social impact of dam failure in Dahe Reservoir is huge, and it declines the values of the whole society.

5 CONCLUSIONS

This study revealed the social impact's formation mechanism of dam failure risk by analyzing the social impact's temporal changes, the constitution, and the direct impact on different members of the engineering community that extends to the whole society, and the social impact on reducing the risk of dam failure.

The study expounded the social impact's diffusion hierarchy of dam break risk, as well as the allocation of the nine roles during dam failure risk, namely direct encounters, responsible people, handlers, supervisors, related encounters, disseminators, sympathizers, followers, and avoiders.

In the light of network information communication, the social impact assessment index of dam failure risk was studied systematically, including personnel loss, publicity, news, and duration. Taking Dahe Reservoir as an example, the practical application showed that the assessment results had great credibility and accuracy, in addition to being objective and reasonable.

This assessment method provided a platform for the comparative analysis of the social impact of different dam failure events based on grade evaluation and easy operability. The data obtained was objective and easy to collect, thereby proving the validity of the assessment results.

ACKNOWLEDGMENTS

This work was supported by the Nonprofit Industry-Specific Research Project of the Ministry of Water Resources of China (201501033), the Research Funds for Youth of Nanjing Hydraulic Research Institute of China (Y715008, Y716005, Y716008, Y716012, Y717004), the National Natural Science Foundation of China (51509156), and the National Key R&D Program of China (2016YFC0401601).

REFERENCES

Cheng Duqiu, Wei Zhengwei. Analysis on social impact of hydraulic conservancy project [J]. Hydraulic science and technology, 2014, (5):60.

Du Bingtao, Yuan Yong-bo. Study on the loss model of dam area after dam break [J]. Journal of Water Resources and Architectural Engineering, 2012, 10(1):64–68.

Huang Junhua. Research on social risk and its regulation in social media communication [J]. News, 2014, (5):21–22, 17.

Hu Suping. Residual risk and risk assessment of dam-failure [J]. China Water Resources, 2010, (8):21–23.

Li Qiang, Shi Lingling, Ye Pengfei, et al. Exploration on the social impact evaluation index system suitable for China's national conditions [J]. Hebei Academic Journal, 2010, 30(1):106–112.

Qi Hongxia, Cai Libin. A review of studies on the social impact of large foreign events [J]. Tourism Tribune, 2014, (5):116–128.

Sheng Yibing. The transformation mechanism from social risk to public crisis [J]. China Management Informationization, 2015, (4):218–219.

Wang Mingchun. Discussion on the social impact of water conservancy projects such as reservoirs and dams [J]. Jilin agriculture, 2010, (11):214–215.

Wang Renzhong, Li Lei, Sheng Jinbao. On criterion of social and environmental risk of reservoir dams [J]. Journal of safety and environment, 2006, (1): 8–11.

Wan Xuanao, Xie Gengyun. Research on the impact of network public opinion on public policy [J]. Editor'sFriend, 2015, 5–11.

Zhang Lijuan. Research on the Countermeasures of local government network public opinion guidance [J]. China Management Informationization, 2015, (18):203–204.

Zhang Xiping, Wei Jiuchang, Liang Liang. Social impact assessment and response to crisis events [J]. Journal of Social Sciences, 2011, (6):90–95.

Zhang Xiuhua. Nature of the engineering community [J]. Journal of Dialectics of Nature, 2008, 30(6):43–47.

Hydraulic Engineering V – Hong, Liu & Xie (Eds)

Experimental measurement of diffusion coefficients under wave action by the tracer method

Liangduo Shen
School of Port and Transportation Engineering, Zhejiang Ocean University, China

Yichun Ding
Zhoushan Port Engineering Planning and Design Institute Co. Ltd., China

Gaofeng Jiang
School of Port and Transportation Engineering, Zhejiang Ocean University, China

Yateng Song & Hua Shi
Zhoushan Harbor Administration Bureau, China

ABSTRACT: The aim of this study was to quantitatively analyze the diffusion coefficient of material under wave action by the tracer method. An experiment was carried out to record the diffusion process of marine material using a CCD camera. A smooth relative concentration contour and centerline series of a tracer image was obtained using a powerful image processing function provided by MATLAB. The image showed the relationship between concentration contour series and diffusion coefficient of material by means of theoretical analysis. On the basis of this result, the experimental measurement of the diffusion coefficient under the combined action of wave and current can be quickly realized.

1 INTRODUCTION

Although economy is more active and developed in the coast, coastal areas and coastal waters are the most complex areas because of the presence of various dynamic factors. The law of various hydrodynamic and material diffusion processes near the shore has become very complex because of the evolution and breakdown of the waves on the shore (Burchard H. et al. 2008, YanWang & ZhiliZou 2015). To realize the harmonious and sustainable development of coastal economic environment, make profit, and avoid disadvantages, it is necessary to understand the law of material diffusion in coastal water. On the one hand, we can develop appropriate sewerage programs according to these characteristics and laws to quickly remove the sewage from the coast and achieve rapid purification. On the other hand, they can be used to predict the corresponding disaster losses for assessing and preventing disaster. Furthermore, people can take advantages of the law to guide the delivery of marine aquaculture material, in order to estimate the changing process of substance concentration with time in the aquaculture area to better guide fisheries production (Yang Xianle & Zheng Zonglin 2007). However, determining the diffusion coefficient of material is key for understanding the law of material diffusion. The traditional diffusion coefficient under the combined action of wave and current is usually calculated by the empirical formula that must influence the accuracy of the calculation results.

In this study, we use contactless measurement, through the use of a high-speed CCD camera (CCD—charge-coupled device, which can convert light into electricity, and then, through the analog-to-digital converter chip, can convert electrical signals into digital signals; the latter can be transmitted to the computer and then compressed to form the resulting image). The change in the information on the observed area of the tracer material with time can be recorded, and the change in the characteristics of the tracer material's concentration contours, contour lines, and centerline with time can be obtained by image processing analysis. Then, their relationship with the corresponding diffusion coefficient can be established by means of theoretical analysis, resulting in a reliable transverse and longitudinal diffusion coefficient (Zou Zhili et al. 2009). On the basis of this, the process of material transport with time can be further calculated through applying the diffusion coefficient to the convective diffusion equation. This can greatly improve the reliability and real-time nature of large-scale ocean monitoring, and significantly reduce monitoring costs. At present, domestic studies involving analysis of coastal pollutant image are neither more nor have enough depth. The existing research generally uses C, C ++, VB, and other computer language programs to deal with the image, and the statement is generally long and prone to grammatical errors. Sun (2004) gave the initial contour line and centerline of the ink on the basis of the application of noncontact measurement method in hydraulic model experiment by image analysis, but he did not further smooth and trim the grayscale mutation, photo deformation, and the centerline with multiple branches and so on brought by environment, equipment, and other factors.

This paper uses a built-in function provided by MATLAB's powerful toolbox to obtain a smooth isotonic line, concentration centerline, and contour lines based on image morphological processing, surface fitting and other techniques. Furthermore, it aims to obtain the concentration distribution of material transport with time and its diffusion coefficient, which provides the necessary data for further analysis of material diffusion.

2 EXPERIMENTAL SETUP

Experiment on the diffusion process of coastal pollutants was conducted in the multifunctional integrated pool of Dalian University of Technology. The pool is 55 m long and 34 m wide and has a maximum water depth of 0.7 m. The experimental coastal topography adopted slopes 1:100 and 1:40, whose depths of the front are 18 and 45 cm, respectively. The contaminants in the experiment used the pure blue ink, and their motion and diffusion processes were recorded by two CCDs placed at a distance of 12 m above the topography. The photo interval for each frame is 0.125 s. To get a clear picture, in the experiment, the light source of the CCD working environment adopted natural light and the surface of the measurement area was made of white cement to form a white camera background. Pouring the ink directly into the water from a cup forms a diffusion movement in the form of instantaneous point source. In the experiment, the ink was diluted to 20%.

2.1 *Ink image processing*

To analyze the concentration and movement trend of the CCD ink image, we developed a detailed process based on MATLAB.

2.1.1 *Obtaining the focused ink image area*

The ink is in motion in the ink image obtained by the CCD, and the ink area needs to be detected. To achieve this, we used the background difference method, which is a common method of motion image detection that uses the difference between the current image containing the moving object and the background image to detect the motion area. Equation (1) gives the specific expression:

$$D_k(x,y) = \left| f_k(x,y) - B_k(x,y) \right| \quad (1)$$

where $B_k(x,y)$ and $f_k(x,y)$ are gray values of the background image and the current image, respectively, $D_k(x,y)$ is the absolute value of the difference between the two. The results are shown in Figure 1(a), (b), and (c). As illustrated, the ink part and the environment in the processed image (Figure 1(c)) achieve a good separation, and are highlighted.

(a) Background frame (b) Current frame (c) difference image

Figure 1. Experimental data and results of background subtraction.

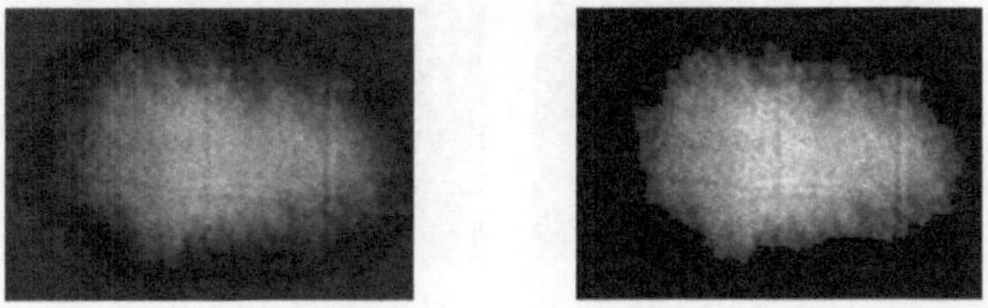

Figure 2. Comparison of the image obtained before and after homomorphic filtering.

2.1.2 *Denoising and enhancing*

In the process of generation, acquisition, and transmission, CCD images were influenced by the lighting source performance, imaging system performance, channel bandwidth, noise, and many other factors, which results in low contrast, decreased clarity, and the presence of interference noise frequently. To improve the image quality, it is necessary to enhance the sharpness of the image. In general, the edge and noise of the image correspond to the high-frequency components of the Fourier-transformed image, whereas the low-frequency component mainly determines the overall gray-level display of the image in the smooth area, so the low-pass filtered image lacks some sharp details compared with the original image. To enhance the details of the image and to retain the low-frequency components of the image, in this study, a homomorphic filtering enhancement method based on the illuminance–reflectance model is adopted to remove the interference and enhance image clarity. As shown in Figure 2(a) and (b), the image obtained after homomorphic filtering is more vivid than the previous image in the gray level.

2.1.3 *Concentration outline*

The concentration outline of ink image reflects the range and shape of material transport. The homogeneous filtered ink image is well separated from the background; at this time, the image can be further subjected to edge detection processing. Concentration outline can be obtained by image binarization, that is, using the grayscale histogram of the image and dividing it according to a given threshold. The points smaller than the threshold are taken as background, that is, taken as the gray value as 0 (black). On the contrary, in areas where the pollutants pass, the gray value is taken as 1 (white). The boundary line of the area with the gray value of 1 is the concentration outline.

The binarized image obtained at this time contains an area with an excessively small grayscale value of 0, which needs to be further refined. After fitting the narrow gap and slender bend, removing the hole, and filling the gap of the contours, the intact binary image with a single region can be obtained, as shown in Figure 3(a). The outline, as shown in Figure 3(b), can be obtained by binary image boundary detection.

2.1.4 *Concentration centerline*

The ink image concentration centerline reflects the ink movement trend and main trajectory. The skeleton, including centerlines and branches, can be well obtained by the thinning of the binary image. The so-called thinning is to continue to convert the boundary line into the inner domain point, so that the boundary gradually contract to the domain until the region

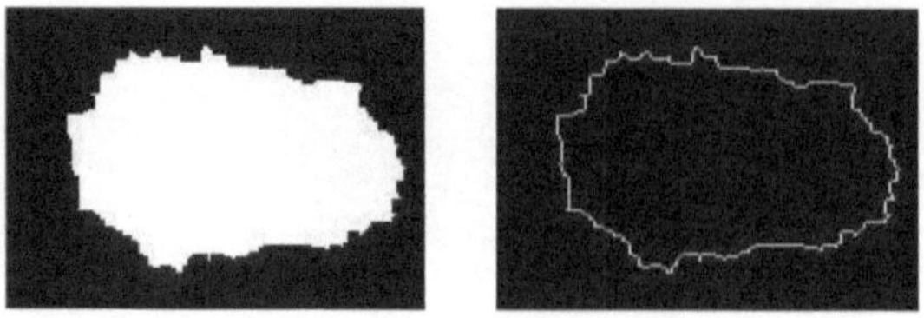

Figure 3. Binary image and corresponding outline.

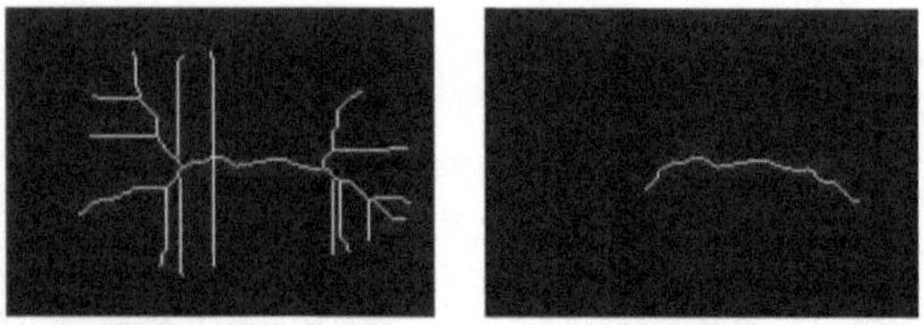

Figure 4. Thinning skeleton of the binary image and the corresponding image after pruning.

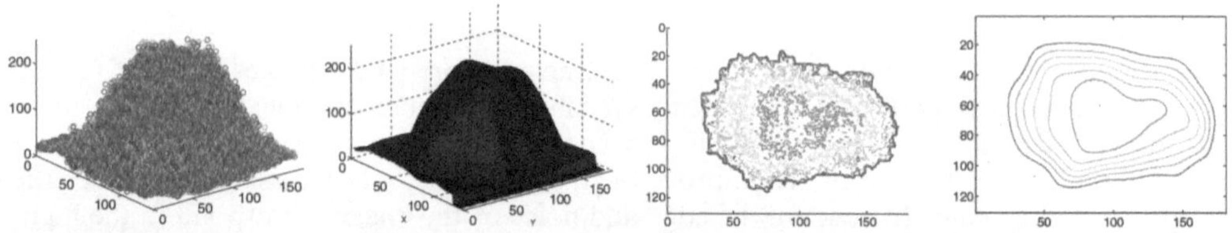

(a) Before surface fitting (b) After surface fitting (c) Concentration contours before fitting (d) Concentration contours after fitti

Figure 5. Surface fitting and concentration contours of the ink image.

shrinks into a number of lines (lines only contain a pixel) and finally produces the skeleton of the image. However, because the ink image diffusion edge is not smooth, which inevitably leads to burr, branch and other information in the refinement process will destroy the one-way consistency of the centerline, as shown in Figure 4(a). To overcome this problem, here we use the pruning algorithm based on the curve tracing technique proposed by Wu et al. (2006), in which the skeleton after pruning is made up of only one trunk centerline through pruning the resulting image skeleton, as shown in Figure 4(b).

2.1.5 *Grayscale fitting and concentration contour*
Due to external environment such as light, dust, signal interference, and CCD sensor, the grayscale of ink image is not in an ideal smooth statement, and few grayscales have a greater mutation, as shown in Figure 5(a). This will make concentration contour obtained from the surface non-smooth, as shown in Figure 5(c). To obtain smooth concentration contours, the spline function is used for surface fitting to obtain a smooth gray distribution, as shown in Figure 5(b). On the basis of this, concentration contours can be obtained as shown in Figure 5(d).

2.1.6 *Space transformation*
The concentration outlines, centerlines, and contours obtained in the previous steps are all based on the image pixel coordinates. However, the actual space coordinates are needed. Because the CCD lens is not exactly perpendicular to the water, the image obtained by it has a certain angle deformation, as shown in Figure 6. Therefore, there is a certain nonlinear transformation relationship between the coordinates on the image and the real physical coordinates.

The method developed by Holland et al. (1997) is used to convert the image pixel coordinates and the actual spatial coordinates. The final map of the concentration outlines, centerlines, and contours in the corresponding physical coordinates is shown in Figure 7.

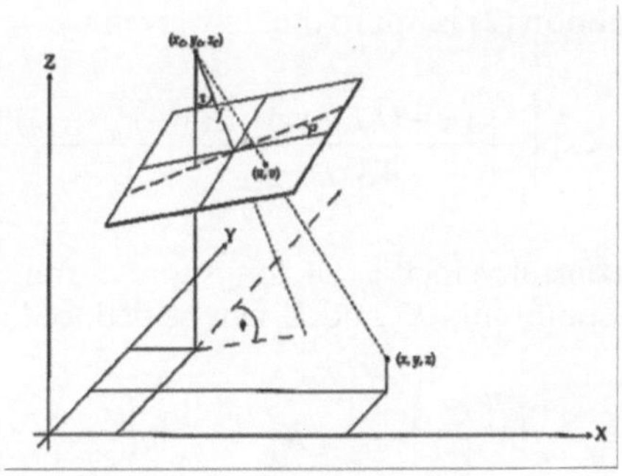

Figure 6. Collinear relationship between the camera image domain and real physical coordinate.

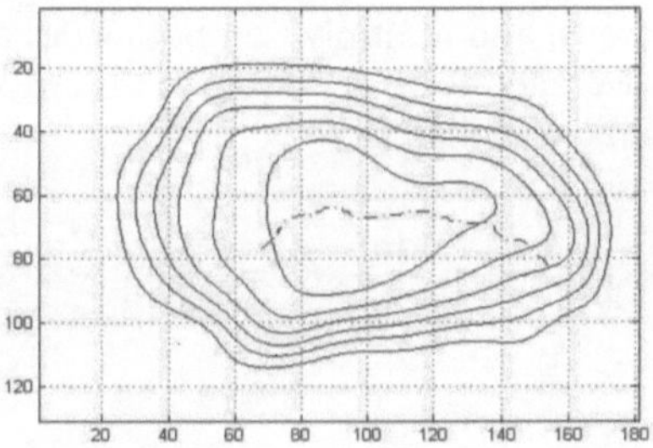

Figure 7. Final contour map and centerline.

After carrying out the above process for every picture in the experimental image series, the corresponding concentration contour series can be obtained.

3 A METHOD OF CALCULATING MIXING COEFFICIENTS

The above concentration contour series reflects the horizontal distribution of the concentration of the ink. To lead the dispersion in the two horizontal directions, a Cartesian coordinate system was used: y points in the longshore direction, x points in the offshore direction, and z points in the upward direction. A three-dimensional convection-diffusion equation of the concentration was applied as follows:

$$\frac{\partial c}{\partial t}+u\frac{\partial c}{\partial x}+v\frac{\partial c}{\partial y}+w\frac{\partial c}{\partial z}=D_{tx}\frac{\partial^2 c}{\partial x^2}+D_{ty}\frac{\partial^2 c}{\partial y^2}+D_{tz}\frac{\partial^2 c}{\partial z^2} \tag{2}$$

where t is the time; u, v, and w are the velocities in the x, y, and z directions, respectively; and D_{tx}, D_{ty}, and D_{tz} are the coefficients of the turbulent diffusion ignoring the molecular diffusion in the x, y, and z directions, respectively.

Dividing the velocities u, v, w, and c into two parts: $u=\bar{U}+\hat{U}+\tilde{u}, v=\bar{V}+\hat{V}+\tilde{v}$, $w=W+\tilde{w}, c=\bar{C}+\hat{C}+\tilde{c}$ (letters with "–" represent depth-averaged part, while letters with "∧" are the deflection part, and letters with "~" represent the wave-related variable), after an integral course in the vertical direction with the upper equation, a two-dimensional horizontal diffusion equation was derived:

$$\frac{\partial \bar{C}}{\partial t}+\bar{U}\frac{\partial \bar{C}}{\partial x}+\bar{V}\frac{\partial \bar{C}}{\partial y}=K_x\frac{\partial^2 \bar{C}}{\partial x^2}+K_y\frac{\partial^2 \bar{C}}{\partial y^2} \tag{3}$$

where K_x and K_y are cross-shore and longshore diffusion coefficients, respectively.

Then, the solution of equation (2) is approximately given by

$$\frac{C_t}{C_c} = \exp\left[-\left(\frac{(x-\bar{U}_0 t - x_0)^2}{4K_x t} + \frac{(y-\bar{V}_0 t - y_0)^2}{4K_y t}\right)\right] \tag{4}$$

where $\bar{U}_0$ and $\bar{V}_0$ are the central velocities of longshore currents and C_c is the maximum concentration of the central point ink. K_x and K_y can be deduced as

$$K_x = -\frac{1}{4}\ln\left(\frac{C_t}{C_c}\right)^{-1}\frac{x'^2}{t},\ K_y = -\frac{1}{4}\ln\left(\frac{C_t}{C_c}\right)^{-1}\frac{y'^2}{t} \tag{5}$$

where C_t/C_c is a constant for the concentration contour of the point ink. Variables x' and y' in (4) are the intercepts of the concentration contour on the axis of coordinates. We replace x' and y' with a and b, respectively, and multiply t on both sides and averaged on the scale of wave periods in (4). The time-averaged scale of wave periods cross-shore and longshore diffusion coefficients K_x and K_y can be obtained as

$$\bar{K}_x = -\frac{1}{4}\ln\left(\frac{C_t}{C_c}\right)^{-1}\frac{d\overline{a^2}}{dt},\ \bar{K}_y = -\frac{1}{4}\ln\left(\frac{C_t}{C_c}\right)^{-1}\frac{d\overline{b^2}}{dt} \tag{6}$$

4 CONCLUSION

This study analyzed the diffusion characteristics of material by the tracer method. Through a built-in function provided by MATLAB's powerful toolbox, the ink image contours were smoothed, the narrow gaps were fused, and the interference caused by the environment and equipment noise were reduced by image morphological processing techniques. The burr defects of concentration contours caused by image grayscale mutation were overcome by using the surface fitting technique, thereby making it possible to obtain smooth concentration outlines, centerlines, and contours. Meanwhile, image pixel coordinates were converted to actual spatial coordinates using the method proposed by Holland et al. On the basis of this result, the relationship between the contours of the tracer image and the diffusion coefficient could be established by means of theoretical analysis. Therefore, the experimental measurement of diffusion coefficients under the combined action of wave and current by the tracer method was realized.

REFERENCES

Burchard H., Craig P.D., Gemmrich J.R. et al. Observational and numerical modeling methods for quantifying coastal ocean turbulence and mixing [J]. Progress in Oceanography. 2008, 76: 399–442.

Holland K.T., Holman R.A., Lippmann T.C., et al. Practical use of video imagery in nearshore oceanographic field studies [J]. Ocean. Eng, 1997, 22(1): 81–92.

Sun Hequan. Application of non-contact measurement method based on image analysis in hydraulic model experiment [D]. Dalian University of Technology, 2004.

Wu qingyang, Su Xianyu and Song Fang. A Pruning Algorithm for the Skeleton of Line-structure-light Pattern [J]. Journal of Sichuan University, 2006, 38(5): 148–151.

Yang Xianle, Zheng Zonglin. The application status and strategy of fishery medicine in China [J]. Journal of Shanghai Fisheries University, 2007, (16): 374–380.

YanWang, ZhiliZou. An experimental and numerical study of bimodal velocity profile of longshore currents over mild-slope barred beaches [J]. Ocean Engineering, 2015, 106: 415–423.

Zou Zhili, Li Liang, Sun Hequan et al. Experimental study on the depth averaged mixing coefficient in longshore currents [J]. Acta Oceanologica Sinica. 2009, (3): 137–148.

Hydraulic Engineering V – Hong, Liu & Xie (Eds)
© 2018 Taylor & Francis Group, London, ISBN 978-0-8153-7471-8

Author index